THE PLANETARY SCIENTIST'S COMPANION

THE PLANETARY SCIENTIST'S COMPANION

Katharina Lodders
Bruce Fegley, Jr.

New York   Oxford

Oxford University Press

1998

Oxford University Press

Oxford   New York
Athens   Auckland   Bangkok   Bogotá   Buenos Aires   Calcutta
Cape Town   Chennai   Dar es Salaam   Delhi   Florence   Hong Kong   Istanbul
Karachi   Kuala Lumpur   Madrid   Melbourne   Mexico City   Mumbai
Nairobi   Paris   São Paulo   Singapore   Taipei   Tokyo   Toronto   Warsaw

and associated companies in
Berlin   Ibadan

Published by Oxford University Press, Inc.
198 Madison Avenue, New York, New York 10016

Oxford is a registered trademark of Oxford University Press

Library of Congress Cataloging-in-Publication Data
Lodders, Katharina
The planetary scientist's companion /
Katharina Lodders, Bruce Fegley.
p.   cm.
Includes index.
ISBN 0-19-511694-1
1. Planetology—Handbooks, manuals, etc.
2. Cosmochemistry—Handbooks, manuals, etc.
3. Geochemistry—Handbooks, manuals, etc.
I. Fegley, Bruce.   II. Title.
QB601.L84   1998
523.2—dc21   9748465

1 3 5 7 9 8 6 4 2
Printed in the United States of America
on acid-free paper

To

our parents

and grandparents

Acknowledgments

For permission to reproduce data we are indebted to the following:
Academic Press, San Diego, for material from: C. A. Barth, 1985, The photochemistry of the atmosphere of Mars, in The photochemistry of atmospheres (J. S. Levine, ed.) p. 348. J. S. Kargel & J. S. Lewis, 1993, The composition and evolution of Earth, Icarus 105, pp. 8-9. G.F. Lindal, G.E Wood, H.B. Hotz, D.N. Sweetnam, V.R. Eshleman, & G.L. Tyler, 1983, The atmosphere of Titan: An analysis of the Voyager 1 radio occultation data, Icarus 53, p. 355-357. G. S. Zhdanov, 1965, Crystal Physics, pp. 192-193.

Elsevier Science Ltd., Oxford and Amsterdam, for material from: E. Anders & M. Ebihara, 1982, Solar-system abundances of the elements, Geochim. Cosmochim. Acta 46, p. 2364. E. Anders & N. Grevesse, 1989, Abundances of the elements: meteoritic and solar, Geochim. Cosmochim. Acta 53, p. 198-200. A.G.W. Cameron, 1968, A new table of abundances of the elements in the solar system, in Origin and distribution of the elements (L. H. Ahrens, ed.), pp. 127-128. G.J. Consolmagno & M.J. Drake, 1997, Composition and evolution of the eucrite parent body: evidence from rare earth elements, Geochim. Cosmochim. Acta 41, p. 1278. A.W. Hofmann, 1988, Chemical differentiation of the Earth: the relationship between mantle, continental crust, and oceanic crust, Earth Planet. Sci. Lett. 90, 299. J. Jones, 1984, The composition of the mantle of the eucrite parent body and the origin of eucrites, Geochim. Cosmochim. Acta 48, p. 645. J.H. Jones & M.J. Drake, 1989, A three-component model for the bulk composition of the Moon, Geochim. Cosmochim. Acta 53, p. 522. A.K. Kennedy, G.E. Lofgren & G.J. Wasserburg, 1993, An experimental study of trace element partitioning between olivine, orthopyroxene and melt in chondrules: equilibrium values and kinetic effects, Earth Planet. Sci. Lett. 115, p. 185-186. A. Seiff et al., 1986, Models of the structure of the atmosphere of Venus from the surface to 100 kilometers altitude, in Advances in space research, Vol. 5, The Venus international reference atmosphere, (A.J. Kliore, V.I. Moroz & G.M. Keating, eds.), pp. 25-26, 28. Y.H. Li, 1991, Distribution patterns of the elements in the ocean: a synthesis, Geochim. Cosmochim. Acta 55, pp. 3224-4225. W.F. McDonough & S.S. Sun, 1995, The composition of the Earth, Chem. Geol. 120, p. 238. J.W. Morgan & E. Anders 1979, Chemical composition of Mars, Geochim. Cosmochim. Acta 43, pp. 1605,1607. J.W. Morgan, H. Higuchi, H. Takahashi & J. Hertogen, 1978, A "chondritic" eucrite parent body: inference from trace elements, Geochim. Cosmochim. Acta 42, p.34. H. Nagasawa, H.D. Schreiber & R.V. Morris, 1980, Experimental mineral/liquid partition coefficients of the rare earth elements (REE), Sc, and Sr for perovskite, spinel and melilite, Earth Planet. Sci. Lett. 46, p. 434. H.S.C. O'Neill, 1991, The origin of the moon and the early history of the earth: a chemical model: Part I: The Moon, Geochim. Cosmochim. Acta 55, p. 1138. A.E. Ringwood, 1991, Phase transformations and their bearing on the constitution and dynamics of the mantle, Geochim. Cosmochim. Acta 55, p. 2087. A.E. Ringwood, S. Seifert & H. Wänke, 1986, A komatiite component in Apollo 16 highland breccias: implications for the nickel-cobalt systematics and bulk composition of the Moon, Earth Planet. Sci. Lett. 81, p. 111. D.M. Shaw, J. Dostal & R.R. Keays, 1976, Additional estimates of continental surface precambrian shield composition in Canada, Geochim. Cosmochim. Acta 40, p. 79-80. D. Stöffler, K. Keil & E.R.D. Scott, 1991, Shock metamorphism of ordinary chondrites, Geochim. Cosmochim. Acta 55, p. 3860. S.S. Sun, 1982, Chemical composition and origin of the earth's primitive mantle, Geochim. Cosmochim. Acta 46, p. 180. S.R. Taylor, 1982, Lunar and terrestrial crusts: a contrast on origin and evolution, Phys. Earth. Planet. Inter. 29, pp. 235-236. W.R. van Schmus & J.A. Wood, 1967, Geochim. Cosmochim. Acta 31, p. 757. B. L. Weaver & J. Tarney, 1984, Major and trace element composition of the continental lithosphere, Phys. Chem. of the Earth (H.N. Pollack & V.R. Murthy, eds.), Vol. 15, p. 43. K.H. Wedepohl, 1995, The composition of the continental crust, Geochim. Cosmochim. Acta 59, pp. 1219-1220. E.J. Whittaker, & R. Muntus, 1970, Ionic radii for use in geochemistry, Geochim. Cosmochim. Acta, 34, p. 952-953.

The Astronomical Journal for material from: G.F. Lindal, 1992, The atmosphere of Neptune: An analysis of radio occultation data acquired with Voyager 2, Astron. J. 103, p. 975.

The Geological Society of America for material from: A.R. Palmer, 1983, The decade of North American geology, 1983 geologic time scale, Geology 11, p. 504.

We also thank the following authors for permissions to use data from their work and for helpful comments: E. Anders, J.A. Burns, R.N. Clayton, B. Edvardsson, B.G. Marsden, H. Palme, Yu. A. Surkov, R.B. Symonds, S.R Taylor, S. Turck-Chiäze, H. Wänke, J.T. Wasson, K.H. Wedepohl.

Preface

Dear friend of planetary sciences,

You certainly have found yourself in situations where you wanted to do a quick calculation or remind yourself about some planetary or cosmochemical quantity but then spent a fair amount of time hunting for the necessary data through a large pile of books and scientific papers. Although there are several handbooks for physics, astronomy, and chemistry, no one handbook contains combined data for the interdisciplinary fields of planetary science and cosmochemistry.

To remedy this situation, we conceived this small book containing physical and chemical data often used in planetary science. The result is a "data journey" through the solar system and beyond. Data for the sun, the terrestrial and outer planets, and their satellites are presented, followed by data for the smaller objects: the asteroids, Centaurs, Kuiper belt objects, and comets. A larger chapter dealing with meteorites, the debris of asteroids, is also included. You will also find information about the newly discovered extrasolar planets and nearby stars in chapter 17, Beyond the Solar System. All chapters start with a small summary describing the planet or smaller bodies, followed by the related tables. Also included in the book are tables listing fundamental constants and unit conversion factors, a small glossary, and an index.

We have included a large number of references to the tables, which allows you to trace data to their original sources. Sometimes data or model results from various authors differ; in these cases we did not always try to make our best pick, but we included the results side by side for comparison.

This handbook cannot, and is not intended to, replace textbooks on planetary science and cosmochemistry. But we think that it can serve as a valuable addition for reference, because it contains a lot of data widely dispersed in scientific journals and books.

In these days of CD-ROMs and the Internet, a small reference book still can find its place on the desk and may be even more convenient for "just looking up that number." Still, we plan to eventually supplement the book with a CD-ROM, which would allow the user to access the data in machine-readable form.

Although we carefully double-checked all numbers, it is almost impossible to avoid errors in a book mainly consisting of tables. Certainly, there will also be changes in some quantities in the future, depending on observational progress. Therefore, we would appreciate feedback and communications from the readers, as well as suggestions on what else to include in the book.

Many of our students and colleagues in the planetary and meteorite community have made suggestions about the content of the book and helped by pointing out useful data and references; our thanks to all of them. We also thank Joyce Berry and Lisa Stallings from Oxford University Press for guiding us through the book creation process.

St. Louis, Missouri   Katharina Lodders
September 1997   Bruce Fegley, Jr.

Contents

17 Beyond the solar system

THE PLANETARY SCIENTIST'S COMPANION

TECHNICAL DATA

Table 1.1 The Greek Alphabet

Letter			Letter		
Uppercase	Lowercase	Name	Uppercase	Lowercase	Name
A	α	alpha	N	ν	nu
B	β	beta	Ξ	ξ	xi
Γ	γ	gamma	O	o	omicron
Δ	δ	delta	Π	π	pi
E	ε	epsilon	P	ρ	rho
Z	ζ	zeta	Σ	σ	sigma
H	η	eta	T	τ	tau
Θ	θ	theta	Y	υ	upsilon
I	ι	iota	Φ	φ	phi
K	κ	kappa	X	χ	chi
Λ	λ	lambda	Ψ	ψ	psi
M	μ	mu	Ω	ω	omega

Table 1.2 Prefixes Used With the SI System

Symbol	Prefix	Factor	Symbol	Prefix	Factor
d	deci	10^{-1}	da	deka	10^{+1}
c	centi	10^{-2}	h	hecto	10^{+2}
m	milli	10^{-3}	k	kilo	10^{+3}
μ	micro	10^{-6}	M	mega	10^{+6}
n	nano	10^{-9}	G	giga	10^{+9}
p	pico	10^{-12}	T	tera	10^{+12}
f	femto	10^{-15}	P	peta	10^{+15}
a	atto	10^{-18}	E	exa	10^{+18}
z	zepto	10^{-21}	Z	zetta	10^{+21}
y	yocto	10^{-24}	Y	yotta	10^{+24}

Table 1.3 Basic SI and cgs Units

Quantity	Symbol	Unit Name SI	Unit Symbol SI	Unit Symbol cgs
Time	t	second	s	s
Mass	M, m	kilogram	kg	g
Amount of substance	n	mole	mol	mol
Length	x, z	meter	m	cm
Thermodynamic temperature	T	kelvin	K	K
Electric current	I	ampere	A	$esu\ s^{-1}$
Luminous intensity	I_v	candela	cd	

Table 1.4 Derived SI Units

Quantity	Symbol or Derivation	Special Name	Symbol SI	Unit Equivalent SI	Unit cgs
Plane angle		radian	rad	$m/m = 1$	
Solid angle		steradian	sr	$m^2/m^2 = 1$	
Area	$A = x^2$			m^2	cm^2
Volume	$V = x^3$			m^3	cm^3
Speed, velocity	$v = x/t$			$m\,s^{-1}$	$cm\,s^{-1}$
Acceleration	$a = x/t^2$			$m\,s^{-2}$	$cm\,s^{-2}$
Force	F	newton	N	$kg\,m\,s^{-2}$	dyn
Pressure, stress	$P = F/A$	pascal	Pa	$N\,m^{-2}$	$dyn\,cm^{-2}$
Energy, work, heat	$E = F \times x$	joule	J	$N\,m = kg\,m^2s^{-2}$	erg
Impulse, momentum	$m \times v$			$N\,s, kg\,m\,s^{-1}$	
Power	$P = E/t$	watt	W	$J\,s^{-1}$	$erg\,s^{-1}$
Frequency	$v = 1/t$	hertz	Hz	s^{-1}	
Electric Charge	Q	coulomb	C	$A\,s$	esu
Electric charge density	ρ_e			$C\,m^{-3}$	$esu\,cm^{-3}$
Electric current density	J_e			$A\,m^{-2}$	$esu\,s^{-1}\,cm^{-2}$
Electric potential, emf	V	volt	V	$J\,C^{-1} = W\,A^{-1}$	$erg\,esu^{-1}$
Electric field	E			$V\,m^{-1} = N\,C^{-1}$	$dyn\,esu^{-1}$
Resistance	R	ohm	Ω	$V\,A^{-1}$	$s\,cm^{-1}$
Electrical resistivity	ρ			$\Omega\,m^{-1}$	s
Electrical conductance	G	siemens	S	$A\,V^{-1} = \Omega^{-1}$	$cm\,s^{-1}$

continued

Table 1.4 *(continued)*

Quantity	Symbol or Derivation	Special Name	Symbol SI	Unit Equivalent SI	Unit cgs
Electrical conductivity	σ			$S\,m^{-1}$	s^{-1}
Electric permittivity	ε			$F\,m^{-1}$	esu
Inductance	L	henry	H	$Wb\,A^{-1}$	$s^2\,cm^{-1}$
Magnetic flux	Φ	weber	Wb	$V\,s$	maxwell, Mx
Magnetic field	H			$A\,m^{-1}$	oersted, Oe
Magnetic flux density	B	tesla	T	$Wb\,m^{-2} = N\,A^{-1}m^{-1}$	Gauss, G
Capacitance	C	farad	F	$C\,V^{-1}$	cm
Luminous flux		lumen	lm	cd sr	
Luminance		lux	lx	$lm\,m^{-2}$	
Power flux, flux unit (f.u.)		jansky	Jy	$10^{-26}\,W\,m^{-2}\,Hz^{-1}$	erg $cm^{-2}s^{-1}Hz^{-1}$
Radioactivity		becquerel	Bq	s^{-1}	
		(1 curie = 37 GBq)			

Gaussian cgs units are given for electrical and magnetic quantities. The electrostatic unit (esu) is also referred to as a statcoulomb, a statampere = esu^{-1}, and a statvolt = erg esu^{-1}. The maxwell = gauss cm^{-2}. A good description of electrical and magnetic concepts, units, and conversion factors is given by Purcell, E. M., 1965, *Electricity and magnetism*, McGraw-Hill, New York, pp. 459. Some conversions between Gaussian cgs units and SI units are (c = the speed of light in cm s^{-1}):

$$1\ coulomb = 0.1c\ esu$$

$$1\ ampere = 0.1c\ esu\ s^{-1}$$

$$1\ volt = 10^8 c^{-1}\ erg\ esu^{-1}$$

$$1\ ohm = 10^9 c^{-2}\ s\ cm^{-1}$$

$$1\ F\,m^{-1} = 10^{-11} c^2\ esu$$

$$1\ A\,m^{-1} = 4\pi 10^{-3}\ oersted$$

$$1\ weber = 10^8\ maxwell = 10^8\ gauss\ cm^{-2}$$

$$1\ gauss = 10^{-4}\ Tesla$$

$$1\ farad = 10^{-9} c^2\ cm$$

Table 1.5 Frequently Used Constants

Name	Symbol	Value & Unit
Universal Constants		
Speed of light in vacuum	c	2.99792458×10^8 m s^{-1} $2.99792458 \times 10^{10}$ cm s^{-1}
Permeability of vacuum	$\mu_0 = 4\pi \times 10^{-7}$	$12.566370614.. \times 10^{-7}$ N A^{-2}
Permittivity of vacuum	$\varepsilon_0 = 1/(\mu_0 c^2)$	$8.854187817 \times 10^{-12}$ F m^{-1}
Universal constant of gravitation	G	$6.67259(85) \times 10^{-11}$ m^3 kg^{-1} s^{-2} $6.67259(85) \times 10^{-8}$ dyn cm^2 g^{-2}
Planck constant	h	$6.6260755(40) \times 10^{-34}$ J s $6.6260755(40) \times 10^{-27}$ erg s
Physicochemical Constants		
Avogadro constant	N_A	$6.0221367(36) \times 10^{23}$ mol^{-1}
Unified atomic mass unit (m^{12}C/12)	u	$1.6605402(10) \times 10^{-27}$ kg $931.49432(28) \times c^{-2}$ MeV
Faraday constant	F	$96485.309(29)$ C mol^{-1}
Molar gas constant	R	$8.314510(70)$ J mol^{-1} K^{-1}
Boltzmann constant	$k = R/N_A$	$1.380658(12) \times 10^{-23}$ J K^{-1} $1.380658(12) \times 10^{-6}$ erg K^{-1}
Molar volume (ideal gas) at P = 101325 Pa		
and T = 273.15 K	V_m	$2.241410(19) \times 10^{-2}$ m^3 mol^{-1}
and T = 298.15 K	V_m	2.445294×10^{-2} m^3 mol^{-1}
Loschmidt constant (273.15 K, 101325Pa)	$n_0 = N_A/V_m$	$2.686763(23) \times 10^{25}$ m^{-3}
Energy in electron volt	eV	$1.60217733(49) \times 10^{-19}$ J
Radiation Constants		
Stefan-Boltzmann constant	σ	$5.67051(19) \times 10^{-8}$ W m^{-2} K^{-4} $5.67051(19) \times 10^{-5}$ erg cm^{-2} s^{-1} K^{-4}
Radiation density constant ($8\pi^5 k^4/15c^3 h^3$)	$a = 4\sigma/c$	$7.56591(19) \times 10^{-16}$ J m^{-3} K^{-4} 7.56591×10^{-15} erg cm^{-3} K^{-4}
First radiation constant (emittance)	$c_1 = 2\pi h c^2$	$3.7417749(22) \times 10^{-16}$ W m^2
Second radiation constant	$c_2 = hc/k$	$0.01438769(12)$ m K
Wien displacement law constant	$b = \lambda_{max} T$	$2.897756(24) \times 10^{-3}$ m K

continued

Table 1.5 *(continued)*

Name	Symbol	Value & Unit
Electromagnetic Constants		
Coulomb law constant	k_C	8.9875518×10^9 N m^{2}C^{-2}
Elementary charge	e	$1.60217733(49) \times 10^{-19}$ C
		4.803206×10^{-10} esu
Magnetic flux quantum ($h/(2e)$)	Φ_0	$2.06783461(61) \times 10^{-15}$ Wb
Quantized Hall conductance	e^2/h	$3.87404614(17) \times 10^{-5}$ S
Quantized Hall resistance	R_H	$25812.8056(12)$ Ω
Bohr magneton	μ_B	$9.2740154(31) \times 10^{-24}$ J T^{-1}
Nuclear magneton	μ_N	$5.0507866(17) \times 10^{-27}$ J T^{-1}
Atomic & Particle Constants		
Rydberg constant	R_∞	$10973731.534(13)$ m^{-1}
Fine-structure constant	α	$7.29735308(33) \times 10^{-3}$
Electron rest mass	m_e	$9.1093897(54) \times 10^{-31}$ kg
		$5.48579903(13) \times 10^{-4}$ u
Electron molar mass	$M(e)$	$5.48579903(13) \times 10^{-7}$ kg mol^{-1}
Electron specific charge	$-e/m_e$	$1.75881962(53) \times 10^{11}$ C kg^{-1}
Electron classical radius	r_e	$2.81794092(38) \times 10^{-15}$ m
Proton rest mass	m_p	$1.6726231(10) \times 10^{-27}$ kg
		$1.007276470(12)$ u
Proton molar mass	$M(P)$	$1.007276470(12) \times 10^{-3}$ kg mol^{-1}
Proton specific charge	e/m_p	$9.5788309(29) \times 10^7$ C kg^{-1}
Neutron rest mass	m_n	$1.6749286(10) \times 10^{-27}$ kg
		$1.008664904(14)$ u
Neutron molar mass	$M(n)$	$1.008664904(14) \times 10^{-3}$ kg mol^{-1}
Deuteron mass	m_d	$3.3435860(20) \times 10^{-27}$ kg
		$2.013553214(24)$ u
Deuteron molar mass	$M(d)$	$2.013553214(24) \times 10^{-3}$ kg mol^{-1}
Astronomical Constants		
Julian day	d	24 h $= 86400$ s
Julian year	yr	365.25 d $= 31557600$ s
Julian century	Cy	36525 d

continued

Table 1.5 *(continued)*

Name	Symbol	Value & Unit
Sidereal second	s	0.9972696 s
Mean sidereal day	d	$23^h56^m04^s.09054 = 86164.09054$ s
Sidereal year (referred to fixed stars)	yr	365.25636 d = 31558149.5 s
Tropical year (equinox to equinox)	yr	365.2421897 d = 31556925.2 s
Anomalistic year (perihelion to perihelion)	yr	365.25964 d
Gregorian calendar year	yr	365.2425 d
Julian year	yr	365.2500 d
Astronomical unit	AU	$1.4959787061 \times 10^{11}$ m
Lightyear	lyr	9.460530×10^{15} m 63239.74 AU
Parsec	pc	3.085678×10^{16} m 3.261633 lyr
Megaparsec	Mpc	3.085678×10^{22} m
Light time for 1 AU		499.0047835s
Earth mass	$M_\oplus$	5.9736×10^{24} kg
Mean Earth radius	$R_\oplus$	6371.01 km
Solar constant (at 1 AU)	S	1367.6 W m^{-2}
Solar mass	$M_\odot$	1.98910×10^{30} kg
Solar radius	$R_\odot$	695950 km
Solar effective temperature	$T_{eff\,\odot}$	5778 K
Solar absolute luminosity	$L_\odot$	3.8268×10^{26} W
Solar absolute bolometric magnitude	$M_{bol\,\odot}$	4.75
Conversions		
plane angle degree	°	$1° = (\pi/180)$ rad
plane angle minute	'	$1' = (1/60)° = (\pi/10800)$ rad
plane angle second	"	$1" = (1/60)' = (\pi/648000)$ rad
	π	3.14159265...
	ln x	$2.3026 \log_{10} x$

Note: IUPAC 1986 recommended values. Digits in parentheses indicate the standard deviation uncertainty in the last digits of the given value.

Source: Cohen, E. R., & Taylor, B. N., 1987, The 1986 adjustment of the fundamental physical constants, *Rev. Modern Phys.* 59, 1121–1148.

1.1 Conversion Factors

Table 1.6 Length

Length	mm	cm	in	m	yd	feet
1 mm =	1	0.1	0.03937	10^{-3}	1.094×10^{-3}	3.281×10^{-3}
1 cm =	10	1	0.3937	0.01	0.01094	0.03281
1 in =	25.4	2.54	1	0.0254	0.02778	0.08333
1 m =	1000	100	39.37	1	1.0936	3.281
1 yd =	914.4	91.44	36.0	0.9144	1	3
1 ft =	304.8	30.48	12	0.3048	0.3333	1
1 km =	10^6	10^5	3.937×10^4	1000	1093.6	3280.8
1 mi =	1.609×10^6	1.609×10^5	6.336×10^4	1609.344	1760	5280
1 AU =	1.496×10^{14}	1.496×10^{13}	5.890×10^{12}	1.496×10^{11}	1.636×10^{11}	4.908×10^{11}
1 lyr =	9.461×10^{18}	9.461×10^{17}	3.725×10^{17}	9.461×10^{15}	1.035×10^{16}	3.105×10^{16}
1 pc =	3.086×10^{19}	3.086×10^{18}	1.215×10^{18}	3.086×10^{16}	3.375×10^{16}	1.013×10^{17}

	km	mi	AU	lyr	pc
1 mm =	10^{-6}	6.215×10^{-7}	6.686×10^{-15}	1.057×10^{-19}	3.241×10^{-20}
1 cm =	10^{-5}	6.215×10^{-6}	6.686×10^{-14}	1.057×10^{-18}	3.241×10^{-19}
1 in =	2.54×10^{-5}	1.578×10^{-5}	1.698×10^{-13}	2.685×10^{-18}	8.232×10^{-19}
1 m =	10^{-3}	6.214×10^{-4}	6.686×10^{-12}	1.057×10^{-16}	3.241×10^{-17}
1 yd =	9.144×10^{-4}	5.682×10^{-4}	6.112×10^{-12}	9.663×10^{-17}	2.963×10^{-17}
1 ft =	3.048×10^{-4}	1.894×10^{-4}	2.037×10^{-12}	3.221×10^{-17}	9.876×10^{-18}
1 km =	1	0.6214	6.686×10^{-9}	1.056×10^{-13}	3.238×10^{-14}
1 mi =	1.609344	1	1.076×10^{-8}	1.701×10^{-13}	5.214×10^{-14}
1 AU =	1.496×10^8	9.296×10^7	1	1.581×10^{-5}	4.848×10^{-6}
1 lyr =	9.461×10^{12}	5.879×10^{12}	6.325×10^4	1	0.3066
1 pc =	3.086×10^{13}	1.918×10^{13}	2.0628×10^5	3.2616	1

1 statute mile = 1.609344 km = 5280 feet
1 nautical mile = 1.8531 km = 6080 feet

1 Ångström (Å) = 10^{-8} cm = 10^{-10} m
1 fermi (f) = 10^{-13} cm = 10^{-15} m

Table 1.7 Area

	cm^2	in^2	ft^2	m^2	ha	km^2
1 cm^2 =	1	0.1550	1.076×10^{-3}	1×10^{-4}	1×10^{-8}	1×10^{-10}
1 in^2 =	6.452	1	6.944×10^{-3}	6.452×10^{-4}	6.452×10^{-8}	6.452×10^{-10}
ft^2 =	929.0	144	1	0.09290	9.290×10^{-6}	9.290×10^{-8}
1m^2 =	1×10^4	1550	10.76	1	1×10^{-4}	1×10^{-6}
1 ha =	1×10^8	1.500×10^7	1.076×10^5	1×10^4	1	0.01
1 km^2 =	1×10^{10}	1.550×10^9	1.076×10^7	1×10^6	100	1

1 acre = 43560 ft^2
1 statute mile^2 = 2.5900 km^2
1 barn = 10^{-28} m^2

Table 1.8 Volume

	cm^3	in^3	l = dm^3	ft^3	m^3	km^3
1 cm^3 =	1	0.06102	0.001	3.531×10^{-5}	1×10^{-6}	1×10^{-15}
1 in^3 =	16.387	1	0.01639	5.787×10^{-4}	1.639×10^{-5}	1.639×10^{-14}
1 l = 1 dm^3 =	1000	61.02	1	0.03531	0.001	1×10^{-2}
ft^3 =	2.832×10^4	1728	28.32	1	0.02832	2.832×10^{-11}
1m^3 =	1×10^6	6.102×10^4	1000	35.31	1	1×10^{-9}
1 km^3 =	1×10^{15}	6.102×10^{13}	1×10^{12}	3.531×10^{10}	1×10^9	1

1 liter = volume of 1 kg water at its maximum density (T = 4°C).

Table 1.9 Pressure

Pressure	Pa	bar	atm	mm Hg (Torr)	dyn cm^{-2}	psi (lb in^{-2})
1 Pa =	1	10^{-5}	9.869×10^{-6}	7.501×10^{-3}	10	1.4504×10^{-4}
1 bar =	10^{5}	1	0.9869	750.1	10^{6}	14.504
1 atm =	1.0133×10^{5}	1.0133	1	760.0	1.013×10^{6}	14.6959
1 mm Hg =	133.3	1.333×10^{-3}	1.316×10^{-3}	1	1333	0.01934
1 dyn cm^{-2} =	0.1	10^{-6}	9.869×10^{-7}	7.501×10^{-4}	1	1.4504×10^{-5}
1 psi (lb in^{-2}) =	6.8948×10^{3}	6.8948×10^{-2}	0.06805	51.7151	6.8948×10^{4}	1

1 cm amagat = 2.69×10^{19} molecules cm^{-3}

Table 1.10 Energy

Energy	J	cal (g)	erg	eV
1 J =	1	0.23901	10^{7}	6.242×10^{18}
1 cal =	4.184	1	4.184×10^{7}	2.612×10^{19}
1 erg =	10^{-7}	2.39006×10^{-8}	1	6.242×10^{11}
1 eV =	1.602×10^{-19}	3.829×10^{-20}	1.602×10^{-12}	1
1 BTU =	1054.35	251.99576	1.05435×10^{10}	6.581×10^{21}
1 watt-hr =	3600	860.421	3.60×10^{10}	2.247×10^{22}
1 Ton TNT =	4.2×10^{9}	1.00×10^{9}	4.2×10^{16}	2.62×10^{28}

	BTU	watt-hr	1 Ton TNT
1 J =	9.4845×10^{-4}	2.778×10^{-4}	2.38×10^{-10}
1 cal =	3.9683×10^{-3}	1.1622×10^{-3}	9.96×10^{-10}
1 erg =	9.4845×10^{-11}	2.778×10^{-11}	2.38×10^{-17}
1 eV =	1.519×10^{-22}	4.450×10^{-23}	3.81×10^{-29}
1 BTU =	1	0.29288	2.51×10^{-7}
1 watt-hr =	3.4144	1	8.57×10^{-7}
1 Ton TNT =	3.98×10^{6}	1.17×10^{6}	1

Photon energy associated with wavelength λ: hc/λ = 1.98648×10^{-23} J (λ in cm).

Temperature Conversions

$$K = 273.15 + {}^\circ C = (5 \times {}^\circ F/9) + 255.22$$

$${}^\circ C = ({}^\circ F - 32) \times 5/9 = K - 273.15$$

$${}^\circ F = (9 \times K/5) - 459.4 = (9 \times {}^\circ C/5) + 32$$

Temperature associated with 1 eV = 11604.8 K

Table 1.11 Time

Time	s	min	h	day	year
1 s =	1	0.01667	2.778×10^{-4}	1.157×10^{-5}	3.169×10^{-8}
1 min =	60	1	0.01667	6.944×10^{-4}	1.901×10^{-6}
1 h =	3.6×10^3	60	1	0.04167	1.141×10^{-4}
1 day =	8.44×10^3	1440	24	1	2.738×10^{-3}
1 year =	3.156×10^7	5.2597×10^5	8766	365.26	1

Table 1.12 Concentration (by Mass)

Concentration	g/g	mass%	g/kg	mg/kg	ppm	µg/kg	ppb	ng/kg	ppt
1 mass% =	10^{-2}	1	10	10^4	10^4	10^7	10^7	10^{10}	10^{10}
1 ppm =	10^{-6}	10^{-4}	10^{-3}	1	1	10^3	10^3	10^6	10^6
1 ppb =	10^{-9}	10^{-7}	10^{-6}	10^{-3}	10^{-3}	1	1	10^3	10^3
1 ppt =	10^{-12}	10^{-10}	10^{-9}	10^{-6}	10^{-6}	10^{-3}	10^{-3}	1	1

1 ppm = 1 µg/g; 1 ppb = 1 ng/g; 1 mg/g = 1 g/kg.

Density Conversions

$$1 \text{ kg m}^{-3} = 10^{-3} \text{ g cm}^{-3}$$

1.2 Mathematical Formulae

Solution of Quadratic Equations

$$x^2 + ax + b = 0$$

$$x = -\frac{1}{2}a \pm \sqrt{\left(\frac{a}{2}\right)^2 - b}$$

Solution of Cubic Equations

$$x^3 + ax^2 + bx + c = 0$$

Set $p = b - \frac{1}{3}a^2$ $q = 2\left(\frac{a}{3}\right)^3 - \frac{1}{3}ab + c$

$$y = \left(-\frac{q}{2} \pm \sqrt{\left(\frac{p}{3}\right)^3 + \left(\frac{q}{2}\right)^2}\right)^{\frac{1}{3}} + \left(-\frac{q}{2} \mp \sqrt{\left(\frac{p}{3}\right)^3 + \left(\frac{q}{2}\right)^2}\right)^{\frac{1}{3}}$$

then $x = y - \frac{1}{3}a.$

If $\left(\frac{p}{3}\right)^3 + \left(\frac{q}{2}\right)^2 < 0;$

set $\sin 3\varepsilon = 4q/(-\frac{4}{3}p)^{\frac{3}{2}}$ *and* $r = \left(-\frac{4}{3}p\right)^{\frac{1}{2}}$

so that $y_1 = r\sin\varepsilon$
$y_2 = r\sin(60^\circ - \varepsilon)$
$y_3 = -r\sin(60^\circ + \varepsilon).$

Some Statistical Formulae

x_i	value of random observation	$\bar{x}$	sample mean
σ	standard deviation	σ^2	variance
N	number of observations		

Sample mean $\bar{x} = \frac{1}{N}\Sigma x_i$

Sample variance $\sigma^2 = \frac{1}{N-1}\Sigma(x_i - \bar{x})^2$

Standard deviation (mean deviation of the observations from the sample average)

$$\sigma = \sqrt{\frac{1}{N-1}\Sigma(x_i - \bar{x})^2}$$

Standard deviation of the mean

$$\bar{\sigma} = \sqrt{\frac{1}{N(N-1)}\Sigma(x_i - \bar{x})^2} = \frac{\sigma}{\sqrt{N}}$$

1σ includes 68.3% of all observations
2σ includes 95.4% of all observations
3σ includes 99.7% of all observations

If N is small, it may be necessary to apply a factor f so that $\bar{\sigma}' = \bar{\sigma} \times f$.

Correction values f for the standard deviation of the mean with low numbers of observations

Number of observations N	3	4	5	6	8	10	20
Confidence interval 68.3%	1.32	1.20	1.15	1.11	1.08	1.06	1.03
Confidence interval 99.7%	19.2	9.2	6.6	5.5	4.5	4.1	3.4

Small number statistics

$\sigma = k \times R$, where N is number of observations and R range of values

N	2	3	4	5	6	7	8	9	10
k	0.886	0.591	0.486	0.430	0.395	0.370	0.351	0.337	0.325

Source: Wilson, E. B., Jr., 1952, *An introduction to scientific research*, McGraw Hill, New York, pp. 373.

Error Propagation

For $x = f(u,v,...)$, the most probable value for x is $\bar{x} = f(\bar{u}, \bar{v}, ...)$.

The variance in x is $\sigma_x^2 = \sigma_u^2 \left(\frac{\partial x}{\partial u}\right)^2 + \sigma_v^2 \left(\frac{\partial x}{\partial v}\right)^2 + 2\sigma_{uv}^2 \left(\frac{\partial x}{\partial u}\right)\left(\frac{\partial x}{\partial v}\right) + \cdots$

where the covariances are

$$\sigma_u^2 = \lim_{N\to\infty} \frac{1}{N} \Sigma (u_i - \bar{u})^2; \quad \sigma_v^2 = \lim_{N\to\infty} \frac{1}{N} \Sigma (v_i - \bar{v})^2$$

$$\sigma_{uv}^2 = \lim_{N\to\infty} \frac{1}{N} \Sigma ((u_i - \bar{u})(v_i - \bar{v}))$$

If u and v are not correlated, $\sigma_{uv}^2 = 0$.

Examples for simple functions

A, B, C   mean values
a, b, c   standard deviations of the mean
$E(A) = A \pm a$, $E(B) = B \pm b$, $E(C) = C \pm c$

$E(A+B) = A+B \pm \sqrt{a^2 + b^2}$

$E(A+B+C) = A + B + C \pm \sqrt{a^2 + b^2 + c^2}$

$E(A-B) = A-B \pm \sqrt{a^2 + b^2}$

$E(AB) = AB \pm \sqrt{B^2 a^2 + A^2 b^2}$

$E(ABC) = ABC \pm \sqrt{B^2 C^2 a^2 + A^2 C^2 b^2 + A^2 B^2 c^2}$

$E(A/B) = A/B \pm (\sqrt{B^2 a^2 + A^2 b^2})/B^2$

$E(A^u B^v C^w) = A^u B^v C^w \sqrt{\left(u\frac{a}{A}\right)^2 + \left(v\frac{b}{B}\right)^2 + \left(w\frac{c}{C}\right)^2}$

Some Simple Geometric Formulae

Circumference of circle	$C_C = 2\pi r$
Area of circle	$A_C = \pi r^2$
Surface area of sphere	$A_{sph} = 4\pi r^2$
Volume of sphere	$V_{sph} = 4/3\ \pi r^3$

Coordinate Transformations

Cartesian coordinates	x, y, z
Spherical coordinates	r, θ, ϕ
Cylindrical polar coordinates	ρ, ϕ, z

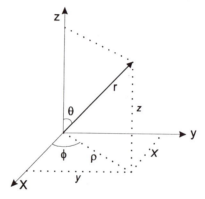

Figure 1.1 Relationships among different coordinate systems

$x = r \sin \theta \cos \phi = \rho \cos \phi$

$y = r \sin \theta \sin \phi = \rho \sin \phi$

$z = r \cos \theta = z$

$r = (x^2 + y^2 + z^2)^{1/2} = (\rho^2 + z^2)^{1/2}$

$\phi = \arctan (y/x)$

$\theta = \arccos [z/(x^2 + y^2 + z^2)^{1/2}] = \arctan (\rho/z)$ *for* $0 \leq \theta \leq \pi$

$\rho = (x^2 + y^2)^{1/2} = r \sin \theta$ *for* $0 \leq \phi < 2\pi$

The Conic Functions

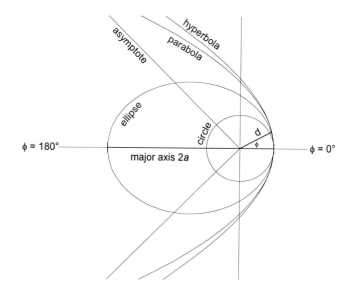

Figure 1.2 The conic functions

General form $\rho = d\,(1 + e)/(1 + e \cos \phi)$

ρ, ϕ planar polar coordinates

e linear eccentricity

d closest distance to origin ($\rho = $ d for $\phi = 0°$)

Special cases

Circle $e = 0$
              circle with radius d

Ellipse $0 < e < 1$,
              $d = a(1 - e)$ where $2a$ is the major (longest)
              axis of the ellipse

Parabola $e = 1$

Hyperbola $e > 1$
              asymptotes ($\rho \rightarrow \infty$) for $\cos \phi = -1/e$

The Celestial Sphere

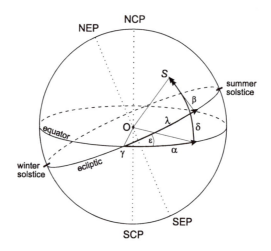

Figure 1.3 The celestial sphere

Ecliptic   apparent path of the sun across the celestial sphere during the course of a year

Equator   celestial equator given by intersection of celestial sphere and projection of Earth's equator

NCP	north celestial pole	NEP	north pole of ecliptic
SCP	south celestial pole	SEP	south pole of ecliptic

O	observer's position	S	observed object
α	right ascension	δ	declination
λ	celestial longitude	β	celestial latitude

ε obliquity of ecliptic to celestial equator (~23.44° for epoch J1997.5)

γ	vernal equinox	$\alpha = 0^h$	$\delta = 0°$	(March 21)
	summer solstice	$\alpha = 6^h$	$\delta = +23.4°$	(June 21)
	autumnal equinox	$\alpha = 12^h$	$\delta = 0°$	(September 21)
	winter solstice	$\alpha = 18^h$	$\delta = -23.4°$	(December 21)

Astronomical Coordinate System Transformations

Horizontal and Celestial (Equatorial) Systems

a   altitude
A   azimuth angle, from north toward east
α   right ascension
δ   declination
h   local hour angle, h = local sidereal time – α
ρ   observer's latitude

$\cos a \sin A = -\cos \delta \sin h$
$\cos a \cos A = \sin \delta \sin \rho - \cos \delta \cos h \sin \rho$
$\sin a = \sin \delta \sin \rho + \cos \delta \cos h \cos \rho$
$\cos \delta \cos h = \sin a \cos \rho - \cos a \cos A \sin \rho$
$\sin \delta = \sin a \sin \rho + \cos a \cos A \cos \rho$

Ecliptic and Celestial (Equatorial) Systems

λ   celestial longitude   β   celestial latitude
ε   obliquity of ecliptic to celestial equator (~23.44° for J1997.5)
     $\varepsilon = 23°26'21.45" - 46.815" \, T - 0.0006" \, T^2 + 0.00181" \, T^3$
     where T = (t–2000.0)/100 = (JD–2451545.0)/36525 (JD = Julian date)

$\cos \delta \cos \alpha = \cos \beta \cos \lambda$
$\cos \delta \sin \alpha = \cos \beta \sin \lambda \cos \varepsilon - \sin \beta \sin \varepsilon$
$\sin \delta = \cos \beta \sin \lambda \sin \varepsilon + \sin \beta \cos \varepsilon$
$\cos \beta \sin \lambda = \cos \delta \sin \alpha \cos \varepsilon + \sin \delta \sin \varepsilon$
$\sin \beta = \sin \delta \cos \varepsilon - \cos \delta \sin \alpha \sin \varepsilon$

Galactic and Celestial (Equatorial) Systems

l^{II} new galactic longitude (equinox 1950.0)
b^{II} new galactic latitude (equinox 1950.0)

$\cos b^{II} \cos (l^{II} - 33°) = \cos \delta \cos (\alpha - 282.25°)$
$\cos b^{II} \sin (l^{II} - 33°) = \cos \delta \sin (\alpha - 282.25°) \cos 62.6° + \sin \delta \sin 62.6°$
$\sin b^{II} = \sin \delta \cos 62.6° - \cos \delta \sin (\alpha - 282.25°) \sin 62.6°$
$\cos \delta \sin (\alpha - 282.25°) = \cos b^{II} \sin (l^{II} - 33°) \cos 62.6° - \sin b^{II} \sin 62.6°$
$\sin \delta = \cos b^{II} \sin (l^{II} - 33°) \sin 62.6° + \sin b^{II} \cos 62.6°$

| Center of galaxy $l^{II} = 0$, $b^{II} = 0$ | $\alpha = 17^h42^m29.3^s$ | $\delta = -28°59'18"$ |
| North galactic pole $b^{II} = +90°$ | $\alpha = 12^h49^m = 282.25°$ | $\delta = +27°24'$ |

Detailed information about reduction of celestial coordinates is described in
Astronomical Almanac, 1997, U.S. Printing Office, Washington, D.C.

Some Formulae Related to Orbital Descriptions

a   semimajor axis of orbit $a = \left[\dfrac{G(M_* + m_P)P^2}{4\pi^2} \right]^{1/3}$

e   (linear) eccentricity

G   Newtonian constant of gravitation ($G = 6.67259 \times 10^{-11}$ m³kg⁻¹s⁻¹)

i   inclination of orbital plane to ecliptic

L   total specific angular momentum

$$L = \sqrt{a(1 - e^2)\mu}$$

L_z component of angular momentum perpendicular to ecliptic plane

$$L_z = \sqrt{a(1 - e^2)\mu} \; \cos i$$

M_* mass of primary object

m_P mass of revolving object (planet, asteroid, comet)

μ $\mu = GM_*$

P   orbital period. For a Keplerian orbit, the orbital period of a revolving object is:

$$P^2 = \frac{4\pi^2}{G(M_* + m_P)} a^3 \quad \text{and } P = a^{3/2} \text{ (in years for a in AU)}$$

q   perihelion or periaston, closest point of planetary orbit to primary star:

$$q = a(1 - e)$$

Q   aphelion or apastron, most distant point of orbit from primary star:

$$Q = a(1 + e)$$

r   distance between M_* and m_P

v   orbital velocity $v = \sqrt{\dfrac{G(M_* + m_P)}{\frac{2}{r} - \frac{1}{a}}}$

v at perihelion $v = \dfrac{2\pi a}{P} \sqrt{\dfrac{1 + e}{1 - e}}$ v at aphelion $v = \dfrac{2\pi a}{P} \sqrt{\dfrac{1 - e}{1 + e}}$

f(m)   mass function $f(m) = \dfrac{(m_P \sin i)^3}{(M_* + m_P)^2} = \dfrac{P}{2\pi G} v^3_{radial,*}$

Formulae Useful for Atmospheric Modeling

Explanation of Symbols

g_P planetary acceleration (GM_P/R_P^2)

γ adiabatic coefficient (C_P/C_V)

k Boltzmann constant

M_P planetary mass

n mole

N_A Avogadro constant

P_i partial pressure of species i

R_P planetary radius

T absolute temperature

V volume

Y_i mass fraction of species i

G gravitational constant

h height above surface

m particle mass

μ_i molecular weight of species i

N number of particles

P total pressure

R molar gas constant

ρ mass density

v speed, velocity

X_i mole fraction of species i

z height above planetary surface

Mean molecular weight	$\bar{\mu} = \Sigma X_i \mu_i$
Column density	$\sigma = \rho H = P/g_P$
Mass fraction	$Y_i = \rho \frac{\mu_i}{\bar{\mu}} X_i$
Number density	$N/V = \rho N_A/\mu = P_i N_A/RT$
Ideal gas law	$PV = nRT$
Maxwellian velocity distribution	$(v) = 4\pi N(m/2\pi kT)^{3/2} v^2 \exp(-mv^2/2kT)$
Root mean square velocity	$v_{rms} = (3kT/m)^{0.5}$
Most probable thermal velocity	$v_{th} = (2kT/m)^{0.5}$
Escape velocity	$v_{esc} = (2GM_P/R_P)^{0.5}$
Hydrostatic pressure	$P = \rho g_P h$
Pressure scale height	$H = k T R_p^2/(mGM_p) = RT/(\mu g_p)$
Barometric equation	$P = P_o \exp(-z/H)$
	$P/P_o = (T/T_o)^{-\beta}$ $\rho/\rho_o = (T/T_o)^{-(1+\beta)}$
	where $\beta = g_P \bar{\mu}/(RdT/dz)$
Adiabatic sound velocity	$v_S = (\gamma P/\rho)^{0.5}$

1.3 Elemental Data

Table 1.13 Periodic Table of the Elements

1	2	3	4	5	6	7	8	9	10	11	12	13	14	15	16	17	18
1 H 1.008																	2 He 4.003
3 Li 6.941	4 Be 9.012											5 B 10.811	6 C 12.011	7 N 14.007	8 O 15.999	9 F 18.998	10 Ne 20.180
11 Na 22.990	12 Mg 24.305											13 Al 26.982	14 Si 28.086	15 P 30.974	16 S 32.066	17 Cl 35.453	18 Ar 39.948
19 K 39.098	20 Ca 40.078	21 Sc 44.956	22 Ti 47.867	23 V 50.942	24 Cr 51.996	25 Mn 54.938	26 Fe 55.845	27 Co 58.933	28 Ni 58.693	29 Cu 63.546	30 Zn 65.39	31 Ga 69.723	32 Ge 72.61	33 As 74.922	34 Se 78.96	35 Br 79.904	36 Kr 83.80
37 Rb 85.468	38 Sr 87.62	39 Y 88.906	40 Zr 91.224	41 Nb 92.906	42 Mo 95.94	43 Tc (97.907)	44 Ru 101.07	45 Rh 102.906	46 Pd 106.42	47 Ag 107.868	48 Cd 112.411	49 In 114.818	50 Sn 118.710	51 Sb 121.760	52 Te 127.60	53 I 126.904	54 Xe 131.29
55 Cs 132.905	56 Ba 137.327	57 La 138.906	72 Hf 178.49	73 Ta 180.948	74 W 183.84	75 Re 186.207	76 Os 190.23	77 Ir 192.217	78 Pt 195.078	79 Au 196.967	80 Hg 200.59	81 Tl 204.383	82 Pb 207.2	83 Bi 208.980	84 Po (208.982)	85 At (209.987)	86 Rn (222.018)
87 Fr (223.020)	88 Ra (226.025)	89 Ac (227.028)	104 104 (261)	105 105 (262)	106 106 (263)	107 107 (262)	108 108 (265)	109 109 (266)	110 110 (269)	111 111 (272)	112 112 (277)						

58 Ce 140.116	59 Pr 140.908	60 Nd 144.24	61 Pm (144.913)	62 Sm 150.36	63 Eu 151.964	64 Gd 157.25	65 Tb 158.925	66 Dy 162.50	67 Ho 164.930	68 Er 167.26	69 Tm 168.934	70 Yb 173.04	71 Lu 174.967
90 Th 232.038	91 Pa (231.036)	92 U 238.029	93 Np (237.048)	94 Pu (244.064)	95 Am (243.061)	96 Cm (247.070)	97 Bk (247.070)	98 Cf (251.080)	99 Es (252.083)	100 Fm (257.095)	101 Md (258.098)	102 No (259.101)	103 Lr (262.110)

Table 1.14 Atomic Weights and Isotopic Compositions of the Elements

Z	Symbol	Element	Atomic Weight [a]	A	Nucleosyn. Process [b]	Isotopic Composition (at%) Terrestrial [c]	Solar [c]
1	H	Hydrogen	1.00794	1		99.985 (water)	99.9966
				2	U	0.015 (water)	0.0034
2	He	Helium	4.002602	3	U,h?	0.000137 (air)	0.0142
				4	U,h	99.999863 (air)	99.9858
3	Li	Lithium	6.941	6	X	7.5	...
				7	U,x,h	92.5	...
4	Be	Beryllium	9.012182	9	X	100	...
5	B	Boron	10.811	10	X	19.9	...
				11	X	80.1	...
6	C	Carbon	12.0107	12	He	98.90	...
				13	H,N	1.10	...
7	N	Nitrogen	14.00674	14	H	99.634 (air)	...
				15	H,N	0.366 (air)	...
8	O	Oxygen	15.9994	16	He	99.762 (water)	...
				17	H,N	0.038 (water)	...
				18	He,N	0.200 (water)	...
9	F	Fluorine	18.9984032	19	N	100	...
10	Ne	Neon	20.1797	20	C	90.48 (air)	92.99
				21	C,Ex	0.27 (air)	0.226
				22	He,N	9.25 (air)	6.79
11	Na	Sodium	22.989770	23	C,Ne,Ex	100	...
12	Mg	Magnesium	24.3050	24	N,Ex	78.99	...
				25	Ne,Ex,C	10.00	...
				26	Ne,Ex,C	11.01	...
13	Al	Aluminum	26.981538	27	Ne,Ex	100	...
14	Si	Silicon	28.0855	28	O,Ex	92.23	...
				29	Ne,Ex	4.67	...
				30	Ne,Ex	3.10	...
15	P	Phosphorous	30.973761	31	Ne,Ex	100	...
16	S	Sulfur	32.066	32	O,Ex	95.02	...
				33	Ex	0.75	...
				34	O,Ex	4.21	...
				36	Ex,Ne,S	0.02	...
17	Cl	Chlorine	35.4527	35	Ex	75.77	...
				37	Ex,C,S	24.23	...
18	Ar	Argon	39.948	36	Ex	0.337 (air)	84.2
				38	O,Ex	0.063 (air)	15.8
				40	S,Ne	99.600 (air)	...

continued

Table 1.14 *(continued)*

Z	Symbol	Element	Atomic Weight [a]	A	Nucleosyn. Process [b]	Isotopic Composition (at%) Terrestrial [c]	Solar [c]
19	K	Potassium	39.0983	39	Ex	93.2581	93.2581
				40	S,Ex,Ne	0.0117	0.01167
				41	Ex	6.7302	6.7302
20	Ca	Calcium	40.078	40	Ex	96.941	...
				42	Ex,O	0.647	...
				43	Ex,C,S	0.135	...
				44	Ex,S	2.086	...
				46	Ex,C,Ne	0.004	...
				48	E,Ex	0.187	...
21	Sc	Scandium	44.955910	45	Ex,Ne,E	100	...
22	Ti	Titanium	47.867	46	Ex	8.0	...
				47	Ex	7.3	...
				48	Ex	73.8	...
				49	Ex	5.5	...
				50	E	5.4	...
23	V	Vanadium	50.9415	50	Ex,E	0.250	...
				51	Ex	99.750	...
24	Cr	Chromium	51.9961	50	Ex	4.345	...
				52	Ex	83.789	...
				53	Ex	9.501	...
				54	E	2.365	...
25	Mn	Manganese	54.938049	55	Ex,E	100	...
26	Fe	Iron	55.845	54	Ex	5.8	...
				56	Ex,E	91.72	...
				57	E,Ex	2.2	...
				58	He,E,C	0.28	...
27	Co	Cobalt	58.933200	59	E,C	100	...
28	Ni	Nickel	58.6934	58	E,Ex	68.077	...
				60	E	26.223	...
				61	E,Ex,C	1.140	...
				62	E,Ex,O	3.634	...
				64	Ex	0.926	...
29	Cu	Copper	63.546	63	Ex,C	69.17	...
				65	Ex	30.83	...
30	Zn	Zinc	65.39	64	Ex,E	48.6	...
				66	E	27.9	...
				67	E,S	4.1	...
				68	E,S	18.8	...

continued

Table 1.14 *(continued)*

Z	Symbol	Element	Atomic Weight[a]	A	Nucleosyn. Process[b]	Isotopic Composition (at%) Terrestrial[c]	Solar[c]
30	Zn			70	E,S	0.6	...
31	Ga	Gallium	69.723	69	S,e,r	60.108	...
				71	S,e,r	39.892	...
32	Ge	Germanium	72.61	70	S,e	21.23	...
				72	S,e,r	27.66	...
				73	e,s,r	7.73	...
				74	e,s,r	35.94	...
				76	E	7.44	...
33	As	Arsenic	74.92160	75	R,s	100	...
34	Se	Selenium	78.96	74	P	0.89	...
				76	S,p	9.36	...
				77	R,s	7.63	...
				78	R,s	23.78	...
				80	R,s	49.61	...
				82	R	8.73	...
35	Br	Bromine	79.904	79	R,s	50.69	...
				81	R,s	49.31	...
36	Kr	Krypton	83.80	78	P	0.35 (air)	0.339
				80	S,p	2.25 (air)	2.22
				82	S	11.6 (air)	11.45
				83	R,s	11.5 (air)	11.47
				84	R,S	57.0 (air)	57.11
				86	S,r	17.3 (air)	17.42
37	Rb	Rubidium	85.4678	85	R,s	72.165	...
				87	S	27.835	...
38	Sr	Strontium	87.62	84	P	0.56	...
				86	S	9.86	...
				87	S	7.00	...
				88	S,r	82.58	...
39	Y	Yttrium	88.90585	89	S	100	...
40	Zr	Zirconium	91.224	90	S	51.45	...
				91	S	11.22	...
				92	S	17.15	...
				94	S	17.38	...
				96	R	2.80	...
41	Nb	Niobium	92.90638	93	S	100	...
42	Mo	Molybdenum	95.94	92	P	14.84	...
				94	P	9.25	...

continued

Table 1.14 *(continued)*

Z	Symbol	Element	Atomic Weight [a]	A	Nucleosyn. Process [b]	Isotopic Composition (at%) Terrestrial [c]	Solar [c]
42	Mo			95	R,s	15.92	...
				96	S	16.68	...
				97	R,s	9.55	...
				98	R,S	24.13	...
				100	R,s	9.63	...
43	Tc	Technetium*	(97.9072)	98	...	...	...
44	Ru	Ruthenium	101.07	96	P	5.52	...
				98	P	1.88	...
				99	R,s	12.7	...
				100	S	12.6	...
				101	R,s	17.0	...
				102	R,S	31.6	...
				104	R	18.7	...
45	Rh	Rhodium	102.90550	103	R,s	100	...
46	Pd	Palladium	106.42	102	P	1.020	...
				104	S	11.14	...
				105	R,s	22.33	...
				106	R,S	27.33	...
				108	R,S	26.46	...
				110	R	11.72	...
47	Ag	Silver	107.8682	107	R,s	51.839	...
				109	R,s	48.161	...
48	Cd	Cadmium	112.411	106	P	1.25	...
				108	P	0.89	...
				110	S	12.49	...
				111	R,S	12.80	...
				112	S,R	24.13	...
				113	R,S	12.22	＇...
				114	S,R	28.73	...
				116	R	7.49	...
49	In	Indium	114.818	113	p,s,r	4.3	...
				115	R,S	95.7	...
50	Sn	Tin	118.710	112	P	0.97	...
				114	P,s	0.65	...
				115	p,s,r	0.34	...
				116	S,r	14.53	...
				117	R,S	7.68	...
				118	S,r	24.23	...

continued

Table 1.14 *(continued)*

Z	Symbol	Element	Atomic Weight[a]	A	Nucleosyn. Process[b]	Isotopic Composition (at%) Terrestrial[c]	Solar[c]
50	Sn			119	S,R	8.59	...
				120	S,R	32.59	...
				122	R	4.63	...
				124	R	5.79	...
51	Sb	Antimony	121.760	121	R,s	57.36	...
				123	R	42.64	...
52	Te	Tellurium	127.60	120	P	0.096	...
				122	S	2.603	...
				123	S	0.908	...
				124	S	4.816	...
				125	R,s	7.139	...
				126	R,S	18.95	...
				128	R	31.69	...
				130	R	33.80	...
53	I	Iodine	126.90447	127	R	100	...
54	Xe	Xenon	131.29	124	P	0.10 (air)	0.121
				126	P	0.09 (air)	0.108
				128	S	1.91 (air)	2.19
				129	R	26.4 (air)	27.34
				130	S	4.1 (air)	4.35
				131	R	21.2 (air)	21.69
				132	R,s	26.9 (air)	26.50
				134	R	10.4 (air)	9.76
				136	R	8.9 (air)	7.94
55	Cs	Cesium	132.90545	133	R,s	100	...
56	Ba	Barium	137.327	130	P	0.106	...
				132	P	0.101	...
				134	S	2.417	...
				135	R,s	6.592	...
				136	S	7.854	...
				137	S,r	11.23	...
				138	S	71.70	...
57	La	Lanthanum	138.9055	138	P	0.0902	...
				139	S,r	99.9098	...
58	Ce	Cerium	140.116	136	P	0.19	...
				138	P	0.25	...
				140	S,r	88.48	...
				142	R	11.08	...

continued

Table 1.14 *(continued)*

Z	Symbol	Element	Atomic Weight [a]	A	Nucleosyn. Process [b]	Isotopic Composition (at%) Terrestrial [c]	Solar [c]
59	Pr	Praseodymium	140.90765	141	R,S	100	...
60	Nd	Neodymium	144.24	142	S	27.13	...
				143	R,S	12.18	...
				144	S,R	23.80	...
				145	R,s	8.30	...
				146	R,S	17.19	...
				148	R	5.76	...
				150	R	5.64	...
61	Pm	Promethium*	(144.9127)	145	...	...	...
62	Sm	Samarium	150.36	144	P	3.1	...
				147	R,s	15.0	...
				148	S	11.3	...
				149	R,S	13.8	...
				150	S	7.4	...
				152	R,S	26.7	...
				154	R	22.7	...
63	Eu	Europium	151.964	151	R,s	47.8	...
				153	R,s	52.2	...
64	Gd	Gadolinium	157.25	152	P,s	0.20	...
				154	S	2.18	...
				155	R,s	14.8	...
				156	R,s	20.47	...
				157	R,s	15.65	...
				158	R,s	24.84	...
				160	R	21.86	...
65	Tb	Terbium	158.92534	159	R	100	...
66	Dy	Dysprosium	162.50	156	P	0.06	...
				158	P	0.10	...
				160	S	2.34	...
				161	R	18.9	...
				162	R,s	25.5	...
				163	R	24.9	...
				164	R,S	28.2	...
67	Ho	Holmium	164.93032	165	R	100	...
68	Er	Erbium	167.26	162	P	0.14	...
				164	P,S	1.61	...
				166	R,s	33.6	...
				167	R	22.95	...

continued

Table 1.14 *(continued)*

Z	Symbol	Element	Atomic Weight [a]	A	Nucleosyn. Process [b]	Isotopic Composition (at%)	
						Terrestrial [c]	Solar [c]
68	Er			168	R,S	26.8	...
				170	R	14.9	...
69	Tm	Thulium	168.93421	169	R,s	100	...
70	Yb	Ytterbium	173.04	168	P	0.127	...
				170	S	3.05	...
				171	R,s	14.3	...
				172	R,S	21.9	...
				173	R,s	16.12	...
				174	S,R	31.8	...
				176	R	12.7	...
71	Lu	Lutetium	174.967	175	R,s	97.41	...
				176	S	2.59	...
72	Hf	Hafnium	178.49	174	P	0.162	...
				176	S	5.206	...
				177	R,s	18.606	...
				178	R,S	27.297	...
				179	R,s	13.629	...
				180	S,R	35.100	...
73	Ta	Tantalum	180.9479	180	p,s,r	0.012	...
				181	R,S	99.988	...
74	W	Tungsten	183.84	180	P	0.13	...
				182	R,s	26.3	...
				183	R,s	14.3	...
				184	R,s	30.67	...
				186	R	28.6	...
75	Re	Rhenium	186.207	185	R,s	37.40	...
				187	R	62.60	...
76	Os	Osmium	190.23	184	P	0.02	...
				186	S	1.58	...
				187	S	1.6	...
				188	R,s	13.3	...
				189	R	16.1	...
				190	R	26.4	...
				192	R	41.0	...
77	Ir	Iridium	192.217	191	R	37.3	...
				193	R	62.7	...
78	Pt	Platinum	195.078	190	P	0.01	...
				192	S	0.79	...

continued

Table 1.14 *(continued)*

Z	Symbol	Element	Atomic Weight[a]	A	Nucleosyn. Process[b]	Isotopic Composition (at%) Terrestrial[c]	Solar[c]
78	Pt			194	R	32.9	...
				195	R	33.8	...
				196	R	25.3	...
				198	R	7.2	...
79	Au	Gold	196.96655	197	R	100	...
80	Hg	Mercury	200.59	196	P	0.15	...
				198	S	9.97	...
				199	R,S	16.87	...
				200	S,r	23.10	...
				201	S,r	13.18	...
				202	S,r	29.86	...
				204	R	6.87	...
81	Tl	Thallium	204.3833	203	R,S	29.524	...
				205	S,R	70.476	...
82	Pb	Lead	207.2	204	S	1.4 (variable)	1.94
				206	R,S	24.1 (variable)	19.12
				207	R,S	22.1 (variable)	20.62
				208	R,s	52.4 (variable)	58.31
83	Bi	Bismuth	208.98038	209	R,s	100	...
84	Po	Polonium*	(208.9824)	209	...	...	...
85	At	Astatine*	(209.9871)	210	...	...	...
86	Rn	Radon*	(222.0176)	222	...	...	...
87	Fr	Francium*	(223.0197)	223	...	...	...
88	Ra	Radium*	(226.0254)	226	...	...	...
89	Ac	Actinium*	(227.0277)	227	...	...	...
90	Th	Thorium*	232.0381 †	232	RA	100	...
91	Pa	Protactinium*	231.03588 †	231	...	100	...
92	U	Uranium*	238.0289 †		...	...	...
			(234.0409)	234	...	0.0055	...
			(235.0439)	235	RA	0.72	0.7200
			(238.0508)	238	RA	99.2745	99.2745
93	Np	Neptunium*	(237.0482)	237	...	...	...
94	Pu	Plutonium*	(244.0642)	244	...	...	...
95	Am	Americium*	(243.0614)	243	...	...	...
96	Cm	Curium*	(247.0703)	247	...	...	...
97	Bk	Berkelium*	(247.0703)	247	...	...	...
98	Cf	Californium*	(251.0796)	251	...	...	...
99	Es	Einsteinium*	(252.0830)	252	...	...	...

continued

Table 1.14   *(continued)*

Z	Symbol	Element	Atomic Weight [a]	A	Nucleosyn. Process [b]	Isotopic Composition (at%) Terrestrial [c]	Solar [c]
100	Fm	Fermium*	(257.0951)	257	...	...	...
101	Md	Mendelevium*	(258.10)	258	...	...	...
102	No	Nobelium*	(259.1009)	259	...	...	...
103	Lr	Lawrencium*	(262.11)	262	...	...	...
104	NN‡	*	(261.11)	261	...	...	...
105	NN‡	*	(262.114)	262	...	...	...
106	NN‡	*	(263.118)	263	...	...	...
107	NN‡	*	(262.12)	262	...	...	...
108	NN‡	*	...	265	...	...	...
109	NN‡	*	...	266	...	...	...
110	NN‡	*	...	269	...	...	...
111	NN‡	*	...	272	...	...	...
112	NN‡	*	...	277	...	...	...

Sources: IUPAC recommended atomic weights of the elements 1993, Heumann, K. G. (editor in chief), *Pure & Appl. Chem.* 66, 2423–2444, © 1994 IUPAC. IUPAC recommended atomic weights of the elements 1995, Coplen, T. B., *Pure & Appl. Chem.* 68, 2339–2359, © 1996 IUPAC. Recommended isotopic compositions of the elements 1989, deLaeter, J. R. (editor in chief), *Pure & Appl. Chem.* 63, 991–1002, © 1991 IUPAC. Solar isotopic compositions and assignments to nucleosynthetic processes are from Anders, E., & Grevesse, N., 1989, *Geochim. Cosmochim. Acta* 53, 197–214, and references therein.

[a] Atomic weights are scaled to $A(^{12}C) = 12$ and are listed for materials with terrestrial isotopic composition. Values in parenthesis are relative atomic weights for unstable radionuclides.

[b] Nuclear process believed to be responsible for natural production of nuclide. For definition of entries, see following table.

[c] Isotopic compositions are listed for terrestrial and solar matter. Only a few direct determinations are available for solar isotopic compositions and for elements other than those listed, the terrestrial isotopic composition is assumed to be representative for solar values.

* Element has no stable isotopes.

† Th, Pa, and U have characteristic terrestrial isotopic compositions and for these an atomic weight is listed. Relative atomic masses for the nuclides are also indicated.

NN‡ The names of elements 104 to 109 are not yet agreed on. As of February 1997, the suggested IUPAC names are: 104 Rutherfordium, Rf; 105 Dubnium, Db; 106 Seaborgium, Sg; 107 Bohrium, Bh; 108 Hassium, Hs; and 109 Meitnerium, Mt.

Assignments to nucleosysnthetic processes

C:	Carbon burning	P:	p-process
E:	Nuclear statistical equilibrium	R:	r-process
Ex:	Explosive nucleosynthesis	r:	r-process contribution 10–30%
H:	Hydrogen burning	RA:	r-process producing actinides
He:	Helium burning	S:	s-process
N:	Hot or explosive hydrogen burning	s:	s-process contribution 10–30%
Ne:	Neon burning	U:	cosmological nucleosynthesis
O:	Oxygen burning	X:	cosmic-ray spallation

Source: Anders, E., & Grevesse, N., 1989, *Geochim. Cosmochim. Acta* 53, 197–214, and references therein.

Table 1.15 Metallic, Covalent, and Ionic Radii, and Coordination Numbers (CN) of the Elements

Element	Metallic Radius Å	Metallic Radius CN	Covalent Radius Å	Covalent Radius CN	Oxidation State	Ionic Radius Å (1)	Ionic Radius Å (2)	CN
H	...	...	0.32	H_2	H^+	...	−0.38	1
Li	1.52	8 cub	1.33	...	Li^+	0.68	0.590	4
					Li^+	0.82	0.76	6
Be	1.13	12 hex	(1.00)	...	Be^{2+}	0.35	0.27	4
					Be^{2+}	...	0.45	6
B	0.89	12 rhbdr	0.83	...	B^{3+}	0.20	0.11	4
C	...	...	0.77	4 dia	C^{4+}	...	0.15	4
			0.71	6 gr	C^{4+}	...	0.16	6
N	...	...	0.74	N_2	N^{3+}	...	0.16	6
					N^{5+}	...	0.13	6
					N^{3-}	...	1.46	4
O	...	...	0.74	H_2O_2	O^{2-}	1.28	1.36	3
					O^{2-}	1.30	1.38	4
					O^{2-}	= 1.32	= 1.40	6
					O^{2-}	1.34	1.42	8
F	...	...	0.72	F_2	F^-	1.23	1.31	4
					F^-	= 1.25	1.33	6
Na	1.85	8 cub	1.54	...	Na^+	1.10	1.02	6
					Na^+	1.24	1.18	8
					Na^+	1.40	1.24	9
Mg	1.60	12 hex	(1.38)	...	Mg^{2+}	0.80	0.720	6
Al	1.43	12 cub	1.26	...	Al^{3+}	0.47	0.39	4
					Al^{3+}	0.61	0.535	6
Si	...	...	1.17	dia	Si^{4+}	0.34	0.26	4
					Si^{4+}	0.48	0.400	6
P	...	...	1.10	P_4	P^{5+}	0.25	0.17	4
S	...	...	1.04	H_2S_2	S^{2-}	1.72	1.84	6
					S^{6+}	0.20	0.12	4
					S^{6+}	...	0.29	6
Cl	...	...	0.99	Cl_2	Cl^-	1.72	1.81	6
K	2.31	8 cub	...	...	K^+	1.59	1.51	8

continued

Table 1.15 *(continued)*

Element	Metallic Radius Å	CN	Covalent Radius Å	CN	Oxidation State	Ionic Radius Å (1)	Å (2)	CN
					K^+	1.68	1.64	12
Ca	1.97	12 cub	...	...	Ca^{2+}	1.08	1.00	6
					Ca^{2+}	1.20	1.12	8
Sc	1.64	12 cub	...	...	Sc^{3+}	0.83	0.745	6
Ti	1.46	12 hex	...	...	Ti^{3+}	0.75	0.670	6
					Ti^{4+}	0.69	0.605	6
V	1.31	8 cub	...	...	V^{2+}	0.87	0.79	6
					V^{3+}	0.72	0.640	6
					V^{4+}	0.67	0.58	6
					V^{5+}	0.62	0.54	6
Cr	1.24	8 cub	...	...	Cr^{3+}	0.70	0.615	6
					Cr^{6+}	0.38	0.26	4
Mn	1.30	12 cub	...	...	Mn^{2+}	0.75 LS	0.67 LS	6
					Mn^{2+}	0.91 HS	0.830 HS	6
					Mn^{3+}	0.66 LS	0.58 LS	6
					Mn^{3+}	0.73 HS	0.645 HS	6
					Mn^{4+}	0.62	0.530	6
Fe	1.23	8 cub	...	...	Fe^{2+}	0.71 HS	0.63 HS	4
					Fe^{2+}	0.69 LS	0.61 LS	6
					Fe^{2+}	0.86 HS	0.780 HS	6
					Fe^{3+}	0.57 HS	0.49 HS	4
					Fe^{3+}	0.63 LS	0.55 LS	6
					Fe^{3+}	0.73 HS	0.645 HS	6
Co	1.25	12 hex	...	...	Co^{2+}	0.73 LS	0.65 LS	6
					Co^{2+}	0.83 HS	0.745 HS	6
Ni	1.24	12 cub	...	...	Ni^{2+}	0.77	0.690	6
Cu	1.28	12 cub	...	...	Cu^+	0.54	0.46	2
					Cu^{2+}	0.81	0.73	6
Zn	1.39	12 hex	1.31	...	Zn^{2+}	0.68	0.60	4
					Zn^{2+}	0.83	0.740	6
Ga	1.36	8 cub	1.27	...	Ga^{3+}	0.55	0.47	4
					Ga^{3+}	0.70	0.620	6

continued

Table 1.15 *(continued)*

Element	Metallic Radius Å	CN	Covalent Radius Å	CN	Oxidation State	Ionic Radius Å (1)	Å (2)	CN
Ge	1.23	4 dia	1.22	Ge_2H_2	Ge^{4+}	0.48	0.390	4
					Ge^{4+}	0.62	0.530	6
As	1.25	layer	1.21	As_4	As^{5+}	0.58	0.46	6
Se	1.16	chain	1.17	Se_8	Se^{2-}	1.88	1.98	6
					Se^{6+}	0.37	0.50	4
Br	...	...	1.14	Br_2	Br^-	1.88	1.96	6
Rb	2.43	8 cub	...	...	Rb^+	1.68	1.61	8
					Rb^+	1.81	1.72	12
Sr	2.15	12 cub	...	...	Sr^{2+}	1.33	1.26	8
Y	1.81	12 hex	...	...	Y^{3+}	1.10	1.019	8
Zr	1.60	12 hex	...	...	Zr^{4+}	0.92	0.84	8
Nb	1.42	8 cub	...	...	Nb^{5+}	0.72	0.64	6
Mo	1.36	8 cub	...	...	Mo^{4+}	0.73	0.650	6
					Mo^{6+}	0.68	0.61	6
Tc	1.36	12 hex	...	...	Tc^{4+}	0.72	0.645	6
Ru	1.34	12 hex	...	...	Ru^{4+}	0.70	0.620	6
Rh	1.34	12 cub	...	...	Rh^{4+}	0.71	0.60	6
Pd	1.37	12 cub	...	...	Pd^{4+}	0.70	0.615	6
Ag	1.44	12 cub	...	...	Ag^+	1.23	1.15	6
Cd	1.56	12 hex	1.48	...	Cd^{2+}	1.03	0.95	6
In	1.66	12 cub	1.44	...	In^{3+}	0.88	0.800	6
Sn	1.52	6 tetr	1.40	dia	Sn^{4+}	0.77	0.690	6
Sb	1.54	6 rhom	1.41	...	Sb^{5+}	0.69	0.60	6
Te	1.43	chain	1.37	...	Te^{2-}	...	2.21	6
					Te^{6+}	...	0.56	6
I	...	...	1.33	I_2	I^-	2.13	2.20	6
Cs	2.63	8 cub	2.35	...	Cs^+	1.82	1.74	8
					Cs^+	1.96	1.88	12
Ba	2.17	8 cub	1.98	...	Ba^{2+}	1.50	1.42	8
La	1.87	12 hex	1.690	...	La^{3+}	1.13	1.032	6
					La^{3+}	1.26	1.160	8
Ce	1.83	12 cub	...	...	Ce^{3+}	1.09	1.01	6

continued

Table 1.15 (continued)

Element	Metallic Radius Å	CN	Covalent Radius Å	CN	Oxidation State	Ionic Radius Å (1)	Å (2)	CN
					Ce³⁺	1.22	1.143	8
Pr	1.82	12 hex	...	...	Pr³⁺	1.08	0.99	6
					Pr³⁺	1.22	1.126	8
Nd	1.82	12 hex	...	...	Nd³⁺	1.06	0.983	6
					Nd³⁺	1.20	1.109	8
Pm	1.81	12 hex	...	...	Pm³⁺	1.04	0.97	6
					Pm³⁺	...	1.093	8
Sm	1.81	6 rhom	...	...	Sm³⁺	1.04	0.958	6
					Sm³⁺	1.17	1.079	8
Eu	1.98	8 cub	...	...	Eu²⁺	1.25	1.17	6
					Eu²⁺	1.33	1.25	8
					Eu³⁺	1.03	0.947	6
					Eu³⁺	1.15	1.066	8
Gd	1.79	12 hex	...	...	Gd³⁺	1.02	0.938	6
					Gd³⁺	1.14	1.053	8
Tb	1.77	12 hex	...	...	Tb³⁺	1.00	0.923	6
					Tb³⁺	1.12	1.040	8
Dy	1.77	12 hex	...	...	Dy³⁺	0.99	0.912	6
					Dy³⁺	1.11	1.027	8
Ho	1.76	12 hex	...	...	Ho³⁺	0.98	0.901	6
					Ho³⁺	1.10	1.015	8
Er	1.75	12 hex	...	...	Er³⁺	0.97	0.890	6
					Er³⁺	1.08	1.004	8
Tm	1.74	12 hex	...	...	Tm³⁺	0.96	0.880	6
					Tm³⁺	1.07	0.994	8
Yb	1.93	12 cub	...	...	Yb²⁺	...	1.14	8
					Yb³⁺	0.95	0.868	6
					Yb³⁺	1.06	0.985	8
Lu	1.74	12 hex	...	...	Lu³⁺	0.94	0.861	6
					Lu³⁺	1.05	0.977	8
Hf	1.59	12 hex	...	...	Hf⁴⁺	0.91	0.83	8
Ta	1.43	8 cub	...	...	Ta⁵⁺	0.72	0.64	6

continued

Table 1.15 *(continued)*

Element	Metallic Radius Å	CN	Covalent Radius Å	CN	Oxidation State	Ionic Radius Å (1)	Å (2)	CN
W	1.37	8 cub	...	...	W^{4+}	0.73	0.66	6
					W^{6+}	0.50	0.42	4
					W^{6+}	0.68	0.60	6
Re	1.37	12 hex	...	...	Re^{4+}	0.71	0.63	6
Os	1.35	12 cub	...	...	Os^{4+}	0.71	0.630	6
Ir	1.35	12 hex	...	...	Ir^{4+}	0.71	0.625	6
Pt	1.38	12 cub	...	...	Pt^{4+}	0.71	0.625	6
Au	1.44	12 cub	...	...	Au^{3+}	0.78	0.68	4sq
Hg	1.536	6 rhom	...	...	Hg^{2+}	1.10	1.02	6
Tl	1.71	12 hex	...	...	Tl^+	1.68	1.59	8
					Tl^+	1.84	1.70	12
					Tl^{3+}	0.97	0.885	6
Pb	1.75	12 cub	...	...	Pb^{2+}	1.26	1.19	6
					Pb^{2+}	1.37	1.29	8
Bi	1.75	6 rhom	...	...	Bi^{3+}	1.10	1.03	6
Th	1.80	12 cub	...	...	Th^{4+}	1.12	1.05	8
U	1.53	12 hex	...	...	U^{4+}	1.08	1.00	8
					U^{6+}	0.81	0.73	6

Notes: Metal crystal structures: cub = cubic, dia = diamond, gr = graphite, hex = hexagonal, tetr = tetragonal, rhbdr = rhombohedral, rhom = rhombic. HS = high spin state (unpaired *d* electrons). LS = low spin (paired *d* electrons).

Sources: Metallic and covalent radii: Zhdanov, G. S., 1965, *Crystal physics*, Academic Press, p. 192–193. Wells, A. F., 1975, *Structural inorganic chemistry*, Oxford Univ. Press, pp. 1095. *Ionic radii:* (1) Scaled to 6-coordinate O^{2-} radius of 1.32 Å and 6-coordinate F^- radius of 1.25 Å. Whittaker, E. J. W., & Muntus, R., 1970, *Geochim. Cosmochim. Acta* 34, 945–956. (2) Scaled to 6-coordinate O^{2-} radius of 1.40 Å. Shannon, R. D., 1974, *Acta Crystallogr.* A32, 751–767.

Table 1.16 Some Radioactive Nuclides, Their Stable Daughters, and Half-lives

Parent ⇨ Daughter	Half-life (years)	Parent ⇨ Daughter	Half-life (years)	Parent ⇨ Daughter	Half-life (years)
^{10}Be ⇨ ^{10}B	1.6×10^6	^{98}Tc ⇨ ^{98}Ru	4.1×10^6	^{186m}Re ⇨ ^{186}Os	2.0×10^5
^{14}C ⇨ ^{14}N	5715	^{99}Tc ⇨ ^{99}Ru	2.13×10^5	^{187}Re ⇨ ^{187}Os	4.23×10^{10}
^{22}Na ⇨ ^{22}Ne	2.605	^{107}Pd ⇨ ^{107}Ag	6.5×10^6	^{190}Pt ⇨ ^{186}Os	6.5×10^{11}
^{26}Al ⇨ ^{26}Mg	7.16×10^6	^{113}Cd ⇨ ^{113}In	9×10^{15}	^{202}Pb ⇨ ^{202}Hg	5.3×10^4
^{36}Cl ⇨ ^{36}Ar	3.01×10^5	^{115}In ⇨ ^{115}Sn	4.4×10^{14}	^{205}Pb ⇨ ^{205}Tl	1.51×10^7
^{40}K ⇨ ^{40}Ar	1.193×10^9	^{126}Sn ⇨ ^{126}Te	1×10^5	^{208}Bi ⇨ ^{208}Pb	3.68×10^5
^{40}K ⇨ ^{40}Ca	1.397×10^9	^{123}Te ⇨ ^{123}Sb	1.24×10^{13}	^{210m}Bi ⇨ ^{206}Pb	3.5×10^6
^{41}Ca ⇨ ^{41}K	(1.02–1.3)×10^5	^{130}Te ⇨ ^{130}Xe	2.51×10^{21}	^{209}Po ⇨ ^{205}Tl	102
^{50}V ⇨ ^{50}Ti	>1.4×10^{17}	^{129}I ⇨ ^{129}Xe	1.57×10^7	^{210}At ⇨ ^{206}Pb	8.1 hours
^{53}Mn ⇨ ^{53}Cr	3.7×10^6	^{135}Cs ⇨ ^{135}Ba	2.3×10^6	^{222}Rn ⇨ ^{206}Pb *	3.823 days
^{60}Fe ⇨ ^{60}Ni	1.5×10^5	^{137}La ⇨ ^{137}Ba	6×10^4	^{226}Ra ⇨ ^{206}Pb *	1599
^{59}Ni ⇨ ^{59}Co	7.5×10^4	^{138}La ⇨ ^{138}Ce	3.10×10^{11}	^{232}Th ⇨ ^{208}Pb *	1.401×10^{10}
^{79}Se ⇨ ^{79}Br	6.5×10^4	^{138}La ⇨ ^{138}Ba	1.57×10^{11}	^{234}U ⇨ ^{206}Pb *†	2.46×10^5
^{81}Kr ⇨ ^{81}Br	2.1×10^5	^{144}Nd ⇨ ^{144}Sm	2.1×10^{15}	^{235}U ⇨ ^{207}Pb *†	7.0381×10^8
^{87}Rb ⇨ ^{87}Sr	4.88×10^{10}	^{145}Pm ⇨ ^{145}Nd	17.7	^{236}U ⇨ ^{232}Th *†	2.3416×10^7
^{93}Zr ⇨ ^{93}Nb	1.5×10^6	^{146}Sm ⇨ ^{142}Nd	1.03×10^8	^{238}U ⇨ ^{206}Pb *†	4.4683×10^9
^{91}Nb ⇨ ^{91}Zr	7×10^2	^{147}Sm ⇨ ^{143}Nd	1.06×10^{11}	^{237}Np †	2.14×10^6
^{92}Nb ⇨ ^{92}Zr	3.7×10^7	^{148}Sm ⇨ ^{144}Nd	7×10^{15}	^{242}Pu †	3.763×10^5
^{94}Nb ⇨ ^{94}Mo	2.03×10^4	^{176}Lu ⇨ ^{176}Hf	3.59×10^{10}	^{244}Pu ⇨ $^{131-136}$Xe †	8.26×10^7
^{93}Mo ⇨ ^{93}Nb	3500	^{174}Hf ⇨ ^{170}Yb	2.0×10^{15}	^{247}Cm †	1.56×10^7
^{97}Tc ⇨ ^{97}Mo	2.6×10^6	^{182}Hf ⇨ ^{182}W	9×10^6		

* decay through a series of intermediate daughter products
† multiple fission products

Sources: Blum, J. D., 1995, in *Global earth physics* (Ahrens, T. J., ed.) Vol. 1, AGU, Washington, D. C., pp. 271–280. IUPAC commission on atomic weights and isotopic abundances, 1994, Heumann, K. G. (chairman), *Pure & Appl. Chem.* 66, 2423–2444. Holden, N. E., 1989, *Pure & Appl. Chem.* 61, 1483–1504. Holden, N. E., 1990, *Pure & Appl. Chem.* 62, 941–958.

Stable Isotopes: Notation and Reference Standards

The isotopic composition of an element can be described by its isotope ratios, for example; D/H, $^{13}C/^{12}C$, or $^{17}O/^{16}O$ and $^{18}O/^{16}O$, for H, C, and O, respectively. Small deviations in isotopic compositions of a sample relative to a standard are described using the δ-notation:

δ-Notation (in per mil): $\delta^0/_{00} = \left(\dfrac{\text{isotope ratio in sample}}{\text{isotope ratio in standard}} - 1 \right) \times 1000$

Conversion of δ-reference scales: The δ-values are not additive. The relation for converting different scales is illustrated for oxygen. A sample relative to reference scale "X" is converted to the VSMOW-scale by:

$$\delta^{18}O_{\underset{\text{VSMOW}}{\text{sample}}} = \delta^{18}O_{\underset{X}{\text{sample}}} + \delta^{18}O_{\underset{\text{VSMOW}}{X}} + 10^{-3} \times \delta^{18}O_{\underset{X}{\text{sample}}} \times \delta^{18}O_{\underset{\text{VSMOW}}{X}}$$

Table 1.17 Stable Isotope Reference Standards

Standard	Value	Normalized Value in ‰	Notes
H VSMOW	$D/H = 1.5576 \times 10^{-4}$	$\delta D_{VSMOW} = 0.00$	Vienna Standard Mean Ocean Water
SLAP	$D/H = 0.8909 \times 10^{-4}$	$\delta D_{SLAP/VSMOW} = -428.0$	Standard Light Antarctic Precipitation
C VPDB	$^{13}C/^{12}C = 0.0112375$	$\delta^{13}C_{VPDB} = 0.00$	Vienna Peedee Belemnite
NBS-19	$^{13}C/^{12}C = 0.0112594$	$\delta^{13}C_{NBS-19/VPDB} = +1.95$	Calcite, NBS-19
N Air	$^{14}N/^{15}N = 272.0$	$\delta^{15}N_{Air} = 0.00$	Air; NBS-14
O VSMOW	$^{18}O/^{16}O = 2.0052 \times 10^{-3}$	$\delta^{18}O_{VSMOW} = 0.00$	Vienna Standard Mean Ocean Water
VSMOW	$^{17}O/^{16}O = 3.7288 \times 10^{-4}$	$\delta^{17}O_{VSMOW} = 0.00$	
SLAP	$^{18}O/^{16}O = 1.8939 \times 10^{-3}$	$\delta^{18}O_{SLAP/VSMOW} = -55.5$	Standard Light Antarctic Precipitation
VPDB	$^{18}O/^{16}O = 2.0672 \times 10^{-3}$	$\delta^{18}O_{VPDB/VSMOW} = +30.91$	Vienna Peedee Belemnite
NBS-19	$^{18}O/^{16}O = 2.06265 \times 10^{-3}$	$\delta^{18}O_{NBS-19/VSMOW} = +28.65$ $\delta^{18}O_{NBS-19/VPDB} = -2.2$	Calcite, NBS-19
S CDT	$^{34}S/^{32}S = 0.044994$		Cañyon Diablo Troilite

Sources: Coplen, T. B., 1994, *Pure Appl. Chem.* 66, 273–276. Coplen, T. B., 1996, *Geochim. Cosmochim. Acta* 60, 3359–3360. O'Neil, J. R., 1986, in *Stable isotopes* (Valley, J. W., Taylor, H. P., & O'Neil, J. R., eds.), Reviews in Mineralogy, Vol. 16., Mineralogical Society of America, pp. 561–570.

1.4 Minerals and Compounds

Table 1.18 Physical Properties of Some Minerals and Compounds

Formula	Mineral Name	MW (g mol^{-1})	mp. (°C)	ρ (STP) (g cm^{-3})	Gravimetric Factor[a]
Metals					
Co	Cobalt	58.9332	1495	8.836	—
Cr	Chromium	51.9961	1857	7.191	—
Fe	Iron	55.845	1536	7.875	—
Mn	Manganese	54.9380	1246	7.470	—
Ni	Nickel	58.6934	1455	8.912	—
P (white)	Phosphorous	30.9738	44	1.820	—
Si	Silicon	28.0855	1412	2.330	—
Oxides					
Al$_2$O$_3$	corundum	101.961	2054	3.987	1.8894
CaO	lime, calcia	56.078	2927	3.345	1.3992
CoO	cobalt oxide	74.933	1830	6.438	1.2715
Cr$_2$O$_3$	eskolaite	151.991	2330	5.225	1.4616
Cu$_2$O	cuprite	143.092	1244	6.105	1.1259
CuO	tenorite	79.546	1124	6.509	1.2518
Fe$_{0.947}$O	wüstite	68.885	1377	5.747	1.3025
FeO	ferrous oxide	71.845	1377	5.987	1.2865
Fe$_2$O$_3$	hematite	159.689	1622	5.275	1.4297
Fe$_3$O$_4$	magnetite	231.533	1597 dec.	5.200	1.3820
K$_2$O	potassium oxide	94.196	881 dec	2.333	1.2046
MgO	periclase	40.304	2832	3.584	1.6583
MnO	manganosite	70.937	1781	5.365	1.2912
Na$_2$O	sodium oxide	61.979	1132 subl.	2.395	1.3480
NiO	bunsenite	74.693	1984	6.809	1.2726
P$_2$O$_5$	phosphorous oxide	141.945	580–585	2.390	2.2914
Rb$_2$O	rubidium oxide	186.935	400 dec.	3.7	1.0936
SiO$_2$	quartz, silica	60.084	1723	2.648	2.1393
TiO$_2$	rutile	79.866	1857 dec.	4.245	1.6685
V$_2$O$_3$	karelianite	149.881	1067	5.022	1.4711

continued

Table 1.18 *(continued)*

Formula	Mineral Name	MW (g mol⁻¹)	mp. (°C)	ρ (STP) (g cm⁻³)	Gravimetric Factorᵃ
ZnO	zincite	81.391	1975	5.675	1.2447
$CaTiO_3$	perovskite	135.944	1915	4.044	3.3920
$FeCr_2O_4$	chromite	223.835	...	5.086	4.0081
$FeTiO_3$	ilmenite	151.711	1200	4.788	2.7166
$MgAl_2O_4$	spinel	142.266	2135	3.583	5.8534
Hydroxides					
$Al(OH)_3$	gibbsite	78.004	107 dec.	2.441	2.8910
$\alpha\text{-}FeO(OH)$	goethite	88.852	dec.	4.269	1.5910
$Mg(OH)_2$	brucite	58.320	350 dec.	2.368	2.3995
Carbonates					
$CaCO_3$	calcite	100.087	886 dec.	2.710	2.4973
$CaMg(CO_3)_2$	dolomite	184.402	800 dec.	2.866	4.6011 (Ca) 7.5870 (Mg)
$FeCO_3$	siderite	115.854	580 dec.	3.943	2.0746
K_2CO_3		138.206	897 dec.	2.3	1.7674
$MgCO_3$	magnesite	84.314	405 dec.	3.010	3.4690
$MnCO_3$	rhodochrosite	114.947	610 dec.	3.700	2.0923
Na_2CO_3		105.989	854 dec.	2.532	2.3051
Phosphates					
$Ca_3(PO_4)_2$	whitlockite	310.178	1670	3.1	2.5798
Silicates					
$CaAl_2Si_2O_8$	anorthite	278.208	1557	2.765	6.9416 (Ca)
$CaMgSi_2O_6$	diopside	216.551	1395	3.275	5.4032 (Ca)
$CaSiO_3$	wollastonite	116.162	1548 *	2.909	2.8983
$Ca_2Mg_5Si_8O_{22}(OH)_2$	tremolite	812.369	1000 dec.	2.977	10.1348 (Ca)
$FeSiO_3$	ferrosilite	131.929	1146	3.998	2.3624
Fe_2SiO_4	fayalite	203.774	1217	4.393	1.8245
$KAlSi_3O_8$	sanidine	278.332	~1200	2.570	7.1188 (K)
$Mg_3Al_2Si_3O_{12}$	pyrope (garnet)	403.128	1297	3.559	5.5287 (Mg)
$MgSiO_3$	enstatite	100.389	1557	3.194	4.1304

continued

Table 1.18 *(continued)*

Formula	Mineral Name	MW (g mol⁻¹)	mp. (°C)	ρ (STP) (g cm⁻³)	Gravimetric Factor[a]
Mg_2SiO_4	forsterite	140.693	1890	3.214	2.8943
$Mg_3Si_4O_{10}(OH)_2$	talc	379.266	800 dec.	2.784	5.2015 (Mg)
$NaAlSi_3O_8$	analbite	262.223	1118	2.611	11.4060 (Na)
$ZrSiO_4$	zircon	183.307	1676 dec.	4.669	2.0094
Sulfates					
$CaSO_4$	anhydrite	136.143	1450	2.964	3.3970
$CaSO_4\cdot2H_2O$	gypsum	172.173	~90 dec.	2.305	4.2959
K_2SO_4	arcanite	174.261	1069	2.661	2.2285
Na_2SO_4	thenardite	142.044	882	2.663	3.0892
Sulfides					
CaS	oldhamite	72.145	2450	2.602	1.8001
CoS	sycoporite	91.000	>1116	5.5	1.5441
Cr_2S_3	chromium sulfide	200.192	1350 dec.	3.8	1.9251
$CuFeS_2$	chalcopyrite	183.525	557	4.088	2.8881(Cu)
FeS	troilite	87.912	1190	4.830	1.5742
FeS_2	pyrite	119.978	742 dec.	5.012	2.1484
FeS_2	marcasite	119.978	450 dec.	4.881	2.1484
Fe_7S_8	pyrrhotite	647.449	...	4.625	1.6562
Fe_8S_9	pyrrhotite	735.361	...	...	1.6460
MgS	niningerite	56.372	>2000 dec.	2.84	2.3193
MnS	alabandite	87.005	dec.	4.055	1.5837
NiS	millerite	90.760	797	5.374	1.5464
Ni_3S_2	heazlewoodite	240.213	790	5.867	1.3642
PbS	galena	239.277	1114	7.597	1.1548
ZnS	sphalerite	97.459	1020 dec.	4.088	1.4904
Halides					
CaF_2	fluorite	78.075	1418	3.179	1.9481
NaCl	halite	58.443	801	2.163	2.5421

[a] gravimetric factor to convert grams of major cation to gram formula
* phase transition to pseudowollastonite at 1125°C; pseudowollastonite mp. is listed

Table 1.19 Melting and Boiling Points of Some Icy Substances

Compound	mp (K)	ΔH_{fus} (J mol^{-1})	bp (K)	ΔH_{vap} (J mol^{-1})
H_2	13.8	120	20.3	897
He	—	—	4.2	83
Ne	24.5	339	27.1	1711
Ar	83.8	1119	87.3	6432
Kr	115.8	1366	119.9	9084
Xe	161.4	1812	165.1	12617
H_2O	273.1	5940	373.1	40600
CH_4	90.6	930	111.6	8170
CO	68.1	835	81.8	6040
CO_2	215.6 *	...	194.67 subl.	25230 subl.
HCHO	181	...	253	24700
CH_3OH	175.4	7540	337	39150
CH_3SH	152	5900	278	28800
N_2	63.14	720	77.32	5580
NH_3	195.36	5655	239.68	23350
$NH_3 \cdot H_2O$	194.2	6560	259.3 †	...
NH_4SH	...	...	317.4 †	...
HCN	259	8400	299	28800
S	388.36	1721	717.824 ‡	10840 (for S_2 ‡)
H_2S	187.61	2380	212.77	18670
SO_2	197.64	7400	263.08	24900
OCS	134.31	4730	222.87	18500

* at 5.3 bar
† Temperature at which dissociation pressure reaches 1 bar.
‡ Sulfur vapor is composed of a mixture of different sulfur species (S_n with n = 1 to 8). The temperature at which S_2 vapor reaches 1 bar is 882.1 K.

Table 1.20 Vapor Pressure Over Low-temperature Solids and Liquids[a]

Compound	Phase	a	b	Range (K)	Sources
H_2O	s	7.610	−2681.18	183–273.1	S47
H_2O	liq	6.079	−2261.10	273.1–373.1	S47
CH_4	s	4.283	−475.6	65–90.6	YE87
CH_4	liq	4.092	−459.8	90.6–190	YE87
$CH_4 \cdot 7H_2O$	s	4.8788	−948.67	<194.45	D73
CO	s	5.243	−411.2	50–68.1	G85
CO	liq	3.993	−326	68.1–130	G85
$CO \cdot 7H_2O$ *	s	5.002	−763	<153	M61
CO_2	s	7.025	−1336	130–195	G69
CO_2	liq	6.045	−1201	135–300	G69
$CO_2 \cdot 6H_2O$	s	5.4303	−1184.626	<218.15	D73
HCHO	liq	5.099	−1290.8	181–253	S47
CH_3OH	liq	6.066	−2044.8	175.4–337	S47
CH_3SH	liq	5.411	−1504.2	152–278	S47
N_2	s	4.798	−360.2	54.78–63.14	G69
N_2	liq	3.944	−305.0	63.14–78.01	G69
$N_2 \cdot 6H_2O$	s	4.6905	−688.9	<147	M69
$N_2 \cdot 7H_2O$	s	5.002	−763	<153	M61
NH_3	s	6.900	−1588	160–195.4	HG78
NH_3	liq	5.201	−1248	195.4–300	HG78
$NH_3 \cdot H_2O$ †	s	8.0519	−2088.0	<193	WL73
$2NH_3 \cdot H_2O$ ‡	s	4.5531	−1291.2	<193	WL73
NH_4HCO_3 §	s	8.248	−2923.2	<354	W68
$NH_4CO_2NH_2$ #	s	7.988	−2741.9	<343	E46
NH_4CN	s	7.814	−2370.4	<309	S47
NH_4SH	s	7.60	−2411.2	280–317	K37
HCN	s	6.747	−1944.5	<259	S47
HCN	liq	5.037	−1504.5	259–299	S47
O_2	liq	4.271	−383.6	54.75–90.19	S47
$O_2 \cdot 6H_2O$	s	4.679	−717	<153	M69
O_3	liq	3.912	−632.4	81.1–161.3	S47
S	s	9.106	−5308.08	≤388.36	B51,N34,WM29
S	liq	4.9948	−3596.71	388.36–Pt_{cr}	RKG73,WM29

continued

Table 1.20   (continued)

Compound	Phase	a	b	Range (K)	Sources
H_2S	s	5.610	−1171.2	<187.61	G69
H_2S	liq	4.780	−1015.5	187.61–213.2	G69
OCS	liq	4.710	−1046.4	134.31–222.87	G69
SO_2	liq	5.046	−1416.6	297.64–263	G69
Ne	s	4.150	−111.5	<24.5	S47
Ar	s	5.254	−445.0	<83.8	S47
Ar·$6H_2O$	s	4.288	−639.5	<149.15	D73
Kr	s	4.500	−545.9	<115.8	S47
Kr·$6H_2O$	s	6.347	−1417.7	<223.35	D73
Xe	s	5.003	−825.2	<161.4	S47
Xe·$6H_2O$	s	4.625	−1215.2	<262.75	D73

[a] Tabulated data from the literature sources were fitted to log P_{vap} = a + b/T(K).
*   The vapor pressure equation for CO·$7H_2O$ is taken as equal to that of N_2·$7H_2O$.
† $P_{vap}(NH_3)$ from NH_3·H_2O (s) = NH_3 (g) + H_2O (g)
‡ $P_{vap}(NH_3)$ from 2 NH_3·H_2O (s) = NH_3·H_2O (s) + NH_3 (g)
§   vapor composed of NH_3 + H_2O + CO_2
#   vapor composed of CO_2 + 2 NH_3
Pt_{cr} The critical point of sulfur is at T_{cr} = 1313 K and P_{cr} = 182.0 bar [RKG73].

Sources: [B51] Bradley, R. S., 1951, *Proc. Roy. Soc.* A205, 553–563. [D73] Davidson, Jr., P. W., 1973, in *Water. A comprehensive treatise*, (Franks, F. ed.), Plenum Press, New York, pp. 115–234. [E46] Egan, E. P., Potts, Jr., J. E., & Potts, G. P., 1946, *Ind. Eng. Chem.* 38, 454–456. [G69] Giauque Scientific Papers Foundation, 1969, *Low temperature, chemical and magneto thermodynamics. The scientific papers of William F. Giauque*, Dover Publ. Inc., New York, pp. 641. [G85] Goodwin, R. D., 1985, *J. Phys. Chem. Ref. Data* 14, 849–932. [HG78] Haar, L., & Gallagher, J. S., 1978, *J. Phys. Chem. Ref. Data* 7, 635–792. [K37] Kelley, K. K., 1937, *U.S. Bureau of Mines Bull.* 406, pp. 153. [M61] Miller, S. L., 1961, *Proc. Natl. Acad. Sci. U.S.* 47, 1798–1808. [M69] Miller, S. L., 1969, *Science* 165, 489–490. [N34] Neumann, K., 1934, *Z. physikal. Chem.* A171, 416–420. [RKG73] Rau, H., Kutty, T. R. N., & Guedes de Carvalho, J. R. F., 1973, *J. Chem. Thermodyn.* 5, 291–302. [S47] Stull, D. R., 1947, *Ind. Eng. Chem.* 39, 517–540. [W68] Wagman, D. D., Evans, W. H., Parker, V. B., Halow, I., Bailey, S. M., & Schumm, R. H., 1968, NBS Technical Note 270-3, pp. 264. [WL73] Weidenschilling, S. J., & Lewis, J. S., 1973, *Icarus* 20, 465–476. [WM29] West, W. A., & Menzies, A. W. C., 1929, *J. Phys. Chem.* 33, 1880–1892. [YE87] Younglove, B. A., & Ely, J. F., 1987, *J. Phys. Chem. Ref. Data* 16, 577–798.

Table 1.21 Thermodynamic Properties of Some Substances at 298.15 K

Chemical Formula	Phase or Name	Formula Weight	S° $(J\ mol^{-1}\ K^{-1})$	$\Delta_f H^{\circ}$ $(kJ\ mol^{-1})$	$\Delta_f G^{\circ}$ $(kJ\ mol^{-1})$
Ag	metal, fcc	107.8682	42.55±0.20	0	0
	gas	107.8682	172.997±0.004	284.9±0.8	246.0±0.8
Ag^+	aq, m=1	107.8677	73.45±0.40	105.79±0.08	77.1±0.1
Ag_2S	acanthite	247.803	142.9±0.3	−32.0±1.0	−39.7±1.0
AgCl	chlorargyrite	143.321	96.2±0.2	−127.1±0.1	−109.8±0.1
AgBr	bromargyrite	187.772	107.1±0.4	−100.4±0.2	−97.0±0.2
AgI	iodargyrite	234.773	115.5±1.7	−61.8±1.7	−66.2±1.8
Al	metal, fcc	26.98154	28.30±0.10	0	0
	gas	26.98154	64.553±0.003	329.7±4.2	289.1±4.2
Al^{3+}	aq, m=1	26.9799	−325±10	−538.4±1.5	−489.4±1.4
$3Al_2O_3 \cdot 2SiO_2$	mullite	426.053	275.0±5.0	−6819.2±10.0	−6441.8±10.0
Al_2SiO_5	kyanite	162.046	82.8±0.5	−2593.8±2.0	−2443.1±2.0
	andalusite	162.046	91.4±0.5	−2589.9±2.0	−2441.8±2.0
	sillimanite	162.046	95.4±0.5	−2586.1±2.0	−2439.1±2.0
$Al_2Si_2O_5(OH)_4$	kaolinite	258.161	200.4±0.5	−4119.0±1.5	−3797.5±1.5
$Al_2SiO_4F_2$	topaz	184.043	105.4±0.2	−3084.5±4.7	−2910.6±4.7
AlN	c, hexagonal	40.988	20.14±0.21	−318.0±2.5	−287.0±2.5
$AlPO_4$	berlinite	121.953	90.8±0.2	−1733.8±5.0	−1617.9±5.0
Al_2O_3	corundum	101.9614	50.92±0.10	−1675.7±1.3	−1582.3±1.3
$Al(OH)_4^-$	aq, m=1	95.012	102.9	−1502.5	−1305.3
AlO(OH)	boehmite	59.988	37.2±0.1	−996.4±2.2	−918.4±2.2
AlO(OH)	diaspore	59.988	35.3±0.2	−1001.3±2.2	−922.7±2.1
$Al(OH)_3$	gibbsite	78.004	68.4±0.1	−1293.1±1.2	−1154.9±1.2
Ar	gas	39.948	154.846±0.003	0	0
As	c, rhomb	74.922	35.69±0.84	0	0
	gas	74.922	174.1	301.8±2.3	260.5±2.3
As_2	gas	149.843	240.8	221.0±7.1	170.5±7.1
As_4	gas	299.686	327.3	153.3±1.7	98.3±1.7
AsH_3	arsine gas	77.945	222.8	66.44	68.93
As_2O_3	arsenolite	197.841	107.4±0.1	−657.0±1.7	−576.0±1.9
As_2O_3	claudetite	197.841	113.3±0.1	−654.8±1.7	−575.6±1.1
AsS	realgar	106.988	63.5±0.6	−30.9±5.0	−29.6±5.0
As_2S_3	orpiment	246.043	163.6±1.6	−91.6±4.2	−90.4±4.2
Au	metal, fcc	196.967	47.49±0.21	0	0
	gas	196.967	180.503	366.1	326.3
B	c, rhombic	10.812	5.90±0.08	0	0

continued

Table 1.21 *(continued)*

Chemical Formula	Phase or Name	Formula Weight	$S°$ ($J\ mol^{-1}\ K^{-1}$)	$\Delta_f H°$ ($kJ\ mol^{-1}$)	$\Delta_f G°$ ($kJ\ mol^{-1}$)
B	gas	10.81	153.435±0.035	560±12	516±12
B_2O_3	c, hexagonal	69.622	53.97±0.30	−1273.5±1.4	−1194.4±1.4
H_3BO_3	c, triclinic	61.834	88.7±0.4	−1094.0±0.8	−968.5±0.4
	aq, m=1	61.834	162.3	−1072.3	−968.8
$H_4BO_4^-$	aq, m=1	78.842	102.5	−1344.0	−1153.2
$Na_2B_4O_7 \cdot 10H_2O$	borax	381.376	586.0±2.3	−6288.6±8.5	−5516.2±8.5
Ba	c, bcc	137.328	62.35±0.30	0	0
	gas	137.328	170.24±0.01	179.1±5.0	146.9±5.0
Ba^{2+}	aq, m=1	137.327	8.40±0.50	−532.5±1.0	−555.4±1.0
$BaCO_3$	witherite	197.337	112.1±2.1	−1210.9±2.2	−1132.2±2.2
BaO	c, cubic	153.327	72.1±0.4	−548.1±2.1	−520.4±2.1
$BaSO_4$	barite	233.392	132.2±0.8	−1473.6±1.0	−1362.5±1.3
Be	metal, hcp	9.01218	9.50±0.10	0	0
	gas	9.01218	136.274±0.020	324±5	286±5
$BeAl_2O_4$	chyrsoberyl	126.973	66.3±0.1	−2298.5±2.8	−2176.2±2.8
Be_2SiO_4	phenakite	110.108	63.4±0.3	−2143.1±4.0	−2028.4±4.0
$Be_3Al_2(Si_6O_{18})$	beryl	537.503	346.7±4.7	−9006.6±7.0	−8500.5±6.4
BeO	bromellite	25.0116	13.77±0.04	−609.4±2.5	−580.1±2.5
Bi	metal, rhomb	208.980	56.74±0.42	0	0
	gas	208.980	186.90	209.6±2.1	170.8±2.1
Bi_2	gas	417.961	273.63	220.1±2.1	172.4±2.1
Bi_2O_3	bismite	465.959	151.5±2.1	−573.9±1.3	−493.5±1.5
Bi_2S_3	bismuthinite	514.161	200.4±3.3	−135.2±2.4	−132.4±2.6
Br_2	liq	159.808	152.21±0.30	0	0
	gas	159.808	245.468±0.005	30.91±0.11	3.1±0.3
Br	gas	79.904	175.017±0.003	111.86±0.06	82.37±0.06
Br^-	aq, m=1	79.9046	82.4	−121.55	−103.96
HBr	gas	80.912	198.699±0.033	−36.44±0.17	−53.51±0.17
	aq, m=1	80.912	82.4	−121.55	−103.96
C	graphite	12.011	5.74±0.10	0	0
	diamond	12.011	2.38±0.02	1.9±0.0	2.9±0.1
C	gas	12.011	158.100±0.003	716.67±0.46	671.24±0.46
CH_4	methane	16.043	186.26±0.21	−74.87±0.34	−50.77±0.40
C_2H_2	acetylene	26.037	200.94	226.73	209.20
C_2H_4	ethylene	28.053	219.33	52.47±0.29	68.42±0.29
C_2H_6	ethane	30.0694	229.60	−84.68	−32.82

continued

Table 1.21 (continued)

Chemical Formula	Phase or Name	Formula Weight	$S°$ ($J\ mol^{-1}\ K^{-1}$)	$\Delta_f H°$ ($kJ\ mol^{-1}$)	$\Delta_f G°$ ($kJ\ mol^{-1}$)
HCN	gas	27.025	201.828±0.040	135.1±8.4	124.7±8.1
CO	gas	28.0102	197.660±0.004	−110.53±0.17	−137.16±0.17
CO_2	gas	44.0096	213.785±0.010	−393.52±0.05	−394.39±0.05
	aq, m=1	44.0096	117.6	−413.8	−386.0
CO_3^{2-}	aq, m=1	60.0102	−50.0±1.0	−675.2±0.1	−527.0±0.3
HCO_3^-	aq, m=1	61.018	98.4±0.5	−689.9±0.20	−586.8±0.3
H_2CO_3	aq, m=1	62.025	184.7±0.9	−699.7±0.1	−623.2±0.1
H_2CO	formaldehyde, g.	30.026	218.95±0.40	−116.0±6.3	−109.9±6.3
CH_3OH	methanol, gas	32.042	239.81	−200.66	−161.96
C_2H_5OH	ethanol, gas	46.069	282.70	−235.10	−168.49
Ca	metal, fcc	40.078	42.54±0.30	0	0
	gas	40.078	154.886±0.004	177.8±0.8	144.0±0.8
Ca^{2+}	aq, m=1	40.077	−56.2±1.0	−543.0±1.0	−553.6
$CaCO_3$	calcite	100.087	91.7±0.2	−1207.4±1.3	−1128.5±1.4
	aragonite	100.087	88.0±0.2	−1207.4±1.4	−1127.4±1.5
$CaMg(CO_3)_2$	dolomite	184.402	155.2±0.3	−2324.5±1.5	−2161.3±1.7
$CaMg_3(CO_3)_4$	huntite	353.030	299.5±0.9	−4529.6±1.6	−4203.1±1.6
$Ca_4(AlSiO_4)_6CO_3$	meionite	934.711	715.2±1.0	−13881.4±6.2	−13131.8±6.2
$CaAl_2SiO_6$	Ca-Al px.	218.124	141.0±2.0	−3306.3±2.5	−2139.6±2.6
$CaAl_2Si_2O_8$	anorthite	278.208	199.3±0.3	−4234.0±2.0	−4007.9±2.0
$CaSiO_3$	wollastonite	116.162	81.7±0.1	−1634.8±1.4	−1549.0±1.4
	pseudowolla-stonite	116.162	87.2±0.9	−1627.6±1.4	−1543.5±1.4
Ca_2SiO_4	larnite	172.240	127.6±0.8	−2306.7±1.5	−2191.2±1.5
	Ca-olivine	172.239	120.5±0.8	−2316.5±2.5	−2198.9±2.5
$Ca_3Si_2O_7$	rankinite	288.402	210.6±2.9	−3949.0±10.0	−3748.1±10.0
$CaMgSiO_4$	monticellite	156.467	108.1±0.3	−2251.0±3.0	−2132.8±3.1
$CaMgSi_2O_6$	diopside	216.551	142.7±0.3	−3201.5±2.0	−3026.8±2.0
$Ca_2MgSi_2O_7$	akermanite	272.629	212.5±0.4	−3864.8±2.0	−3667.5±2.0
$Ca_3Mg(SiO_4)_2$	merwinite	328.707	253.1±2.1	−4536.2±3.0	−4307.7±3.0
$Ca_2Al_2SiO_7$	gehlenite	274.201	210.1±0.6	−3985.0±5.0	−3785.5±5.0
$Ca_3Al_2Si_3O_{12}$	grossular	450.448	260.1±0.5	−6640.0±3.2	−6278.5±3.2
$CaTiSiO_5$	sphene	196.028	129.2±0.8	−2596.6±3.0	−2454.6±3.2
$CaFeSi_2O_6$	hedenbergite	248.091	174.2±0.3	−2839.9±3.0	−2676.3±3.0
$Ca_3Fe_2Si_3O_{12}$	andradite	508.175	316.4±2.0	−5771.0±5.9	−5427.0±5.9
$Ca_2Mg_5Si_8O_{22}(OH)_2$	tremolite	812.369	548.9±1.3	−12303.0±7.0	−11574.6±7.0

continued

Table 1.21 *(continued)*

Chemical Formula	Phase or Name	Formula Weight	$S°$ (J mol^{-1} K^{-1})	$\Delta_f H°$ (kJ mol^{-1})	$\Delta_f G°$ (kJ mol^{-1})
$Ca_3(PO_4)_2$	whitlockite	310.178	236.0±0.8	−4120.8±5.0	−3883.6±5.0
$Ca_5(PO_4)_3OH$	hydroxyapatite	502.314	390.4±1.7	−6738.5±5.0	−6337.1±5.0
$Ca_5(PO_4)_3F$	fluorapatite	504.305	387.9±1.7	−6872.0±5.0	−6489.7±5.0
CaO	lime, calcia	56.078	38.1±0.4	−635.1±0.9	−603.1±0.9
$Ca(OH)_2$	portlandite	74.093	83.4±0.4	−986.1±1.3	−898.0±1.3
CaS	oldhamite	72.145	56.7±1.3	−474.9±2.1	−469.5±2.1
$CaSO_4$	anhydrite	136.143	107.4±0.2	−1434.4±4.2	−1321.8±4.3
$CaSO_4 \cdot 2H_2O$	gypsum	172.173	193.8±0.3	−2023.0±4.3	−1797.0±4.4
$CaCl_2$	hydrophilite	110.984	104.6±1.3	−795.8±0.7	−747.7±0.8
CaF_2	fluorite	78.075	68.9±0.3	−1228.0±2.0	−1175.3±2.0
$CaTiO_3$	perovskite	135.944	93.6±0.4	−1660.6±1.7	−1574.8±1.8
$CaMoO_4$	powellite	200.017	122.6±0.8	−1541.4±0.9	−1434.7±0.9
$CaWO_4$	scheelite	287.917	126.4±0.8	−1645.2±0.9	−1538.0±0.9
Cd	metal, hexag	112.412	51.80±0.15	0	0
	gas	112.412	167.749±0.004	111.80±0.2	77.2±0.2
Cd^{2+}	aq, m=1	112.411	−2.8±1.5	−75.92±0.6	−77.6±0.6
$CdCO_3$	octavite	172.421	92.5±5.5	−750.6±2.5	−669.4±2.6
CdO	monteponite	128.411	54.8±1.5	258.35±0.4	−228.7±0.6
CdS	greenockite	144.478	72.2±0.3	−149.6±1.3	−146.1±1.3
Ce	metal, fcc	140.116	72.0±4.0	0	0
	gas	140.116	191.77	420.7±2.1	385.0±2.1
Ce^{3+}	aq, m=1	140.114	−205.	−696.2	−672.0
Ce^{4+}	aq, m=1	140.114	−301.	−537.2	−503.8
CeO_2	cerianite	172.115	62.3±0.1	−1088.7±1.5	−1025.4±1.9
Ce_2O_3	c, hexagonal, α	328.230	150.62±4.18	−1796.2±8.4	−1707.9±8.4
Cl_2	gas	70.9058	223.081±0.010	0	0
Cl	gas	35.4529	165.190±0.004	121.302±0.008	105.306±0.01
Cl^-	aq, m=1	35.4534	56.60±0.20	−167.080±0.1	−131.2±0.1
HCl	gas	36.4608	186.902±0.005	−92.31±0.10	−95.3±0.1
	aq, m=1	36.4608	56.5	−167.2	−131.2
ClO	gas	51.452	224.96	101.63	98.40
ClO_4^-	aq, m=1	99.4512	182.0	−129.33	−8.52
Co	metal, hcp	58.9332	30.04±0.42	0	0
	gas	58.9332	179.52±0.01	426.7	382.1
Co_2SiO_4	Co olivine	209.950	142.6±0.2	−1412.0±2.0	−1308.7±2.0
CoO	c, cubic	74.933	52.8±0.3	−237.9±1.3	−214.1±1.3

continued

Table 1.21 *(continued)*

Chemical Formula	Phase or Name	Formula Weight	$S°$ ($J\ mol^{-1}\ K^{-1}$)	$\Delta_f H°$ ($kJ\ mol^{-1}$)	$\Delta_f G°$ ($kJ\ mol^{-1}$)
Co_3O_4	c, cubic	240.797	109.3±0.3	−918.8±2.0	−802.2±2.0
CoS_2	cattierite	123.066	74.8±0.2	−150.9±4.9	−145.1±4.9
Co_3S_4	linnaeite	305.066	176.0±0.4	−347.5±7.3	−334.9±7.3
Cr	metal, bcc	51.9962	23.62±0.21	0	0
	gas	51.9962	174.31±0.40	397.5±4.2	352.6±4.2
CrN	carlsbergite	66.003	37.7±2.1	−117.2±8.4	−92.8±8.4
Cr_2O_3	eskolaite	151.991	81.2±1.3	−1134.7±8.4	−1053.1±8.4
CrO_4^{2-}	aq, m=1	115.995	50.21	−881.15	−727.75
$Cr_2O_7^{2-}$	aq, m=1	215.989	261.9	−1490.3	−1301.1
$FeCr_2O_4$	chromite	223.835	146.0±1.7	−1445.5±5.0	−1344.5±6.0
Cs	metal, bcc	132.9055	85.10±0.30	0	0
	gas	132.9055	175.599±0.003	76.5±1.0	49.5±1.0
Cs^+	aq, m=1	132.9049	132.1±0.5	−258.0±0.5	−291.5±1.0
Cs_2O	c, hexagonal	281.810	146.9±0.4	−346.0±1.2	−308.4±1.2
Cu	metal, fcc	63.546	33.15±0.08	0	0
	gas	63.546	166.397±0.004	337.6±1.2	297.9±1.2
Cu^+	aq, m=1	63.546	40.60±0.40	71.7±0.1	50.0±0.1
Cu^{2+}	aq, m=1	63.545	−98.0±4.0	64.9±1.0	65.1±0.1
$Cu_2(OH)_2CO_3$	malachite	221.116	166.3±2.5	−1054.0±2.1	−890.2±2.2
$Cu_3(OH)_2(CO_3)_2$	azurite	344.671	254.4±3.8	−1632.2±2.0	−1391.4±2.2
CuO	tenorite	79.546	42.6±0.2	−156.1±2.0	−128.3±2.0
Cu_2O	cuprite	143.092	92.4±0.3	−170.6±0.1	−147.8±0.1
CuS	covellite	95.613	67.4±0.1	−54.6±0.3	−55.3±0.3
Cu_2S	chalcocite	159.159	116.2±0.2	−83.9±1.1	−89.2±1.1
$CuFeS_2$	chalcopyrite	183.525	124.9±0.2	−194.9±1.6	−195.1±1.6
Cu_5FeS_4	bornite	501.843	398.5±0.8	−371.6±2.1	−394.7±2.1
$CuFe_2S_3$	cubanite	271.436	205.0±0.8	...	−564.5±3.4
$CuSO_4$	chalcocyanite	159.611	109.5±0.6	−771.4±1.2	−662.3±1.4
$CuSO_4 \cdot 5H_2O$	chalcanthite	249.687	301.2±0.6	−2279.7±3.4	−1880.0±3.6
CuCl	nantokite	98.999	86.2±2.0	−137.2±10.0	−119.9±10.0
CuI	marshite	190.451	96.6	−67.8±5.0	−69.4±5.0
Dy	metal, hcp	162.503	74.89±0.84	0	0
	gas	162.503	199.63	290.4	254.4
Dy^{3+}	aq, m=1	162.501	−231.0	−699.0	−665.0
Dy_2O_3	c, cubic	373.004	149.79±0.85	−1863.13±3.93	−1771.4±4.0
Er	metal, hcp	167.263	73.18±0.15	0	0

continued

Table 1.21 (continued)

Chemical Formula	Phase or Name	Formula Weight	$S°$ (J mol^{-1} K^{-1})	$\Delta_f H°$ (kJ mol^{-1})	$\Delta_f G°$ (kJ mol^{-1})
Er	gas	167.263	195.59	317.1	280.7
Er^{3+}	aq, m=1	167.261	−244.3	−705.4	−669.1
Er$_2$O$_3$	c, cubic	382.524	155.64±0.85	−1897.86±1.92	−1808.9±2.0
Eu	metal, cubic	151.964	80.79±0.16	0	0
Eu	gas	151.964	188.795	175.3	142.2
EuO	c, cubic	167.964	62.76±0.85	−592.0±8.4	−556.1±8.4
Eu$_2$O$_3$	c, cubic	351.926	146.44±8.50	−1651.4±8.4	−1555.2±8.4
F$_2$	gas	37.9968	202.791±0.005	0	0
F	gas	18.9984	158.750±0.004	79.39±0.30	62.29±0.30
F$^-$	aq, m=1	18.9990	−13.8±0.8	−335.35±0.65	−281.5±0.7
HF	gas	20.0063	73.779±0.003	−273.30±0.70	−275.4±0.7
	aq, m=1	20.0063	88.7	−320.08	−296.82
Fe	metal, bcc	55.845	27.09±0.13	0	0
	gas	55.845	180.49±0.04	415.5±1.3	369.8±1.3
Fe^{2+}	aq, m=1	55.844	−107.1±2.0	−91.1±3.0	−90.0±2.0
Fe^{3+}	aq, m=1	55.844	−280.0±13.0	−49.9±5.0	−16.7±2.0
FeAl$_2$O$_4$	hercynite	173.806	117.0±3.0	−1950.5±8.5	−1838.1±10.0
Fe$_3$C	cohenite	179.546	104.4±3.4	24.9±1.3	19.7±1.7
FeCO$_3$	siderite	115.854	95.5±0.2	−755.9±5.5	−682.8±5.5
FeSiO$_3$	ferrosilite	131.929	94.6±0.3	−1195.2±3.0	−1118.0±3.0
Fe$_2$SiO$_4$	fayalite	203.774	151.0±0.2	−1478.2±1.3	−1379.1±1.3
Fe$_3$Al$_2$Si$_3$O$_{12}$	almandine	497.748	342.6±1.4	−5264.7±3.0	−4942.0±3.3
FePO$_4$·2H$_2$O	strengite	186.847	171.3±1.3	−1888.2±0.9	−1657.5±1.0
Fe$_{0.947}$O	wüstite	68.885	56.6±0.4	−266.3±0.8	−244.9±0.8
'FeO'	stoichiometric	71.845	60.6±1.7	−272.0±2.1	−251.4±2.2
Fe$_2$O$_3$	hematite	159.689	87.4±0.2	−826.2±1.3	−744.4±1.3
Fe$_3$O$_4$	magnetite	231.533	146.1±0.4	−1115.7±2.1	−1012.7±2.1
α-FeO(OH)	goethite	88.852	60.4±0.6	−562.6±2.1	−491.8±2.1
FeS	troilite	87.912	60.3±0.2	−101.0±1.5	−101.3±1.5
Fe$_{0.90}$S	pyrrhotite	82.327	63.2±0.1	−97.6±2.0	−99.6±2.0
Fe$_{0.875}$S	pyrrhotite	80.931	60.7±0.2	−97.5±2.0	−98.9±2.0
FeS$_2$	pyrite	119.978	52.9±0.1	−171.5±1.7	−160.1±1.7
	marcasite	119.978	53.9±0.1	−169.5±2.1	−158.4±2.1
(Fe,Ni)$_9$S$_8$	pentlandite	771.957	...	−837.4±14.6	...
Fe$_{0.9}$Ni$_{2.1}$S$_4$	violarite	301.783	63.80±0.30	−378±11.8	−333±12
FeSO$_4$·7H$_2$O	melanterite	278.017	409.2±1.3	−3014.3±0.6	−2509.5±1.3

continued

Table 1.21 *(continued)*

Chemical Formula	Phase or Name	Formula Weight	S° (J mol⁻¹ K⁻¹)	$\Delta_f H°$ (kJ mol⁻¹)	$\Delta_f G°$ (kJ mol⁻¹)
$FeCl_2$	lawrencite	126.751	118.0±0.4	−341.7±0.4	−302.2±0.4
$FeCl_3$	molysite	162.204	142.3±0.4	−399.5±0.4	−334.0±0.4
$FeTiO_3$	ilmenite	151.711	108.9±0.3	−1232.0±2.5	−1155.5±2.5
Fe_2TiO_4	ulvöspinel	223.555	180.4±2.5	−1493.8±2.0	−1399.9±2.1
$FeWO_4$	ferberite	303.684	131.8±1.7	−1154.8±8.5	−1054.0±8.5
Ga	metal, orthorh	69.723	40.84±0.20	0	0
	gas	69.723	169.04±0.01	272.0±2.1	233.7±2.1
Ga^{3+}	aq, m=1	69.721	−331.	−211.7	−159.0
Ga_2O_3	c, monocl	187.444	84.98±0.42	−1089.10±0.85	−998.34±0.85
Gd	metal, hcp	157.253	31.09±0.21	0	0
	gas	157.253	194.314	397.5	359.8
Gd^{3+}	aq, m=1	157.251	−205.9	−686.	−661.
Gd_2O_3	c, monoclinic	362.504	151.88±0.85	−1819.6±3.6	−1732.3±3.6
Ge	c, cubic	72.612	31.09±0.15	0	0
	gas	72.612	167.900	376.6	335.9
GeH_4	germane gas	76.644	217.13	90.8	113.4
GeO	gas	88.611	224.29	−46.19	−73.19
GeS	c, orthorh	104.679	55.27±0.27	−551.0±0.8	−497.1±0.9
H_2	gas	2.0159	130.680±0.003	0	0
H	gas	1.00795	114.716±0.002	217.999±0.006	203.278±0.006
H^+	gas	1.00740	108.946±0.02	1536.246±0.04	1516.990±0.04
H^+	aq, m=1	1.00740	0	0	0
OH	hydroxyl	17.0074	183.708±0.04	38.99±1.21	34.28±1.21
HO_2	hydroperoxyl	33.007	229.106±0.08	2.1±8.4	14.4±8.4
OH^-	aq, m=1	17.0079	−10.90±0.20	−230.015±0.040	−157.3±0.1
H_2O	liq	18.0153	69.95±0.03	−285.830±0.040	−237.14±0.04
	gas	18.0153	188.835±0.010	−241.826±0.040	−228.58±0.04
H_2O_2	gas	34.015	232.99	−136.11	−105.44
He	gas	4.0026	126.153±0.002	0	0
Hf	metal, hex	178.49	43.56±0.21	0	0
	gas	178.49	186.897±0.2	618.4±6.3	575.6±6.3
HfO_2	c, monocl	210.491	59.3±0.4	−1117.6±1.3	−1061.1±1.3
Hg	liq	200.592	75.90±0.12	0	0
	gas	200.592	174.970±0.005	61.38±0.04	31.88±0.04
Hg^{2+}	aq, m=1	200.591	−36.19±0.80	170.21±0.20	163.5±0.2
Hg_2^{2+}	aq, m=1	401.183	65.74±0.80	166.87±0.50	153.6±0.5

continued

Table 1.21 *(continued)*

Chemical Formula	Phase or Name	Formula Weight	S° ($J\ mol^{-1}\ K^{-1}$)	$\Delta_f H^\circ$ ($kJ\ mol^{-1}$)	$\Delta_f G^\circ$ ($kJ\ mol^{-1}$)
HgO	montroydite	216.591	70.25±0.30	−90.79±0.12	−58.5±0.1
HgS	cinnabar	232.659	82.5±2.1	−54.3±2.1	−40.7±2.7
	metacinnabar	232.659	96.2±4.2	−46.7±1.5	−43.3±0.8
Hg_2Cl_2	calomel	472.090	191.6±0.8	−265.4±0.4	−210.4±0.4
HgI_2	coccinite	454.401	180.0±6.3	−105.4±1.7	−101.7±2.6
Ho	metal, hcp	164.930	75.02±1.67	0	0
	gas	164.930	195.59	300.8	264.8
Ho_2O_3	c, cubic	377.859	158.16±0.32	−1880.70±4.85	−1791.4±5.0
I_2	c, orthorh	253.8089	116.14±0.30	0	0
	gas	253.8089	260.687±0.005	62.42±0.08	19.32±0.08
I	gas	126.9045	180.786±0.002	106.76±0.04	70.17±0.04
I^-	aq, m=1	126.9050	106.45±0.30	−56.78±0.05	−51.7±0.1
HI	gas	127.9124	206.589±0.004	26.36±0.21	1.56±0.21
In	metal, tetragonal	114.818	57.84±0.84	0	0
	gas	114.818	173.79	243.30	208.71
In_2O_3	c, cubic	277.635	104.2	−925.79	−830.68
Ir	metal, fcc	192.217	35.48±0.17	0	0
	gas	192.217	193.578	665.3	617.9
K	metal, bcc	39.0983	64.63±0.20	0	0
	gas	39.0983	160.340±0.003	89.0±0.4	60.5±0.4
K^+	aq, m=1	39.0978	101.20±0.20	−252.14±0.08	−282.5±0.1
$KAlSiO_4$	kaliophilite	158.163	133.3±1.2	−2124.7±3.1	−2008.8±3.1
$KAlSi_2O_6$	leucite	218.247	200.2±1.7	−3037.8±2.7	−2875.1±2.7
$KAlSi_3O_8$	microcline	278.332	214.2±0.4	−3974.6±3.9	−3749.3±3.9
	sanidine	278.332	232.8±0.5	−3965.6±4.1	−3745.8±4.1
$KAl_3Si_3O_{10}(OH)_2$	muscovite (ord)	398.309	287.7±0.6	−5990.0±4.9	−5608.4±4.9
	muscovite (dis)	398.309	306.4±0.6	−5974.8±4.9	−5598.8±4.9
$KMg_3AlSi_3O_{10}(OH)_2$	phlogopite (ord)	417.261	315.9±1.0	−6246.0±6.0	−5860.5±6.0
	phlogopite (dis)	417.261	334.6±1.0	−6226.0±6.0	−5846.0±6.0
$KMg_3AlSi_3O_{10}F_2$	F-phlogopite (ord)	421.243	317.6±2.1	−6375.5±4.0	−6030.1±4.2
	F-phlogopite (dis)	421.243	336.3±2.1	−6355.5±4.0	−6015.7±4.2
K_2O	c, cubic	94.196	94.1±6.3	−363.2±2.1	−322.1±2.8
KOH	c, monocl	56.106	78.9±0.8	−424.7±0.6	−378.9±0.6
KBr	c	119.002	95.9±0.2	−393.8±0.2	−380.4±0.2
KCl	sylvite	74.551	82.6±0.2	−436.5±0.2	−408.6±0.2

continued

Table 1.21 *(continued)*

Chemical Formula	Phase or Name	Formula Weight	S° (J mol⁻¹ K⁻¹)	$\Delta_f H°$ (kJ mol⁻¹)	$\Delta_f G°$ (kJ mol⁻¹)
KCl	gas	74.551	239.09	−214.7±0.4	−233.4±0.4
Kr	gas	83.801	164.085±0.003	0	0
La	metal, hex	138.906	56.90±2.51	0	0
	gas	138.906	182.377	431.0	393.56
La₂O₃	c, cubic	325.809	127.32±0.84	−1793.68±1.59	−1705.96±1.60
Li	metal, bcc	6.941	28.99±0.30	0	0
	gas	6.941	138.781±0.025	159.3±1.0	126.6±1.0
Li⁺	aq, m=1	6.9407	12.24±0.15	−278.47±0.08	−292.9±0.1
LiAlSiO₄	eucryptite	126.006	103.8±0.8	−2123.3±2.0	−2009.2±2.0
LiAlSi₂O₆	α-spodumene	186.090	129.3±0.8	−3053.5±2.8	−2880.2±3.0
	β-spodumene	186.090	154.4±1.2	−3025.3±2.8	−2859.5±3.0
LiAlSi₄O₁₀	petalite	306.259	233.2±0.6	−4886.5±6.3	−4610.7±6.3
Li₂SiO₃	c, orthorh	89.966	80.3±1.3	−1649.5±4.2	−1558.7±4.2
Li₂O	c, cubic	29.882	37.6±0.3	−597.9±2.1	−561.2±2.1
Lu	metal, hcp	174.967	50.96±0.84	0	0
	gas	174.967	184.80	427.6	387.8
Lu₂O₃	c, cubic	397.932	109.96±0.85	−1878.20±7.53	−1788.85±6.28
Mg	metal, hcp	24.305	32.54±0.20	0	0
	gas	24.305	148.648±0.003	147.1±0.8	112.5±0.8
Mg²⁺	aq, m=1	24.304	−137±4	−467.0±0.6	−455.4±0.6
MgAl₂O₄	spinel	142.266	88.7±4.0	−2299.1±2.0	−2176.6±2.3
MgCO₃	magnesite	84.314	65.1±0.1	−1113.3±1.3	−1029.5±1.4
MgCO₃·3H₂O	nesquehonite	138.360	195.6±0.6	−1977.3±0.3	−1723.8±0.5
4MgCO₃·Mg(OH)₂·4H₂O	hydromagnesite	467.638	503.7±1.6	−6514.9±1.1	−5864.2±1.1
Mg₂(OH)₂CO₃·3H₂O	artinite	196.680	232.9±0.7	−2920.6±0.7	−2568.4±0.8
MgSiO₃	enstatite	100.389	66.3±0.1	−1545.6±1.5	−1458.3±1.6
	clinoenstatite	100.389	67.9±0.4	−1545.0±1.5	−1458.1±1.6
	perovskite	100.389	63.6±3.0	−1445.1±5.0	−1357.0±5.0
	ilmenite	100.389	60.4±3.0	−1486.6±5.0	−1397.5±5.0
Mg₂SiO₄	forsterite	140.693	94.1±0.1	−2173.0±2.0	−2053.6±2.0
Mg₃Al₂Si₃O₁₂	pyrope	403.128	266.3±0.8	−6285.0±4.0	−5934.5±4.0
Mg₇Si₈O₂₂(OH)₂	anthophyllite	780.822	534.5±3.5	−12070.0±8.0	−11343.4±8.5
Mg₃Si₂O₅(OH)₄	chrysotile, antigorite	277.113	221.3±0.8	−4360.0±3.0	−4032.4±3.1
Mg₃Si₄O₁₀(OH)₂	talc	379.266	260.8±0.6	−5900.0±2.0	−5520.2±2.1
Mg₃(PO₄)₂	farringtonite	262.858	188.3	−3745.1±10.5	−3502.8±10.5

continued

Table 1.21 *(continued)*

Chemical Formula	Phase or Name	Formula Weight	S° (J mol⁻¹ K⁻¹)	Δ_fH° (kJ mol⁻¹)	Δ_fG° (kJ mol⁻¹)
MgO	periclase	40.3045	26.95±0.15	−601.60±0.30	−569.3±0.3
Mg(OH)₂	brucite	58.320	63.2±0.1	−924.5±0.4	−833.5±0.4
MgS	niningerite	56.372	50.3±0.4	−345.7±4.2	−341.4±4.2
MgSO₄	c	120.369	91.4±0.8	−1284.9±0.6	−1170.5±0.8
MgSO₄·7H₂O	epsomite	246.477	372.0±4.0	−3388.7±0.1	−2871.2±0.9
MgF₂	sellaite	62.3019	57.2±0.5	−1124.2±1.2	−1071.1±1.2
MgCl₂	chlormagnesite	95.211	89.6±0.8	−641.3±0.7	−591.8±0.7
MgTiO₃	geikelite	120.170	74.6±0.2	−1572.8±1.2	−1484.4±1.2
MgCr₂O₄	picrochromite	192.295	106.0±0.8	−1783.6±0.9	−1669.1±0.9
MgFe₂O₄	magnesioferrite	199.993	121.8±2.0	−1441.5±3.0	−1329.6±3.1
Mn	metal, alpha	54.938	32.01±0.08	0	0
	gas	54.938	173.716	283.3±4.2	241.0±4.2
Mn²⁺	aq, m=1	54.937	−73.60±1.0	−220.8±0.5	−228.1±0.5
MnCO₃	rhodochrosite	114.947	98.0±0.1	−892.9±0.5	−819.1±0.6
MnSiO₃	rhodonite	131.022	100.5±1.0	−1321.6±2.0	−1244.7±2.2
	pyroxmangite	131.022	99.4±2.0	−1322.3±2.0	−1245.0±2.4
Mn₂SiO₄	tephroite	201.959	155.9±0.5	−1731.5±3.0	−1631.0±3.0
Mn₇SiO₁₂	braunite	604.645	416.4±0.8	−4260.0±4.0	−3944.7±4.0
MnO₄⁻	aq, m=1	118.936	191.2	−541.4	−447.2
MnO	manganosite	70.937	59.7±0.4	−385.2±0.5	−362.9±0.5
MnO₂	pyrolusite	86.937	52.8±0.1	−520.0±0.7	−465.0±0.7
Mn₂O₃	bixbyite	157.874	113.7±0.2	−959.0±1.0	−882.1±1.0
Mn₃O₄	hausmannite	228.812	164.1±0.2	−1384.5±1.4	−1282.5±1.4
MnS	alabandite	87.005	80.3±0.8	−213.9±0.8	−218.7±0.9
MnS₂	hauserite	119.071	99.9±0.1	−223.8±10.0	−224.6±10.0
MnCl₂	scacchite	125.844	118.2±0.2	−481.3±0.8	−440.5±0.8
MnTiO₃	pyrophanite	150.803	104.9±0.2	−1360.1±4.0	−1280.9±4.0
Mo	metal, bcc	95.94	28.66±0.21	0	0
Mo	gas	95.94	181.95±0.04	659.0±3.8	613.3±3.8
MoO	gas	111.940	241.8±8.4	311.0±33.5	278.0±33.5
MoO₂	gas	127.940	277.0±8.4	−8.3±12.6	−21.2±12.6
MoO₃	gas	143.939	284.0±12.6	−346.4±20.9	−330.8±20.9
H₂MoO₄	gas	161.955	355.6±8.4	−851.0±4.2	−787.2±4.2
MoO₄²⁻	aq, m=1	159.940	27.2	−997.9	−836.3
MoO₃	molybdite	143.939	77.7±0.4	−745.2±0.4	−668.1±0.4
MoS₂	molybdenite	160.074	62.6±0.2	−271.8±4.9	−262.8±4.9

continued

Table 1.21 *(continued)*

Chemical Formula	Phase or Name	Formula Weight	$S°$ ($J\ mol^{-1}\ K^{-1}$)	$\Delta_f H°$ ($kJ\ mol^{-1}$)	$\Delta_f G°$ ($kJ\ mol^{-1}$)
N_2	gas	28.0135	191.609	0	0
N	gas	14.00675	153.30±0.02	472.68±0.10	455.54±0.10
NH_3	gas	17.0306	192.77±0.05	−45.90±0.40	−16.37±0.40
NH_4^+	aq, m=1	18.0380	111.17±0.40	−133.26±0.25	−79.4±0.3
NH_4F	c	37.037	71.96	−463.96	−348.68
NH_4Cl	salammoniac	53.491	94.6±0.4	−314.4±0.3	−202.9±0.3
NH_4Br	c	97.943	113.0	−270.83	−175.2
NH_4I	c	144.943	117.0	−201.42	−112.5
NO	gas	30.006	210.76	90.29±0.17	86.60±0.17
NO_2	gas	46.006	240.1±0.1	33.1±0.4	51.2±0.4
NO_3^-	aq, m=1	62.0056	146.70±0.40	−206.85±0.40	−110.8±0.4
HNO_3	aq, m=1	63.013	146.4	−207.36	−111.25
Na	metal, bcc	22.98977	51.10±0.30	0	0
	gas	22.98977	153.667±0.025	107.3±0.7	76.8±0.7
Na^+	aq, m=1	22.98922	58.45±0.15	−240.34±0.06	−261.5±0.1
$NaAlSiO_4$	nepheline	142.055	124.4±1.3	−2090.4±3.9	−1975.8±3.9
	carnegieite	142.055	118.7±0.3	−2104.3±4.0	−1988.0±4.0
$NaAlSi_2O_6$	jadeite	202.139	133.5±1.3	−3029.3±3.6	−2850.6±4.0
$NaAlSi_3O_8$	albite	262.223	207.4±0.4	−3935.0±2.6	−3711.6±2.6
	analbite	262.223	225.6±0.4	−3923.6±2.6	−3705.6±2.6
$Na_2Mg_3Al_2Si_8O_{22}(OH)_2$	glaucophane	783.544	541.2±3.0	−11964.0±9.0	−11230.8±10.0
$NaCa_2Mg_4Al_3Si_6O_{22}(OH)_2$	pargasite	835.827	582.0±4.0	−12719.8±22.0	−11981.5±22.0
$NaAl_3Si_3O_{10}(OH)_2$	paragonite (ord)	382.200	277.1±0.9	−5949.3±3.8	−5568.5±3.9
	paragonite (dis)	382.200	295.8±0.9	−5933.0±3.8	−5555.7±3.9
$NaHCO_3$	nahcolite	84.007	102.1±1.7	−949.0±0.2	−851.2±0.6
Na_2CO_3	c, monocl	105.989	135.0±0.6	−1129.2±0.3	−1045.3±0.4
$Na_2CO_3 \cdot H_2O$	thermonatrite	124.004	168.1±0.8	−1429.7±0.4	−1286.1±0.5
Na_2O	c	61.979	73.3±0.8	−414.8±0.3	−376.0±0.4
NaOH	c	39.997	64.4±0.8	−425.8±0.1	−379.6±0.3
Na_2SO_4	thenardite	142.044	149.6±0.1	−1387.8±0.4	−1269.8±0.4
$Na_2SO_4 \cdot 10H_2O$	mirabilite	322.197	591.9±0.6	−4327.3±4.0	−3645.8±3.4
NaF	villiaumite	41.988	51.5±0.1	−573.6±0.7	−543.4±0.7
Na_3AlF_6	cryolite	209.941	238.5±0.5	−3316.8±6.0	−3152.1±6.0
$NaCa_2Mg_4Al_3Si_6O_{22}F_2$	fluorpargasite	839.809	583.0±5.0	−12800.5±14.0	−12102.2±14.0
NaCl	halite	58.443	72.1±0.2	−411.3±0.1	−384.2±0.1
	gas	58.443	229.79	−181.4±2.1	−201.3±2.1

continued

Table 1.21 *(continued)*

Chemical Formula	Phase or Name	Formula Weight	S^o (J mol^{-1} K^{-1})	$\Delta_f H^o$ (kJ mol^{-1})	$\Delta_f G^o$ (kJ mol^{-1})
Na$_4$(AlSiO$_4$)$_3$Cl	sodalite	484.606	848.1±4.2	−13457.9±15.8	−12703.7±16.6
NaFeSi$_2$O$_6$	acmite	231.003	170.6±0.8	−2584.5±4.0	−2417.2±4.2
Nb	metal, bcc	92.906	36.46±0.4	0	0
	gas	92.906	186.26±0.4	773.0±8.0	688.4±8.0
NbO	c, hex	108.906	46.0±8.4	−419.7±12.6	−391.9±12.6
	gas	108.906	238.98±3.56	198.7±20.9	168.9±20.9
NbO$_2$	c, tetrag	124.905	54.51±0.29	−795.0±8.4	−739.2±8.4
	gas	124.905	272.9±8.4	−200.0±20.9	−209.3±20.9
Nb$_2$O$_5$	c, orthorh	265.810	137.3±1.3	−1899.5±4.2	−1765.8±4.2
Nd	metal, hcp	144.243	71.09±4.18	0	0
	gas	144.243	189.406	327.6	292.4
Nd$_2$O$_3$	c, hexagonal	336.484	158.57±4.20	−1807.9±1.0	−1721.0±1.0
Ne	gas	20.1798	146.328±0.003	0	0
Ni	metal, fcc	58.6934	29.87±0.08	0	0
	gas	58.6934	182.19±0.08	430.1±8.4	384.7±8.4
Ni^{2+}	aq, m=1	58.692	−128.9±2.0	−54.0±0.9	−45.6±0.9
Ni$_2$SiO$_4$	Ni olivine	209.470	128.1±0.2	−1396.5±3.0	−1288.9±3.0
NiO	bunsenite	74.693	38.0±0.2	−239.3±0.4	−211.1±0.4
NiS	millerite	90.756	53.0±0.4	−91.0±3.0	−63.9±3.0
Ni$_3$S$_2$	heazlewoodite	240.213	133.2±0.3	−216.3±3.0	−210.2±3.0
NiSO$_4$	c	154.758	101.3±0.3	−873.2±1.0	−762.7±1.0
NiSO$_4$·6H$_2$O	retgersite (α)	262.850	334.5±0.4	−2683.4±0.5	−2225.1±0.5
NiSO$_4$·7H$_2$O	morenosite	280.865	378.9±0.4	−2976.5±0.5	−2461.9±0.5
NiCl$_2$	c	129.599	98.2±0.2	−304.9±2.0	−258.8±2.0
NiFe$_2$O$_4$	trevorite	234.382	140.9±5.0	−1070.5±2.0	−965.1±2.5
O	gas	15.9994	161.058±0.003	249.17±0.10	231.74±0.10
O$_2$	gas	31.9989	205.152±0.005	0	0
O$_3$	ozone, gas	47.998	238.93	142.7±1.7	163.2±1.7
Os	metal, hcp	190.23	32.64±0.06	0	0
	gas	190.23	192.573	791.0	745.0
P	white	30.97376	41.09±0.25	0	0
	red	30.97376	22.85±0.08	−17.46	−12.03
	gas	30.97376	163.198±0.003	316.4±1.0	280.0±1.0
P$_2$	gas	61.9475	218.13±0.4	143.65±2.1	103.1±2.1
P$_4$	gas	123.8950	280.0±0.4	58.9±2.1	24.4±2.1
PO$_4^{3-}$	aq, m=1	94.973	−222.0±4.2	−1259.6±0.9	−1001.6±0.9

continued

Table 1.21 *(continued)*

Chemical Formula	Phase or Name	Formula Weight	$S°$ ($J\ mol^{-1}\ K^{-1}$)	$\Delta_f H°$ ($kJ\ mol^{-1}$)	$\Delta_f G°$ ($kJ\ mol^{-1}$)
HPO_4^{2-}	aq, m=1	95.9805	−33.5±1.5	−1292.14	−1089.15
$H_2PO_4^-$	aq, m=1	96.9879	90.4	−1296.29	−1130.28
H_3PO_4	aq, m=1	97.9953	158.2	−1288.34	−1142.54
P_2O_5	c, hex	141.945	114.4±0.4	−1504.9±0.5	−1361.6±0.5
Pb	metal, fcc	207.21	64.80±0.30	0	0
	gas	207.21	175.374±0.005	195.2±0.8	162.2±0.8
Pb^{2+}	aq, m=1	207.21	18.5±1.0	0.92±0.25	−24.2±0.2
$PbCO_3$	cerussite	267.219	131.0±3.4	−699.2±1.2	−625.5±1.6
PbO	litharge (red)	223.209	66.5±0.2	−219.0±0.8	−188.9±0.8
	massicot (yel)	223.209	68.7±0.2	−217.3±0.3	−187.9±0.3
PbO_2	plattnerite	239.209	71.8±0.4	−277.4±2.9	−218.3±2.9
Pb_3O_4	minium	685.628	212.0±6.7	−718.7±6.3	−601.6±6.6
PbS	galena	239.277	91.7±0.7	−98.3±2.0	−96.8±2.0
	gas	239.277	251.41±0.21	131.8±6.3	85.7±6.3
$PbSO_4$	anglesite	303.274	148.5±0.6	−919.97±0.40	−813.1±0.5
$PbCl_2$	cotunnite	278.116	136.0±2.1	−359.4±0.3	−314.1±0.7
$PbMoO_4$	wulfenite	367.149	166.1±2.1	−1051.9±0.9	−951.2±1.2
Pd	metal, fcc	106.421	37.82±0.21	0	0
	gas	106.421	167.05	378.2	339.7
Pr	metal, hcp	140.908	73.93±4.18	0	0
	gas	140.908	189.808	355.6	320.9
Pr_2O_3	c, hex	329.814	158.57±4.20	−1809.58±6.69	−1721.0±6.7
Pt	metal, fcc	195.08	41.63±0.21	0	0
	gas	195.08	192.406	565.3	520.5
PtS	cooperite	227.145	55.1±0.1	−82.4±3.4	−76.9±3.4
Rb	metal, bcc	85.4678	76.74±0.30	0	0
	gas	85.4678	170.093±0.025	80.9±0.4	53.1±0.4
Rb^+	aq, m=1	85.4673	121.5	−251.17	−283.98
Re	metal, hcp	186.207	36.53±0.38	0	0
	gas	186.207	188.938	769.9	724.6
Rh	metal, fcc	102.906	31.54±0.21	0	0
	gas	102.906	185.808	556.9	510.8
Rn	gas	222.018	176.235±0.003	0	0
Ru	metal, hcp	101.072	28.53±0.21	0	0
	gas	101.072	186.507	642.7	595.8
S	c, orthorh	32.067	32.054±0.050	0	0

continued

Table 1.21 *(continued)*

Chemical Formula	Phase or Name	Formula Weight	$S°$ (J mol^{-1} K^{-1})	$\Delta_f H°$ (kJ mol^{-1})	$\Delta_f G°$ (kJ mol^{-1})
S	c, monocl	32.067	33.03±0.05	0.360±0.003	0.070±0.003
S	gas	32.067	167.828±0.035	276.98±0.25	236.50±0.25
S^{2-}	aq, m=1	32.068	−14.6±1.0	33.1±1.0	85.8±1.0
S_2	gas	64.133	228.167±0.010	128.60±0.30	79.69±0.30
S_8	gas	256.533	432.5±5.0	101.3±2.0	48.8±2.0
HS	gas	33.075	195.55±0.02	140.4±3.5	111.1±3.5
HS$^-$	aq, m=1	33.075	67.0±0.9	16.3±0.2	44.8±0.3
H_2S	gas	34.0825	205.81±0.05	−20.6±0.5	−33.4±0.6
	aq, m=1	34.0825	121	−39.7	−27.83
SO_2	gas	64.0646	248.223±0.050	−296.81±0.20	−300.1±0.2
SO_3	gas	80.065	256.8±0.8	−395.7±0.7	−371.0±0.7
SO_3^{2-}	aq, m=1	80.066	−29.0±4.2	−635.5±0.9	−486.5±0.9
SO_4^{2-}	aq, m=1	96.0654	18.50±0.40	−909.34±0.40	−744.0±0.4
HSO_4^-	aq, m=1	97.0728	131.8	−887.34	−755.91
H_2SO_4	aq, m=1	98.080	20.1	−909.27	−744.53
OCS	gas	60.077	231.64±0.02	−141.7±2.0	−168.9±2.0
CS_2	gas	76.144	237.88±0.05	116.7±1.0	66.6±1.0
Sb	c, rhomb	121.760	45.52±0.21	0	0
	gas	121.760	180.15	264.6±2.5	224.5±2.5
Sb_2	gas	243.520	254.81	231.2±2.5	182.4±2.5
Sb_4	gas	487.040	350.00	206.5±0.8	156.4±0.8
Sb_2O_3	valentinite	291.518	123.0±2.5	−708.6±2.9	−626.4±3.0
Sb_2S_3	stibnite	339.720	182.0±3.3	−151.4±2.3	−149.9±2.3
Sc	metal, hex	44.956	34.64±0.21	0	0
	gas	44.956	174.79	377.8	336.03
Sc_2O_3	c	137.910	76.99±0.42	−1908.82±2.51	−1819.37±2.52
Se	c, hex	78.963	42.27±0.05	0	0
	gas	78.963	176.72	227.07	187.03
Se_2	gas	157.926	252.0	146.0	96.2
H_2Se	gas	80.979	219.02	29.7	15.9
Si	c	28.0855	18.81±0.08	0	0
	gas	28.0855	167.980±0.004	450±8	405.5±8
SiC	c, α (hex)	40.096	16.48±0.13	−71.55±6.3	−69.14±6.3
	c, β (cubic)	40.096	16.61±0.13	−73.22±6.3	−70.85±6.3
Si_3N_4	nierite	140.284	113.0±16.7	−744.8±29.3	−647.3±29.3
Si_2N_2O	sinoite	100.184	45.35±0.10	−887.5±10	−802.1±10

continued

Table 1.21 *(continued)*

Chemical Formula	Phase or Name	Formula Weight	$S°$ (J mol^{-1} K^{-1})	$\Delta_f H°$ (kJ mol^{-1})	$\Delta_f G°$ (kJ mol^{-1})
SiO	gas	44.085	211.6±0.8	−100.4±8.4	−127.3±8.4
SiO$_2$	quartz	60.0844	41.46±0.20	−910.7±1.0	−856.3±1.0
	cristobalite	60.0844	43.4±0.1	−908.4±2.1	−854.6±2.1
	tridymite	60.0844	43.9±0.4	−907.5±2.4	−853.8±2.4
	coesite	60.0844	38.5±1.0	−907.8±2.1	−852.5±2.1
	stishovite	60.0844	27.8±0.4	−861.3±2.1	−802.8±2.1
	glass	60.0844	48.5±1.0	−901.6±2.1	−849.3±2.1
H$_4$SiO$_4$	aq, m=1	96.115	180.0±4.2	−1460.0±1.7	−1307.5±2.1
SiF$_4$	gas	104.0791	282.76±0.42	−1614.94±0.84	−1572.71±0.84
Sm	metal, hcp	150.363	69.50±2.09	0	0
	gas	150.363	183.042	206.7	172.8
Sm$_2$O$_3$	c, monocl	348.724	151.04±4.20	−1822.97±2.01	−1796.69±2.02
Sn	metal, white	118.711	51.18±0.08	0	0
	metal, gray	118.711	44.14	−2.09	0.009
	gas	118.711	168.492±0.004	301.2±1.5	266.2±1.5
Sn^{2+}	aq, m=1	118.710	−17.	−8.8	−27.2
SnO	c, tetrag	134.710	57.17±0.30	−280.71±0.20	−251.91±0.21
SnO$_2$	cassiterite	150.710	49.04±0.10	−577.63±0.20	−515.8±0.2
SnS	herzenbergite	150.777	77.0±0.8	−106.5±1.5	−104.6±1.5
SnS$_2$	berndtite	182.844	87.5±0.2	−149.8±5.0	−141.5±5.0
Sr	metal, fcc	87.621	55.00±0.30	0	0
	gas	87.621	164.64±0.02	164.0±1.7	131.5±1.7
Sr^{2+}	aq, m=1	87.620	−31.5±2.0	−550.9±0.5	−563.8±0.8
SrCO$_3$	strontianite	147.630	97.1±1.7	−1218.7±1.5	−1137.6±1.5
SrO	c	103.620	55.5±0.4	−591.3±1.0	−560.7±1.0
SrSO$_4$	celestite	183.685	117.0±4.2	−1453.2±4.2	−1339.6±4.4
Ta	metal, bcc	180.948	41.47±0.21	0	0
	gas	180.948	185.22	782.0±2.1	739.1±2.1
Ta$_2$O$_5$	c, orthorh	441.893	143.1±1.3	−2046.0±4.2	−1911.0±4.2
Tb	metal, hcp	158.925	73.30±0.84	0	0
Tb	gas	158.925	203.58	388.7	349.7
Tb$_2$O$_3$	c, cubic	365.849	156.90±4.2	−1865.2±8.4	−1776.6±8.4
Te	c, hex	127.603	49.71±0.20	0	0
	gas	127.603	182.74	196.73	157.08
Te$_2$	gas	255.206	268.14	168.2	118.0
Th	metal, fcc	232.0381	51.8±0.5	0	0

continued

Table 1.21 *(continued)*

Chemical Formula	Phase or Name	Formula Weight	$S°$ (J mol^{-1} K^{-1})	$\Delta_f H°$ (kJ mol^{-1})	$\Delta_f G°$ (kJ mol^{-1})
Th	gas	232.0381	190.17±0.05	602±6	560.7±6
ThO$_2$	thorianite	264.037	65.23±0.20	−1226.4±3.5	−1169.2±3.5
Ti	metal, hex	47.867	30.72±0.10	0	0
	gas	47.867	180.298±0.010	473±3	428.4±3
TiN	osbornite	61.874	30.24±0.21	−337.6±4.2	−308.9±4.2
TiO$_2$	rutile	79.866	50.62±0.30	−944.0±0.8	−888.8±1.0
	anatase	79.866	49.9±0.3	−938.7±2.1	−883.2±2.1
Ti$_2$O$_3$	c, trigonal	143.732	77.3±1.0	−1520.9±8.4	−1433.9±8.4
Tl	c, hcp	204.384	64.18±0.21	0	0
	gas	204.384	180.963	182.21	147.39
Tm	metal, hcp	168.934	74.01	0	0
	gas	168.934	190.113	232.2	197.5
Tm$_2$O$_3$	c, cubic	385.867	139.75±0.85	−1888.66±0.85	−1794.45±0.85
U	metal, orthorh	238.0289	50.20±0.20	0	0
	gas	238.0289	199.79±0.10	533±8	488.4±8
UO$_2$	uraninite	270.0278	77.0±0.2	−1084.9±1.0	−1031.7±1.0
UO$_2^{2+}$	aq, m=1	270.0267	−97.5	−1019.6	−953.5
UO$_3$	c, monocl	286.0272	96.11±0.40	−1223.8±1.2	−1145.7±1.2
U$_3$O$_8$	c	842.0822	282.55±0.50	−3574.8±2.5	−3369.5±2.5
V	metal, bcc	50.9415	28.94±0.42	0	0
	gas	50.9415	182.3±0.8	515.5±8	469.7±8
VO	c, cubic	66.941	39.0±0.8	−431.8±6.3	−404.2±6.3
V$_2$O$_3$	karelianite	149.881	98.1±1.3	−1218.8±6.3	−1139.0±6.3
V$_2$O$_4$	c	165.881	103.5±2.1	−1427.2±6.3	−1318.4±6.3
V$_2$O$_5$	c, orthorh	181.880	130.5±2.1	−1550.6±6.3	−1419.3±6.3
W	metal, bcc	183.841	32.65±0.10	0	0
	gas	183.841	173.95±0.08	851.0±6.3	808.9±6.3
WO$_2$	c, monocl	215.840	50.6±0.3	−589.7±0.9	−533.9±0.9
WO$_3$	c, monocl	231.839	75.9±1.3	−842.9±0.8	−764.1±0.9
WS$_2$	tungstenite	247.974	67.8±0.3	−241.6±2.5	−233.0±2.5
Xe	gas	131.292	169.685±0.003	0	0
Y	metal, hcp	88.906	44.43±0.25	0	0
	gas	88.906	179.48	421.3	381.1
Y$_2$O$_3$	c, cubic	225.810	99.08±4.20	−1905.31±2.26	−1816.6±2.4
Yb	metal, beta	173.04	59.83±0.17	0	0
	gas	173.04	173.126	152.3	118.4

continued

Table 1.21 *(continued)*

Chemical Formula	Phase or Name	Formula Weight	$S°$ ($J\ mol^{-1}\ K^{-1}$)	$\Delta_f H°$ ($kJ\ mol^{-1}$)	$\Delta_f G°$ ($kJ\ mol^{-1}$)
Yb_2O_3	c, cubic	394.078	133.05±0.85	−1814.60±0.85	−1726.84±0.85
Zn	metal, hcp	65.392	41.63±0.15	0	0
	gas	65.392	160.989±0.004	130.42±0.20	94.9±0.2
Zn^{2+}	aq, m=1	65.391	−109.8±0.5	−153.39±0.20	−147.3±0.2
$ZnCO_3$	smithsonite	125.401	81.2±0.2	−817.0±3.1	−735.3±3.1
Zn_2SiO_4	willemite	222.867	131.4±0.8	−1636.7±5.0	−1523.1±5.0
ZnO	zincite	81.391	43.2±0.1	−350.5±0.3	−320.4±0.3
ZnS	sphalerite	97.459	58.7±0.2	−204.1±1.5	−199.6±1.5
	wurtzite	97.459	58.8±0.2	−203.8±1.5	−199.3±1.5
$ZnSO_4$	zinkosite	161.456	110.5±1.3	−980.1±0.8	−868.7±0.9
Zn_2TiO_4	Zn, Ti spinel	242.649	143.1±3.0	−1652.1±2.0	−1538.4±2.0
$ZnMn_2O_4$	hetaerolite	239.266	149.7±0.5	−1337.0±4.0	−1227.8±4.0
$ZnFe_2O_4$	franklinite	241.080	150.7±0.3	−1188.1±4.0	−1082.1±4.0
Zr	metal, hcp	91.224	38.87±0.20	0	0
	gas	91.224	183.03±0.04	610.0±8.4	567.0±8.4
$ZrSiO_4$	zircon	183.307	84.0±1.3	−2034.2±3.1	−1919.7±3.1
ZrO_2	baddeleyite	123.223	50.4±0.3	−1100.6±1.7	−1042.9±1.7

Notes: This table gives selected thermodynamic data that are useful for calculations in planetary science, cosmochemistry, and terrestrial geochemistry. The data are taken from the following references and the selected values are based on the same reference states for consistency with each other.

Sources: **[AC91]** Abramowitz, S. & Chase, M. W., 1991, *Pure & Appl. Chem.* 63, 1449–1454. **[ACI93]** Alcock, C. B., Chase, M. W., & Itkin, V. P., 1993, *J. Phys. Chem. Ref. Data* 22, 1–85. **[ACI94]** Alcock, C. B., Chase, M. W., & Itkin, V. P., 1994, *J. Phys. Chem. Ref. Data* 23, 385–497. **[Cha85]** Chase, M. W., et al., 1985, *JANAF thermochemical tables,* 3rd ed., AIP, Washington, D.C. **[GVA89]** Gurvich, L. V., Veyts, I. V., & Alcock, C. B., 1989-1994, *Thermodynamic properties of individual substances,* 3 vols., Hemisphere Publishing, N.Y. **[Feg81]** Fegley, M. B., Jr., 1981, *J. Am. Ceram. Soc.* 64, C124–C126. **[HD73]** Hultgren, R., Desai, P., et al., 1973, *Selected values of the thermodynamic properties of the elements,* Am. Soc. Metals, Metals Park, OH. **[KWH95]** Komada, N., Westrum, E. F., Jr., Hemingway, B. S., Zolotov, M. Yu., Semenov, Y. V., Khodakovsky, I. L., & Anovitz, L. M., 1995, *J. Chem. Thermo.* 27, 1119–1132. **[RCB96]** Rocabois, P., Chatillon, C., & Bernard, C., 1996, *J. Am. Ceram. Soc.* 79, 1361–1365. **[RH95]** Robie, R. A. & Hemingway, B. S., 1995, *Thermodynamic properties of minerals and related substances at 298.15 K and 1 bar (10^5 pascals) pressure and at higher temperatures,* USGS Bull No. 2131, Washington, D.C. **[RHF78]** Robie, R. A., Hemingway, B. S., & Fisher, J. R., 1978, *Thermodynamic properties of minerals and related substances at 298.15 K and 1 bar (10^5 pascals) pressure and at higher temperatures,* USGS Bull. No. 1452, Washington, D.C. **[Wag82]** Wagman, D. D. et al., 1982, *The NBS tables of chemical thermodynamic properties,* NBS, Washington, D.C.

Oxygen Fugacity Buffers as a Function of Temperature

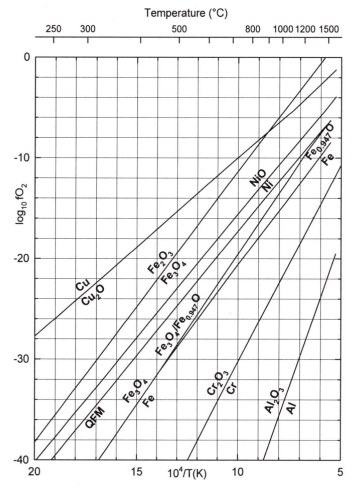

Figure 1.4 The Ellingham diagram for the oxygen fugacity (fO_2) of selected buffers as a function of inverse temperature. In a closed system the fO_2 is regulated by the coexistence of the pure buffer components. Among the buffers shown are the iron-wüstite (Fe-Fe$_{0.947}$O), nickel-nickel oxide (Ni-NiO), hematite-magnetite (Fe$_2$O$_3$-Fe$_3$O$_4$), and the quartz-fayalite-magnetite (SiO$_2$-Fe$_2$SiO$_4$-Fe$_3$O$_4$, QFM) buffers.

1.5 Partition Coefficients

During igneous differentiation processes, elements are distributed among different mineral phases and melt according to their compatibility, which is measured by the partition coefficient or distribution coefficient. The partition coefficients are given by the concentration ratios of an element between two phases, e.g., a solid mineral phase and a liquid melt.

The Nernstian partition coefficient (D) is defined by the ratio of the mass concentration (C) of an element in two phases; e.g., the D for a mineral (min) and coexisting melt (melt) is

$$D^{min/melt} = \frac{C_1}{C_2} = \frac{C_{min}}{C_{melt}} \qquad (1)$$

The molar partition coefficient (k) is defined similarly by the mole fraction (X) ratios of an element between two phases:

$$k^{min/melt} = \frac{X_1}{X_2} = \frac{X_{min}}{X_{melt}} \qquad (2)$$

Partition coefficients are often applied in mass-balance calculations. For a system consisting of i phases, we have the following mass-balance equations for the mass-fractions Y_i of each phase and the concentrations C_i of an element in each phase:

$$\Sigma Y_i = 1 \qquad (3)$$

$$C_{tot} = \Sigma C_i Y_i \qquad (4)$$

For a system containing two phases, the following equations derived from equations (1), (3), and (4) are often useful:

$$\frac{C_1}{C_{tot}} = 1/\left[1 + Y_2 \left(\frac{1}{D} - 1\right)\right] \qquad (5)$$

$$Y_2 = (C_{tot} - D \cdot C_2)/(C_2 - D \cdot C_2) \qquad (6)$$

Partition coefficients are typically a function of temperature and total pressure, and in some systems also depend on oxygen and/or sulfur fugacity (fO_2, fS_2). Partition coefficients are related to thermodynamic properties of the exchange reaction of an element between two phases. For example, the exchange reaction for partitioning of an element between metal and silicate is:

$$M \text{ (in metal)} + \frac{y}{2}O_2 = MO_y \text{ (in silicate)} \tag{7}$$

for which the equilibrium constant K_{eq} is defined as

$$\log K_{eq} = \log \frac{X_{MO_y}}{X_M} + \log \frac{\gamma_{MO_y}}{\gamma_M} - \frac{y}{2} \log fO_2 \tag{8}$$

where the activity a_i is replaced by the product of mole fraction and activity coefficient ($X_i \gamma_i$) of species i in the respective phases. Molar metal/silicate partition coefficients can be parameterized by

$$\log k = a' + b \cdot \log fO_2 + c'/T \tag{9}$$

or, for isothermal cases

$$\log k = a + b \cdot \log fO_2 \tag{10}$$

Comparing the parameters in equations (8) and (10) yields the valence state ($2 \cdot y$) of the oxide in the silicate:

$$2 \cdot y = -b/4 \tag{11}$$

and the relation of the activity coefficients ratio to the fit coefficient a:

$$\log \frac{\gamma_{MO_y}}{\gamma_M} = \log K_{eq} - a \tag{12}$$

Fits similar to equations (9) and (10) can be obtained from Nernstian partition coefficients, and the fit coefficient b for the fO_2 term (or, in the isothermal case, the slope) can be used to derive the valence state of the element in the silicate phase from equation (11). However, equation (12) only applies to fit parameters obtained from molar partition coefficients; Nernstian partition coefficients have to be converted accordingly.

Table 1.22 Some Partition Coefficients Between Minerals and Silicate Melt (weight ratios)

	T (°C)	P (bar)	−log fO$_2$	Buffer [a]	D (crystal/liquid) [b]	Sources
orthopyroxene/liq						
Ca	1150	1	13.1	IW−0.5	0.041	KLW93
Ca	1425	1	9.9	IW−0.5	0.062	KLW93
Sr	1150	1	13.1	IW−0.5	0.0068	KLW93
Sr	1425	1	9.9	IW−0.5	0.00051	KLW93
Ba	1150	1	13.1	IW−0.5	0.0067	KLW93
Ba	1425	1	9.9	IW−0.5	0.00017	KLW93
Ge	1150	1	13.1	IW−0.5	0.025	KLW93
Ge	1425	1	9.9	IW−0.5	0.065	KLW93
P	1330–1340	28×10^3	~6.4	~NNO	0.03	Ul89
Sc	1150	1	13.1	IW−0.5	0.60	KLW93
Sc	1195	1	13.3	IW−1.3	1.4	MW77
Sc	1425	1	9.9	IW−0.5	0.63	KLW93
Sc	1330–1340	28×10^3	~6.4	~NNO	0.33	Ul89
Y	1330–1340	28×10^3	~6.4	~NNO	≤0.01	Ul89
Ti	1150	1	13.1	IW−0.5	0.072	KLW93
Ti	1425	1	9.9	IW−0.5	0.082	KLW93
Ti	1330–1340	28×10^3	~6.4	~NNO	0.10	Ul89
Zr	1150	1	13.1	IW−0.5	0.0040	KLW93
Zr	1425	1	9.9	IW−0.5	0.0033	KLW93
Zr	1330–1340	28×10^3	~6.4	~NNO	0.03	Ul89
Hf	1150	1	13.1	IW−0.5	0.021	KLW93
Hf	1425	1	9.9	IW−0.5	0.0044	KLW93
Hf	1330–1340	28×10^3	~6.4	~NNO	0.14	Ul89
V	1150	1	13.1	IW−0.5	2.6	KLW93
V	1425	1	9.9	IW−0.5	0.61	KLW93
V	1330–1340	28×10^3	~6.4	~NNO	0.90	Ul89
Nb	1150	1	13.1	IW−0.5	0.015	KLW93
Nb	1425	1	9.9	IW−0.5	0.0014	KLW93
Nb	1330–1340	28×10^3	~6.4	~NNO	≤0.01	Ul89
Ta	1330–1340	28×10^3	~6.4	~NNO	≤0.01	Ul89
Cr	1150	1	13.1	IW−0.5	2.0	KLW93
Cr	1425	1	9.9	IW−0.5	0.97	KLW93
Mn	1150	1	13.1	IW−0.5	0.56	KLW93

continued

Table 1.22 *(continued)*

	T (°C)	P (bar)	$-\log fO_2$	Buffer [a]	D (crystal/liquid) [b]	Sources
Mn	1425	1	9.9	IW–0.5	0.39	KLW93
Fe	1150	1	13.1	IW–0.5	0.54	KLW93
Fe	1425	1	9.9	IW–0.5	0.37	KLW93
Co	1150	1	13.1	IW–0.5	1.5	KLW93
Co	1425	1	9.9	IW–0.5	0.71	KLW93
Ni	1150	1	13.1	IW–0.5	1.5	KLW93
Ni	1425	1	9.9	IW–0.5	0.97	KLW93

clinopyroxene/liq

	T (°C)	P (bar)	$-\log fO_2$	Buffer [a]	D (crystal/liquid) [b]	Sources
Sr	1140–1190	1	8–14	~NNO to IW–1	$\log D = -13.62 + 18434/T$	SWS74
P	1330–1340	28×10^3	~6.4	~NNO	0.03	Ul89
Sc	1330–1340	28×10^3	~6.4	~NNO	0.51	Ul89
Y	1330–1340	28×10^3	~6.4	~NNO	0.20	Ul89
Ti	1330–1340	28×10^3	~6.4	~NNO	0.18	Ul89
Zr	1330–1340	28×10^3	~6.4	~NNO	0.03	Ul89
Hf	1330–1340	28×10^3	~6.4	~NNO	0.22	Ul89
V	1330–1340	28×10^3	~6.4	~NNO	1.31	Ul89
Nb	1330–1340	28×10^3	~6.4	~NNO	0.02	Ul89
Ta	1330–1340	28×10^3	~6.4	~NNO	0.02	Ul89
Eu	1140–1190	1	8–14		$\log D = -4.49 + 6350/T$ $+0.04 \log fO_2$	SWS74

diopside/liq

	T (°C)	P (bar)	$-\log fO_2$	Buffer [a]	D (crystal/liquid) [b]	Sources
Sr	1265	1	0.679	air	0.078	GKW74
Ga	1300	1	...	...	0.19	MD87
Ge	1300	1	...	...	1.4	MD87
Mn	1200–1350	1	...	...	$\log D = -5.859 + 8414/T$	AS82
Co	1200–1350	1	...	...	$\log D = -3.789 + 5490/T$	AS82
Ni	1200–1350	1	...	...	$\log D = -2.837 + 4802/T$	AS82
La	1265	1	0.679	air	0.69	GKW74
Ce	1265	1	0.679	air	0.098	GKW74
Nd	1265	1	0.679	air	0.21	GKW74
Sm	1265	1	0.679	air	0.26	GKW74
Eu	1265	1	0.679	air	0.31	GKW74
Eu	1265	1	10–16	IW–4.9 to IW+1.1	$\log D = 0.190$ $+0.0846 \log fO_2$	GKW74
Gd	1265	1	0.679	air	0.30	GKW74

continued

Table 1.22 (continued)

	T (°C)	P (bar)	–log fO₂	Buffer [a]	D (crystal/liquid) [b]	Sources
Dy	1265	1	0.679	air	0.33	GKW74
Er	1265	1	0.679	air	0.30	GKW74
Lu	1265	1	0.679	air	0.28	GKW74
olivine/liq						
Ca	1200	1	12.4	IW–0.5	0.030	KLW93
Ca	1525	1	9.0	IW–0.5	0.022	KLW93
Sr	1200	1	12.4	IW–0.5	0.0034	KLW93
Sr	1250	1	12.5	IW–1.2	0.003	MW77
Sr	1525	1	9.0	IW–0.5	0.0015	KLW93
Ba	1200	1	12.4	IW–0.5	0.000053	KLW93
Ba	1250	1	12.5	IW–1.2	0.005	MW77
Ba	1525	1	9.0	IW–0.5	0.00032	KLW93
Ga	1300	1	...	...	0.024	MD87
Ge	1200	1	12.4	IW–0.5	0.056	KLW93
Ge	1300	1	...	...	0.52	DM87
Ge	1300–1450	1	...	...	0.68	CW83
Ge	1500	20×10³	...	...	0.54	CW83
Ge	1525	1	9.0	IW–0.5	0.097	KLW93
P	1330–1350	1	~6.4	~NNO	≤0.01	Ul89
Sc	1120	1	8.7–10.7	~NNO to NNO–2	0.37	Li76
Sc	1200	1	12.4	IW–0.5	0.53	KLW93
Sc	1250	1	12.5	IW–1.2	0.265	MW77
Sc	1330–1350	1	~6.4	~NNO	0.16	Ul89
Sc	1525	1	9.0	IW–0.5	0.47	KLW93
Y	1200	1	12.4	IW–0.5	0.0070	KLW93
Y	1330–1350	1	~6.4	~NNO	≤0.01	Ul89
Y	1525	1	9.0	IW–0.5	0.0094	KLW93
Ti	1200	1	12.4	IW–0.5	0.0090	KLW93
Ti	1330–1350	1	~6.4	~NNO	0.02	Ul89
Ti	1525	1	9.0	IW–0.5	0.014	KLW93
Zr	1200	1	12.4	IW–0.5	0.0013	KLW93
Zr	1330–1350	1	~6.4	~NNO	0.01	Ul89
Zr	1525	1	9.0	IW–0.5	0.00068	KLW93
Hf	1200	1	12.4	IW–0.5	0.0046	KLW93
Hf	1330–1350	1	~6.4	~NNO	≤0.01	Ul89

continued

Table 1.22 *(continued)*

	T (°C)	P (bar)	–log fO$_2$	Buffer [a]	D (crystal/liquid) [b]	Sources
Hf	1525	1	9.0	IW–0.5	0.0011	KLW93
V	1200	1	12.4	IW–0.5	0.27	KLW93
V	1330–1350	1	~6.4	~NNO	0.06	Ul89
V	1525	1	9.0	IW–0.5	0.30	KLW93
Nb	1330–1350	1	~6.4	~NNO	≤0.01	Ul89
Ta	1330–1350	1	~6.4	~NNO	≤0.01	Ul89
Cr	1200	1	12.4	IW–0.5	0.52	KLW93
Cr	1250	1	12.5	IW–1.2	1.08	MW77
Cr	1525	1	9.0	IW–0.5	0.45	KLW93
Mn	1200	1	12.4	IW–0.5	0.55	KLW93
Mn	1525	1	9.0	IW–0.5	0.50	KLW93
Mn	1200–1350	1	...	...	log D = –25.94+39986/T	AS82
Fe	1200	1	12.4	IW–0.5	0.84	KLW93
Fe	1525	1	9.0	IW–0.5	0.67	KLW93
Co	1200	1	12.4	IW–0.5	2.4	KLW93
Co	1200–1350	1	...	...	log D = –2.989+5328/T	AS82
Ni	1200	1	12.4	IW–0.5	7.6	KLW93
Ni	1525	1	9.0	IW–0.5	9.0	KLW93
Ni	1200–1350	1	...	...	log D = –1.953+4648/T	AS82
La	1200	1	12.4	IW–0.5	0.000028	KLW93
La	1525	1	9.0	IW–0.5	0.000031	KLW93
Ce	1200	1	12.4	IW–0.5	0.000038	KLW93
Ce	1250	1	12.5	IW–1.2	0.12	MW77
Ce	1525	1	9.0	IW–0.5	0.00010	KLW93
Nd	1200	1	12.4	IW–0.5	0.00020	KLW93
Nd	1350	1	10.3	IW+0.5	0.00007	McK86
Nd	1525	1	9.0	IW–0.5	0.00042	KLW93
Sm	1200	1	12.4	IW–0.5	0.00062	KLW93
Sm	1250	1	12.5	IW–1.2	0.010	MW77
Sm	1350	1	9.7	IW+0.5	0.00058	McK86
Sm	1525	1	9.0	IW–0.5	0.0011	KLW93
Eu	1200	1	12.4	IW–0.5	0.00015	KLW93
Eu	1250	1	12.5	IW–1.2	0.010	MW77
Eu	1525	1	9.0	IW–0.5	0.00075	KLW93
Gd	1200	1	12.4	IW–0.5	0.00099	KLW93

continued

Table 1.22 *(continued)*

	T (°C)	P (bar)	−log fO₂	Buffer [a]	D (crystal/liquid) [b]	Sources
Gd	1350	1	9.7	IW+0.5	0.00102	McK86
Gd	1525	1	9.0	IW−0.5	0.0012	KLW93
Dy	1200	1	12.4	IW−0.5	0.0039	KLW93
Dy	1525	1	9.0	IW−0.5	0.0014	KLW93
Er	1200	1	12.4	IW−0.5	0.0087	KLW93
Er	1525	1	9.0	IW−0.5	0.013	KLW93
Yb	1200	1	12.4	IW−0.5	0.017	KLW93
Yb	1250	1	12.5	IW−1.2	0.042	MW77
Yb	1350	1	10.3	IW+0.5	0.0194	McK86
Yb	1525	1	9.0	IW−0.5	0.030	KLW93
Lu	1200	1	12.4	IW−0.5	0.020	KLW93
Lu	1525	1	9.0	IW−0.5	0.039	KLW93
plagioclase/liq						
K	1240	1	12.5	IW−1.1	0.20	MW77
Rb	1240	1	12.5	IW−1.1	0.08	MW77
Sr	1240	1	12.5	IW−1.1	1.87	MW77
Sr	1140−1190	1	8−14	~NNO to IW−1	log D = −4.30+6570/T	SWS74
Sr	1150−1400	1	0.679	air	log D = −2.28+3930/T	DW75
Ba	1240	1	12.5	IW−1.1	0.16	MW77
Ba	1150−1400	1	0.679	air	log D = −3.84+5125/T	DW75
Al	1300	1	0.679	air	1.75	MD87
Ga	1300	1	0.679	air	0.86	MD87
Ge	1300	1	0.679	air	0.51	MD87
Y	1150−1400	1	0.679	air	log D = +0.05−3340/T	DW75
La	1150−1400	1	0.679	air	log D = −2.78+3040/T	DW75
Ce	1240	1	12.5	IW−1.1	0.047	MW77
Ce	1150−1400	1	0.679	air	log D = −2.26+2000/T	DW75
Nd	1150−1400	1	0.679	air	log D = −1.83+1268/T	DW75
Sm	1240	1	12.5	IW−1.1	0.033	MW77
Sm	1150−1400	1	0.679	air	log D = −1.79+1016/T	DW75
Eu	1240	1	12.5	IW−1.1	1.2	MW77
Eu	1150−1400	1	0.679	air	log D = −1.59+680/T	DW75
Eu	1140−1190	1	8−14		log D = −4.22+2460/T−0.15 log fO₂	SWS74
Gd	1150−1400	1	0.679	air	log D = −1.34+104/T	DW75
Dy	1150−1400	1	0.679	air	log D = −0.67−1024/T	DW75

continued

Table 1.22 *(continued)*

	T (°C)	P (bar)	−log fO₂	Buffer [a]	D (crystal/liquid) [b]	Sources
Er	1150–1400	1	0.679	air	log D = −0.05–2006/T	DW75
Yb	1240	1	12.5	IW−1.1	0.035	MW77
Lu	1150–1400	1	0.679	air	log D = −2.35+1390/T	DW75
spinel/liq						
Al	1300	1	0.679	air	3.3	MD87
Ga	1300	1	0.679	air	4.6	MD87
Ge	1300	1	0.679	air	0.11	MD87
Sc	1400	1	8–0.679	IW+1.7 to air	0.0478	NSM80
La	1400	1	8–0.679	IW+1.7 to air	0.010	NSM80
Sm	1400	1	8–0.679	IW+1.7 to air	0.0064	NSM80
Eu	1400	1	8–0.679	IW+1.7 to air	0.0061	NSM80
Tb	1400	1	8–0.679	IW+1.7 to air	0.0078	NSM80
Yb	1400	1	8–0.679	IW+1.7 to air	0.0076	NSM80
Lu	1400	1	8–0.679	IW+1.7 to air	0.0213	NSM80
garnet/liq						
P	1330–1340	28×10³	~6.4	~NNO	0.10	UI89
Sc	1330–1340	28×10³	~6.4	~NNO	2.27	UI89
Y	1330–1340	28×10³	~6.4	~NNO	2.11	UI89
Ti	1330–1340	28×10³	~6.4	~NNO	0.28	UI89
Zr	1330–1340	28×10³	~6.4	~NNO	0.32	UI89
Hf	1330–1340	28×10³	~6.4	~NNO	0.44	UI89
V	1330–1340	28×10³	~6.4	~NNO	1.57	UI89
Nb	1330–1340	28×10³	~6.4	~NNO	0.07	UI89
Ta	1330–1340	28×10³	~6.4	~NNO	0.04	UI89
melilite/liq						
Sr	1500	1	0.679	air	0.66	NSM80
Sc	1500	1	0.679	air	0.69	NSM80
La	1500	1	0.679	air	0.475	NSM80
Sm	1500	1	0.679	air	0.608	NSM80
Eu	1500	1	0.679	air	0.578	NSM80
Eu	1535–1542	1	8–0.679	IW+0.4 to air	1.2	NSM80
Gd	1535–1542	1	8–0.679	IW+0.4 to air	1.2	NSM80
Tb	1500	1	0.679	air	0.486	NSM80
Yb	1500	1	0.679	air	0.222	NSM80

continued

Table 1.22 *(continued)*

	T (°C)	P (bar)	–log fO₂	Buffer ᵃ	D (crystal/liquid) ᵇ	Sources
Lu	1500	1	0.679	air	0.238	NSM80
perovskite/liq						
Sr	1420	1	0.679	air	0.734	NSM80
Sc	1420	1	0.679	air	0.161	NSM80
La	1420	1	0.679	air	2.62	NSM80
Sm	1420	1	0.679	air	2.70	NSM80
Eu	1420	1	0.679	air	2.34	NSM80
Eu	1420	1	8	IW+1.4	5.4	NSM80
Gd	1420	1	0.679	air	2.56	NSM80
Tb	1420	1	0.679	air	1.58	NSM80
Yb	1420	1	0.679	air	0.488	NSM80
Lu	1420	1	0.679	air	0.411	NSM80

ᵃ Corresponding buffer for oxygen fugacity. IW = Fe/Fe$_{0.947}$O (wüstite); NNO = Ni/NiO. NNO–0.5 means oxygen fugacity is 0.5 log units below NNO buffer.

ᵇ In some cases, the log of the partition coefficients are given as function of oxygen fugacity or temperature, or both. The temperature in the fit equations is in Kelvin.

Sources: [AS82] Arutyunyan, L. A., & Sargsyan, G. O., 1982, *Dokl. Akad. Nauk SSSR* 264, 159–164. [CW83] Capobianco, C. J., & Watson, E. B., 1983, *Geochim. Cosmochim. Acta* 46, 235–240. [DW75] Drake, M. J., & Weill, D. F., 1975, *Geochim. Cosmochim. Acta* 39, 689–712. [GKW74] Grutzeck, M., Kridelbaugh, S., & Weill, D., 1974, *Geophys. Res. Lett.* 1, 273–275. [KLW93] Kennedy, A. K., Lofgren, G. E., & Wasserburg, G. J., 1993, *Earth Planet. Sci. Lett.* 115, 177–195. [Li76] Lindstrom, D. J., 1976, cited by Jones, J. H., 1984, *Geochim. Cosmochim. Acta* 48, 641–648. [McK86] McKay, G., A., 1986, *Geochim. Cosmochim. Acta* 50, 69–79. [MW77] McKay, G. A., & Weill, D. F., 1977, *Proc. 8th. Lunar Sci. Conf.* 2339–2355. [NSM80] Nagasawa, H., Schreiber, H. D., & Morris, R. V., 1980, *Earth Planet. Sci. Lett.* 46, 431–327. [MD87] Malvin, D. J., & Drake, M. J., 1987, *Geochim. Cosmochim. Acta* 51, 2117–2178. [SWS74] Sun, C. O., Williams, R. J., & Sun, S. S., 1974, *Geochim. Cosmochim. Acta* 38, 1415–1433. [Ul89] Ulmer, P., 1989, *Annu. Rep. of the Director, Geophysical Laboratory, Carnegie Inst.* 1988/1989, pp. 42–47.

Table 1.23 Some Experimental Metal/Silicate and Sulfide/Silicate Partition Coefficients (weight ratios)

	T (°C)	–log fO$_2$	Buffer	wt% S [a]	X$_S$ [a]	D	Phases	Sources
Ga	1200	11.5	IW+0.4	36.5	0.5	0.13	sul$_{liq}$/sil$_{liq}$	Lo91
Ga	1270	12.5	IW–1.4	19–22	~0.3	1.25	sul$_{liq}$/sil$_{liq}$	JD86
Ga	1190	13.6	IW–1.5	0	0	3.7	met$_{sol}$/sil$_{liq}$	DNR84
Ga	1200	13.8	IW–1.9	0	0	5.4	met$_{sol}$/sil$_{liq}$	Lo91
Ga	1270	12.5	IW–1.4	0	0	7.5	met$_{sol}$/sil$_{liq}$	JD86
Ga	1300	12.2	IW–1.5	0	0	4.1	met$_{sol}$/sil$_{liq}$	DNR84
Ga	1300	12.2	IW–1.5	0	0	15	met$_{sol}$/sil$_{liq}$	SPW89
Pb	1270	12.2	IW–1.1	20	0.3	6.7	sul$_{liq}$/sil$_{liq}$	JD86
P	1250	12.6	IW–1.3	18	0.28	4.2	sul$_{liq}$/sil$_{liq}$	JD86
P	1190	13.6	IW–1.5	0	0	2.8	met$_{sol}$/sil$_{liq}$	ND83
P	1250	12.6	IW–1.3	0	0	7	met$_{sol}$/sil$_{liq}$	JD86
P	1300	12.2	IW–1.5	0	0	8.6	met$_{sol}$/sil$_{liq}$	ND83
P	1300	12.2	IW–1.5	0	0	14	met$_{sol}$/sil$_{liq}$	SPW89
As	1200	11.5	IW+0.4	36.5	0.5	180	sul$_{liq}$/sil$_{liq}$	Lo91
As	1200	13.8	IW–1.9	0	0	970	met$_{sol}$/sil$_{liq}$	Lo91
Sb	1200	11.5	IW+0.4	36.5	0.5	100	sul$_{liq}$/sil$_{liq}$	Lo91
Sb	1200	13.8	IW–1.9	0	0	2150	met$_{sol}$/sil$_{liq}$	Lo91
V	1260	13	IW–1.8	0	0	0.011	met$_{sol}$/sil$_{liq}$	DNC89
V	1260	13	IW–1.8	...	...	0.493	sul$_{liq}$/sil$_{liq}$	DNC89
Cr	1200	11.5	IW+0.4	36.5	0.5	1.0	sul$_{liq}$/sil$_{liq}$	Lo91
Cr	1200	13.8	IW–1.9	26	0.38	0.5	sul$_{liq}$/sil$_{liq}$	Lo91
Cr	1260	13	IW–1.8	...	...	2.96	sul$_{liq}$/sil$_{liq}$	DNC89
Cr	1260	13	IW–1.8	0	0	0.358	met$_{sol}$/sil$_{liq}$	DNC89
Mo	1200	9.3	NNO–1.5	42	0.56	3.7	sul$_{liq}$/sil$_{liq}$	Lo91
Mo	1300	8.8	NNO–2.1	39	0.53	26.5	sul$_{liq}$/sil$_{liq}$	Lo91
Mo	1300	12	IW–1.3	26	0.38	935	sul$_{liq}$/sil$_{liq}$	Lo91
Mo	1250–1270	12.5–13	~IW–1.5	24	0.36	1250	sul$_{liq}$/sil$_{liq}$	JD86

continued

Table 1.23 *(continued)*

	T (°C)	–log fO$_2$	Buffer	wt% S[a]	X$_S$[a]	D	Phases	Sources
Mo	1200	13.8	IW–1.9	0	0	350	met$_{sol}$/sil$_{liq}$	Lo91
Mo	1250–1270	12.5–13	~IW–1.5	0	0	3060	met$_{sol}$/sil$_{liq}$	JD86
Mo	1300	12.2	IW–1.5	0	0	2355	met$_{sol}$/sil$_{liq}$	Ra78
Mo	1300	12.2	IW–1.5	0	0	300	met$_{sol}$/sil$_{liq}$	SPW89
W	1200	9.4	NNO–1.6	42	0.56	0.2	sul$_{liq}$/sil$_{liq}$	Lo91
W	1200	13.8	IW–1.9	26	0.38	2.5	sul$_{liq}$/sil$_{liq}$	Lo91
W	1250–1270	12.3–12.5	~IW–1.2	19–24	~0.33	1	sul$_{liq}$/sil$_{liq}$	JD86
W	1190	13.4	IW–1.9	0	0	25	met$_{sol}$/sil$_{liq}$	ND82
W	1200	13.8	IW–1.9	0	0	35	met$_{sol}$/sil$_{liq}$	Lo91
W	1250–1270	12.3–12.5	~IW–1.2	0	0	36	met$_{sol}$/sil$_{liq}$	JD86
W	1300	12.2	IW–1.5	0	0	16	met$_{sol}$/sil$_{liq}$	Ra78
W	1300	12.2	IW–1.5	0	0	42	met$_{sol}$/sil$_{liq}$	SPW89
Mn	1200	11.5	IW+0.4	36.5	0.5	0.35	sul$_{liq}$/sil$_{liq}$	Lo91
Mn	1250	...	...	37.3	0.51	1.6	sul$_{liq}$/sil$_{liq}$	Sa85
Mn	1200	13.8	IW–1.9	26	0.38	0.08	sul$_{liq}$/sil$_{liq}$	Lo91
Mn	1260	13	IW–2.4	0	0	0.004	met$_{sol}$/sil$_{liq}$	DNC89
Mn	1260	13	IW–2.4	...	...	0.062	sul$_{liq}$/sil$_{liq}$	DNC89
Re	1250–1260	13.0–13.5	~IW–2	23–24	0.35	2000	sul$_{liq}$/sil$_{liq}$	JD86
Re	1250–1260	13.0–13.5	~IW–2	0	0	1.7×10^5	met$_{sol}$/sil$_{liq}$	JD86
Fe	1150	...	...	36.5	0.5	1.2	sul$_{liq}$/sil$_{liq}$	MS76
Fe	1255	...	...	32	0.47	0.15	sul$_{liq}$/sil$_{liq}$	RN78
Co	1150	...	...	36.5	0.5	7	sul$_{liq}$/sil$_{liq}$	MS76
Co	1200	11.5	IW+0.4	36.5	0.5	20	sul$_{liq}$/sil$_{liq}$	Lo91
Co	1200	13.8	IW–1.9	26	0.38	55	sul$_{liq}$/sil$_{liq}$	Lo91
Co	1250	...	...	37.3	0.51	150	sul$_{liq}$/sil$_{liq}$	Sa85
Co	1255	...	...	32	0.47	80	sul$_{liq}$/sil$_{liq}$	RN78
Co	1260	12.6–12.7	~IW–1.5	24–25	0.36	140	sul$_{liq}$/sil$_{liq}$	JD86
Co	1200	13.8	IW–1.9	0	0	180	met$_{sol}$/sil$_{liq}$	Lo91
Co	1260	12.6–12.7	~IW–1.5	0	0	330	met$_{sol}$/sil$_{liq}$	JD86

continued

Table 1.23 *(continued)*

	T (°C)	$-\log fO_2$	Buffer	wt% S [a]	X_S [a]	D	Phases	Sources
Ni	1150	...	...	36.5	0.5	150	sul_{liq}/sil_{liq}	MS76
Ni	1250	...	...	37.3	0.51	390	sul_{liq}/sil_{liq}	Sa85
Ni	1255	...	...	32	0.47	274	sul_{liq}/sil_{liq}	RN78
Ni	1270	12.5	IW–1.4	23	0.34	5000	sul_{liq}/sil_{liq}	JD86
Ni	1270	12.5	IW–1.4	0	0	6650	met_{sol}/sil_{liq}	JD86
Ru	1250	8.5	NNO–1.3	36	0.52	4300	sul_{liq}/sil_{liq}	FCS96
Pd	1200	9.2	NNO–1.4	37.6	0.51	8.8×10^4	sul_{liq}/sil_{liq}	SCF90
Pd	1350	12.2	IW–2	0	0	1.6×10^7	met_{sol}/sil_{liq}	BPS94
Os	1250	8.5	NNO–1.3	36	0.52	2.5×10^4	sul_{liq}/sil_{liq}	FCS96
Ir	1250	8.5	NNO–1.3	36	0.52	2.2×10^4	sul_{liq}/sil_{liq}	FCS96
Ir	1200	9.2	NNO–1.4	37.6	0.51	1.3×10^5	sul_{liq}/sil_{liq}	SCF90
Ir	1270	12.5–13.2	~IW–1.8	22–23	0.34	2.0×10^4	sul_{liq}/sil_{liq}	JD86
Ir	1270	12.5–13.2	~IW–1.8	0	0	1.7×10^6	met_{sol}/sil_{liq}	JD86
Ir	1300	12.7	IW–2	0	0	1×10^{12}	met_{sol}/sil_{liq}	BP95
Pt	1200	9.2	NNO–1.4	37.6	0.51	9.1×10^3	sul_{liq}/sil_{liq}	SCF90
Pt	1250	8.5	NNO–1.3	36	0.52	1.2×10^4	sul_{liq}/sil_{liq}	FCS96
Cu	1150	...	...	36.5	0.5	50	sul_{liq}/sil_{liq}	MS76
Cu	1200	11.5	I+0.4W	36.5	0.5	400	sul_{liq}/sil_{liq}	Lo91
Cu	1255	...	...	32	0.47	245	sul_{liq}/sil_{liq}	RN78
Cu	1245	7.4	~NNO	0	0	90	met_{sol}/sil_{liq}	RB95
Cu	1245	11.9	IW–0.5	0	0	2190	met_{sol}/sil_{liq}	RB95
Cu	1300	12.2	IW–1.5	0	0	45	met_{sol}/sil_{liq}	SPW89
Ag	1250	12.7–13.0	~IW–1.6	23–27	0.37	100	sul_{liq}/sil_{liq}	JD86
Ag	1250	12.7–13.0	~IW–1.6	0	0	1	met_{sol}/sil_{liq}	JD86
Au	1200	9.2	NNO–1.4	37.6	0.51	1000	sul_{liq}/sil_{liq}	SCF90
Au	1270	12.2–13.0	~IW–1.5	20–22	0.32	10000	sul_{liq}/sil_{liq}	JD86
Au	1270	12.2–13.0	~IW–1.5	0	0	13000	met_{sol}/sil_{liq}	JD86

continued

Table 1.23 (continued)

	T (°C)	−log fO$_2$	Buffer	wt% S [a]	X$_S$ [a]	D	Phases	Sources
Zn	1150	...	...	36.5	0.5	0.5	sul$_{liq}$/sil$_{liq}$	MS76
Zn	1200	11.5	IW+0.4	36.5	0.5	1.2	sul$_{liq}$/sil$_{liq}$	Lo91
Zn	1200	13.8	IW−1.9	26	0.38	0.5	sul$_{liq}$/sil$_{liq}$	Lo91

Notes: Data for total pressure of one bar. The "buffer" column gives the oxygen fugacity relative to the IW (iron-wüstite) or NNO (nickel-nickel oxide) buffers in logarithmic units. Phases: met: metal, sul: sulfide bearing metal, sil: silicate, liq: liquid, sol: solid.
[a] weight percent (wt%) sulfur and mole fraction sulfur (X$_S$) in metallic phase

Sources: **[BP95]** Borisov, A., & Palme, H., 1995, *Geochim. Cosmochim. Acta* 59, 481–485. **[BP96]** Borisov, A., & Palme, H., 1996, *Mineral. Petrol.* 56, 297–312. **[BPS94]** Borisov, A., Palme, H., & Spettel, B., 1994, *Geochim. Cosmochim. Acta* 58, 705– 716. **[DNC89]** Drake, M. J., Newsom, H. E., & Capobianco, C. J., 1989, *Geochim. Cosmochim. Acta* 53, 2101–2111. **[DNR84]** Drake, M. J., Newsom, H. E., Reed, J. B., & Enright, M. C., 1984, *Geochim. Cosmochim. Acta* 48, 1609–1615. **[EOD96]** Ertel, W, O'Neill, H. St. C., Dingwell, D. B., & Spettel, B., 1996, *Geochim. Cosmochim. Acta* 60, 1171–1180. **[FCS96]** Fleet, M. E., Crocket, J. H., & Stone, W. E., 1996, *Geochim. Cosmochim. Acta* 60, 2397–2412. **[HBP94]** Holzheid, A., Borisov, A., & Palme, H., 1994, *Geochim. Cosmochim. Acta* 58, 1975–1981. **[Hi91]** Hillgren, V. J., 1991, *Geophys. Res. Lett.* 18, 2077–2080. **[HP96]** Holzheid, A., & Palme, H., 1996, *Geochim. Cosmochim. Acta* 60, 1181–1193. **[JD86]** Jones, J., & Drake, M. J., 1986, *Nature* 322, 221–228. **[Lo91]** Lodders, K., 1991, *Spurenelementverteilung zwischen Sulfid und Silikatschmelze und kosmochemische Anwendungen*, Ph.D. Thesis, Univ. Mainz, Germany, pp. 176. **[MS76]** MacLean, W. H., & Shimazaki, H., 1976, *Econ. Geol.* 76, 1049–1057. **[ND82]** Newsom, H. E., & Drake, M. J., 1982, *Geochim. Cosmochim. Acta* 46, 2483–2489. **[ND83]** Newsom, H., & Drake, M. J., 1983, *Geochim. Cosmochim. Acta* 47, 93–100. **[Ra78]** Rammensee, W., 1978, *Verteilungsgleichgewichte von Spurenelementen zwischen Metallen und Silikaten*, Ph.D. Thesis, Univ. Mainz, Germany, pp. 159. **[RB95]** Ripley, E. M., & Brophy, J. G., 1995, *Geochim. Cosmochim. Acta* 59, 5027–5030. **[RN78]** Rajamani, V., & Naldrett, A. J., 1978, *Econ. Geol.* 73, 82–93. **[RPW83]** Rammensee, W., Palme, H., & Wänke, H., 1983, *Lunar Planet. Sci. Conf.* XIV, 628–624. **[Sa85]** Sargsyan, G. O., 1985, *Geokhimiya* 6, 796–800. **[SCF90]** Stone, W. E., Crocket, J. H., & Fleet, M. E., 1990, *Geochim. Cosmochim. Acta* 54, 2341–2344. **[SPW89]** Schmitt, W., Palme, H., & Wänke, H., 1989, *Geochim. Cosmochim. Acta* 53, 173–185.

Table 1.24 Isothermal Metal/Silicate Partition Coefficients as a Function of Oxygen Fugacity

$$\log D = a + b \log fO_2$$

Element	T (°C)	a	b	$-\log fO_2$ range	Sources
Ga	1190	−11.9	−0.92	13–13.4	DNR84
Ga	1300	−8.8	−0.77	12–12.6	DNR84
Ga	1300	−8.22	−0.77	10.8–13.5	SPW89
Ga	1600	−5.45	−0.68	8.2–10.6	SPW89
Ge	1300	−3.27	−0.52	10.8–13.5	SPW89
Ge	1600	−1.57	−0.50	8.2–10.6	SPW89
P	1190	−15.95	−1.21	11.9–13.5	ND83
P	1300	−17.72	−1.53	11.9–13.5	ND83
P	1300	−13.00	−1.16	10.8–13.5	SPW89
P	1600	−7.02	−0.90	8.2–10.6	SPW89
V	1600	−0.643	−1.014	8–11	RPW83
Ta	1600	−19.65	−1.62	8–11	RPW83
Cr	1600	−6.85	−0.643	8–11	RPW83
Mo	1260	−12.45	−1.28	8.6–12.4	Hi91
Mo	1300	−14.44	−1.46	8–13	Ra78
Mo	1300	−15.09	−1.44	10.8–13.5	SPW89
Mo	1600	−11.68	−1.66	6–10	Ra78
W	1260	−16.69	−1.40	8.6–12.4	Hi91
W	1300	−19.17	−1.67	8–13	Ra78
W	1300	−18.51	−1.65	10.8–13.5	SPW89
W	1600	−12.58	−1.57	6–10	Ra78
W	1600	−8.68	−1.07	8.2–10.6	SPW89
Mn	1600	−6.86	−0.503	8–11	RPW83
Fe	1300	−4.43	−0.44	10.8–13.5	SPW89
Fe	1403	−4.40	−0.48	8.1–12.6	HP96
Co	1260	−5.71	−0.67	8.7–12.3	Hi91
Co	1403	−2.97	−0.49	8.1–12.6	HP96
Ni	1260	−3.16	−0.54	8.6–12.4	Hi91
Ni	1403	−1.79	−0.48	8.1–12.6	HP96
Cu	1245	−0.315	−0.309	7.4–11.9	RB95
Cu	1300	−2.98	−0.38	10.8–13.5	SPW89

Note: See Table 1.23 for sources.

Table 1.25 Metal/Silicate Partition Coefficients as a Function of Oxygen
Fugacity and Temperature

$$\log D = a + b \log fO_2 + c \times 10^4 / T(K)$$

Element	a	b	c	Sources
Mo	7.626	−1	−2.46	EOD96, HBP94
W	4.282	−1	−2.37	EOD96
Fe	−4.48	−0.48	0.01	HP96
Co	−3.53	−0.48	0.26	HP96
Ni	−3.19	−0.49	0.12	HP96
Au	6.7	−0.25	0.375	BP96

Note: See Table 1.23 for sources.

Table 1.26 Isothermal Sulfide/Silicate Partition Coefficients as a Function
of Oxygen Fugacity

$$\log D = a + b \log fO_2$$

Element	T (°C)	a	b	−log fO_2 range	Sources
V	1260	−10.32	−0.76	12.2–16.5	DNC89
Mn	1260	−8.73	−0.58	12.2–16.5	DNC89
Mo	1260	−14.00	−1.39˙	8.6–12.4	Hi91
W	1260	−20.97	−1.67	8.6–12.4	Hi91
Ni	1260	−3.10	−0.53	8.6–12.4	Hi91
Co	1260	−6.18	−0.69	8.7–12.3	Hi91

Note: See Table 1.23 for sources.

2

THE SOLAR SYSTEM

2.1 Solar System Elemental Abundances

Solar system elemental abundances (often referred to as "cosmic" abundances in the older literature) represent the best abundance estimates for our solar system. Abundance data are obtained from solar spectroscopy (solar abundances) and from abundance determinations in primitive meteorites (CI-chondrites). Except for volatile elements (e.g., H, C, N, O, noble gases), which are lost from meteorites and elements that are destroyed in the sun (e.g., Li, Be, B), the agreement between solar and meteoritic abundances is generally fair. Because abundance determinations in meteorites are less uncertain, preference is given to these data when selecting solar system abundances.

Elemental abundances are generally reported on an atomic (N) scale, which is standardized by setting the hydrogen abundance to log N_H = [H] = εH = 12 (the astronomical scale) or to N_{Si} = Si = 10^6 atoms (the cosmochemical scale). Data from older compilations are listed for reference.

Table 2.1 Solar System Abundances on the Atomic Astronomical Scale

	Selected	Selected	[GN93]	[AG89]	[AG89]	[LL78]	[RA76]	[All61]
	Sp. & Met.	Met.*	Sp.	Sp.	Met.	Sp.	Sp. & Met.	Sp. & Met.
H	12	...	12	12	...	12	12	12
He	11	...	11	10.99	...	...	10.8	11.21
Li	3.31	3.31	1.16	1.16	3.31	...	1.0	3.50
Be	1.42	1.15	1.15	1.15	1.42	...	1.15	2.80
B	2.88	2.88	(2.6)	(2.6)	2.88	...	<2.1	2.88
C	8.55	...	8.55	8.56	...	8.67	8.62	8.60
N	7.97	...	7.97	8.05	...	7.99	7.94	8.05
O	8.87	...	8.87	8.93	...	8.92	8.84	8.95
F	4.47	4.47	...	4.56	4.48	...	4.56	6.0
Ne	8.07	...	8.07	8.09	...	...	7.57	8.70
Na	6.31	6.31	...	6.33	6.31	6.32	6.28	6.30
Mg	7.56	7.56	...	7.58	7.58	7.62	7.60	7.4

continued

76

Table 2.1 *(continued)*

	Selected	Selected	[GN93]	[AG89]	[AG89]	[LL78]	[RA76]	[All61]
	Sp. & Met.	Met.*	Sp.	Sp.	Met.	Sp.	Sp. & Met.	Sp. & Met.
Al	6.48	6.48	...	6.47	6.48	6.49	6.52	6.22
Si	7.55	7.55	...	7.55	7.55	7.63	7.65	7.50
P	5.46	5.46	...	5.45	5.57	5.45	5.50	5.40
S	7.20	7.20	...	7.21	7.27	7.23	7.2	7.35
Cl	5.27	5.27	...	5.5	5.27	...	5.5	6.25
Ar	6.60	...	6.60	6.56	...	...	6.0	6.88
K	5.12	5.12	...	5.12	5.13	5.12	5.16	4.82
Ca	6.36	6.36	...	6.36	6.34	6.34	6.35	6.19
Sc	3.09	3.09	3.20	3.10	3.09	...	3.04	2.85
Ti	4.93	4.93	5.04	4.99	4.93	...	5.05	4.89
V	4.00	4.00	...	4.00	4.02	...	4.02	3.82
Cr	5.68	5.68	...	5.67	5.68	...	5.71	5.38
Mn	5.52	5.52	...	5.39	5.53	...	5.42	5.12
Fe	7.50	7.48	7.51	7.67	7.51	...	7.50	6.57
Co	4.91	4.91	...	4.92	4.91	...	4.90	4.75
Ni	6.25	6.25	...	6.25	6.25	...	6.28	5.95
Cu	4.27	4.27	...	4.21	4.27	...	4.06	4.50
Zn	4.65	4.65	...	4.60	4.65	...	4.45	4.28
Ga	3.12	3.12	...	2.88	3.13	...	2.8	2.45
Ge	3.63	3.63	...	3.41	3.63	...	3.50	3.2
As	2.36	2.36	...	...	2.37	...	...	2.11
Se	3.40	3.40	...	...	3.35	...	...	3.33
Br	2.61	2.61	...	...	2.63	...	...	2.65
Kr	3.23	3.23	...	...	3.23	...	...	3.21
Rb	2.40	2.40	...	2.60	2.40	...	2.60	2.35
Sr	2.89	2.89	...	2.90	2.93	...	2.90	2.79
Y	2.22	2.22	...	2.24	2.22	...	2.10	2.45
Zr	2.60	2.60	...	2.60	2.61	...	2.75	2.50
Nb	1.40	1.40	...	1.42	1.40	...	1.9	1.50
Mo	1.95	1.95	...	1.92	1.96	...	2.16	1.88
Ru	1.82	1.82	...	1.84	1.82	...	1.83	1.44
Rh	1.07	1.07	...	1.12	1.09	...	1.40	0.80

continued

Solar System

Table 2.1 *(continued)*

	Selected Sp. & Met.	Selected Met.*	[GN93] Sp.	[AG89] Sp.	[AG89] Met.	[LL78] Sp.	[RA76] Sp. & Met.	[All61] Sp. & Met.
Pd	1.70	1.70	...	1.69	1.70	...	1.5	1.26
Ag	1.24	1.24	...	(0.94)	1.24	...	0.85	0.82
Cd	1.76	1.76	1.77	1.86	1.76	...	1.85	1.45
In	0.82	0.82	...	(1.66)	0.82	...	1.65	0.75
Sn	2.13	2.13	...	2.0	2.14	...	2.0	1.57
Sb	1.02	1.02	...	1.0	1.04	...	1.0	0.95
Te	2.23	2.23	...	...	2.24	...	...	2.05
I	1.51	1.50	...	...	1.51	...	...	1.35
Xe	2.23	2.23	...	...	2.23	...	...	2.06
Cs	1.13	1.13	...	...	1.12	...	<1.9	1.16
Ba	2.21	2.21	...	2.13	2.21	...	2.09	2.08
La	1.20	1.20	...	1.22	1.20	...	1.13	1.1
Ce	1.62	1.62	...	1.55	1.61	...	1.55	1.29
Pr	0.80	0.80	...	0.71	0.78	...	0.66	0.66
Nd	1.48	1.48	...	1.50	1.47	...	1.23	1.36
Sm	0.97	0.97	1.01	1.00	0.97	...	0.72	0.89
Eu	0.55	0.55	...	0.51	0.54	...	0.7	0.48
Gd	1.08	1.08	...	1.12	1.07	...	1.12	1.05
Tb	0.34	0.34	...	(−0.1)	0.33	...	...	0.24
Dy	1.16	1.16	...	1.1	1.15	...	1.06	1.08
Ho	0.50	0.50	...	(0.26)	0.50	...	...	0.39
Er	0.97	0.97	...	0.93	0.95	...	0.76	0.84
Tm	0.14	0.14	...	(0.00)	0.13	...	0.26	0.08
Yb	0.95	0.95	...	1.08	0.95	...	0.9	0.78
Lu	0.13	0.13	...	(0.76)	0.12	...	0.76	0.06
Hf	0.74	0.74	...	0.88	0.73	...	0.8	0.40
Ta	−0.13	−0.13	...	...	−0.13	...	...	0.75
W	0.68	0.68	...	(1.11)	0.68	...	1.7	0.60
Re	0.27	0.27	...	...	0.27	...	≤−0.3	0.90
Os	1.38	1.38	...	1.45	1.38	...	0.7	1.40
Ir	1.36	1.36	...	1.35	1.37	...	0.85	1.20
Pt	1.68	1.68	...	1.8	1.68	...	1.75	1.70

continued

Table 2.1 *(continued)*

	Selected Sp. & Met.	Selected Met.*	[GN93] Sp.	[AG89] Sp.	[AG89] Met.	[LL78] Sp.	[RA76] Sp. & Met.	[All61] Sp. & Met.
Au	0.84	0.84	...	(1.01)	0.83	...	0.75	0.66
Hg	1.16	1.16	...	...	1.09	...	<2.1	0.75
Tl	0.81	0.81	...	(0.9)	0.82	...	0.90	0.55
Pb	2.05	2.05	1.95	1.85	2.05	...	1.93	1.50
Bi	0.69	0.69	...	...	0.71	...	<1.9	0.50
Th	0.07	0.07	0.27	0.12	0.08	...	0.2	0.00
U	−0.50	−0.50	...	(<−0.47)	−0.49	...	<0.60	−0.3

[a] Derived from solar spectroscopic and meteorite analyses.

Data are normalized so that $\log N_H = [H] = 12.00$.

Sp.: Spectroscopic solar photospheric value

Met.: Data from analyses of CI-chondrites

Values in parenthesis are uncertain.

* Solar system values based on meteorites were obtained from the selected CI-chondrite data (see Chapter 16) via: $\log (C/(MW \times 1.0676 \times 10^{-4}))$ where "C" = CI-abundances in ppm and "MW" = molecular weight.

Sources: **[AE82]** Anders, E., & Ebihara, M., 1982, *Geochim. Cosmochim. Acta* 46, 2363–2380. **[AG89]** Anders, E., & Grevesse, N., 1989, *Geochim. Cosmochim. Acta* 53, 197–214. **[All61]** Aller, L. H., 1961, in *Interscience monographs and texts in physics and astronomy* (Marshak, R. E., ed.) Vol. VII, pp. 177–195. **[GN93]** Grevesse, N., & Noels, A., 1993, in *Origin and evolution of the elements* (Prantzos, N., Vangioni-Flam, E., Cassé, M., eds.), Cambridge Univ. Press, pp. 14–25. **[LL8]** Lambert, D. L., 1978, *Mon. Not. R. Astron. Soc.* 182, 249–272 (C,N,O) and Lambert, D. L., & Luck, R. E., 1978, *Mon. Not. R. Astron. Soc.* 183, 79–100. **[RA76]** Ross, J. E., & Aller, L. H., 1976, *Science* 191, 1223–1229.

Table 2.2 Solar System Abundances on the Atomic Cosmochemical Scale [a]

	Selected [b]	[AG89]	[AE82]	[Cam73]	[Cam68]	[SU56]
H	2.82×10^{10}	2.79×10^{10}	2.72×10^{10}	3.18×10^{10}	2.6×10^{10}	4.00×10^{10}
He	2.82×10^{9}	2.72×10^{9}	2.18×10^{9}	2.21×10^{9}	2.1×10^{9}	3.08×10^{9}
Li	57.5	57.1	59.7	49.5	45	100
Be	0.72	0.73	0.78	0.81	0.69	20
B	21.4	21.2	24	350	6.2	24
C	1.00×10^{7}	1.01×10^{7}	1.21×10^{7}	1.18×10^{7}	1.35×10^{7}	3.5×10^{6}
N	2.63×10^{6}	3.13×10^{6}	2.48×10^{6}	3.74×10^{6}	2.44×10^{6}	6.6×10^{6}
O	2.09×10^{7}	2.38×10^{7}	2.01×10^{7}	2.15×10^{7}	2.36×10^{7}	2.15×10^{7}
F	832	843	843	2450	3630	1600
Ne	3.31×10^{6}	3.44×10^{6}	3.76×10^{6}	3.44×10^{6}	2.36×10^{6}	8.6×10^{6}
Na	5.75×10^{4}	5.74×10^{4}	5.70×10^{4}	6.0×10^{4}	6.32×10^{4}	4.28×10^{4}
Mg	1.023×10^{6}	1.074×10^{6}	1.075×10^{6}	1.050×10^{6}	1.050×10^{6}	9.12×10^{5}
Al	8.51×10^{4}	8.49×10^{4}	8.49×10^{4}	8.5×10^{4}	8.51×10^{4}	9.48×10^{4}
Si	1.00×10^{6}	1.00×10^{6}	1.00×10^{6}	1.00×10^{6}	1.00×10^{6}	1.00×10^{6}
P	8130	1.04×10^{4}	1.04×10^{4}	9600	1.27×10^{4}	1.00×10^{4}
S	4.47×10^{5}	5.15×10^{5}	5.15×10^{5}	5.0×10^{5}	5.06×10^{5}	3.75×10^{5}
Cl	5250	5240	5240	5700	1970	8850
Ar	1.12×10^{5}	1.01×10^{5}	1.04×10^{5}	1.172×10^{5}	2.28×10^{5}	1.5×10^{5}
K	3720	3770	3770	4200	3240	3160
Ca	6.46×10^{4}	6.11×10^{4}	6.11×10^{4}	7.21×10^{4}	7.36×10^{4}	4.90×10^{4}
Sc	34.7	34.2	33.8	35	33	28
Ti	2400	2400	2400	2775	2300	2440
V	282	293	295	262	900	220
Cr	1.35×10^{4}	1.35×10^{4}	1.34×10^{4}	1.27×10^{4}	1.24×10^{4}	7800
Mn	9330	9550	9510	9300	8800	6850
Fe	8.91×10^{5}	9.00×10^{5}	9.00×10^{5}	8.3×10^{5}	8.90×10^{5}	6.00×10^{4}
Co	2290	2250	2250	2210	2300	1800
Ni	5.01×10^{4}	4.93×10^{4}	4.93×10^{4}	4.80×10^{4}	4.57×10^{4}	2.74×10^{4}
Cu	524	522	514	540	919	212
Zn	1260	1260	1260	1244	1500	486
Ga	37.2	37.8	37.8	48	45.5	11.4
Ge	120	119	118	115	126	50.5
As	6.46	6.56	6.79	6.6	7.2	4.0
Se	70.8	62.1	62.1	67.2	70.1	67.6

continued

Table 2.2 *(continued)*

	Selected [b]	[AG89]	[AE82]	[Cam73]	[Cam68]	[SU56]
Br	11.5	11.8	11.8	13.5	20.6	13.4
Kr	47.9	45	45.3	46.8	64.4	51.3
Rb	7.08	7.09	7.09	5.88	6.95	6.5
Sr	21.9	23.5	23.8	26.9	58.4	18.9
Y	4.68	4.64	4.64	4.8	4.6	8.9
Zr	11.2	11.4	10.7	28	30	54.5
Nb	0.708	0.698	0.71	1.4	1.15	1.00
Mo	2.51	2.55	2.52	4.0	2.52	2.42
Ru	1.86	1.86	1.86	1.9	1.6	1.49
Rh	0.331	0.344	0.344	0.4	0.33	0.214
Pd	1.41	1.39	1.39	1.3	1.5	0.675
Ag	0.490	0.486	0.529	0.45	0.5	0.26
Cd	1.62	1.61	1.59	1.48	2.12	0.89
In	0.186	0.184	0.184	0.189	0.217	0.11
Sn	3.80	3.82	3.82	3.6	4.22	1.33
Sb	0.295	0.309	0.352	0.316	0.381	0.246
Te	4.79	4.81	4.91	6.42	6.76	4.67
I	0.91	0.90	0.90	1.09	1.41	0.80
Xe	4.79	4.7	4.35	5.38	7.10	4.0
Cs	0.380	0.372	0.372	0.387	0.367	0.456
Ba	4.57	4.49	4.36	4.8	4.7	3.66
La	0.447	0.4460	0.448	0.445	0.36	2.00
Ce	1.175	1.136	1.16	1.18	1.17	2.26
Pr	0.178	0.1669	0.174	0.149	0.17	0.40
Nd	0.851	0.8279	0.836	0.78	0.77	1.44
Sm	0.263	0.2582	0.261	0.226	0.23	0.664
Eu	0.100	0.0973	0.0972	0.085	0.091	0.187
Gd	0.339	0.3300	0.331	0.297	0.34	0.684
Tb	0.062	0.0603	0.0589	0.055	0.052	0.0956
Dy	0.407	0.3942	0.398	0.36	0.36	0.556
Ho	0.089	0.0889	0.0875	0.079	0.090	0.118
Er	0.263	0.2508	0.253	0.225	0.22	0.316
Tm	0.039	0.0378	0.0386	0.034	0.035	0.0318
Yb	0.251	0.2479	0.243	0.216	0.21	0.220

continued

Table 2.2 *(continued)*

	Selected [b]	[AG89]	[AE82]	[Cam73]	[Cam68]	[SU56]
Lu	0.038	0.0367	0.0369	0.036	0.035	0.050
Hf	0.155	0.154	0.176	0.21	0.16	0.438
Ta	0.0209	0.0207	0.0226	0.021	0.022	0.065
W	0.135	0.133	0.137	0.16	0.016	0.49
Re	0.0525	0.0517	0.0507	0.053	0.055	0.135
Os	0.676	0.675	0.717	0.75	0.71	1.00
Ir	0.646	0.661	0.660	0.717	0.43	0.821
Pt	1.35	1.34	1.37	1.4	1.13	1.625
Au	0.195	0.187	0.186	0.202	0.20	0.145
Hg	0.407	0.34	0.52	0.4	0.75	0.284
Tl	0.182	0.184	0.184	0.192	0.182	0.108
Pb	3.16	3.15	3.15	4	2.90	0.47
Bi	0.138	0.144	0.144	0.143	0.164	0.144
Th	0.0331	0.0335	0.0335	0.058	0.034	...
U	0.0089	0.0090	0.0090	0.0262	0.0234	...

[a] Derived from solar spectroscopic and meteorite (CI-chondrite) analyses
  Data are normalized so that $N_{Si} = 10^6$ Atoms.
[b] Selected values are obtained from the selected abundances on the log $N_H = 12$ scale via:
  N(on Si-scale) = antilog[log N(on H-scale) − 1.55], so that Si = 10^6.

Sources: **[AE82]** Anders, E., & Ebihara, M., 1982, *Geochim. Cosmochim. Acta* 46, 2363–2380. **[AG89]** Anders, E., & Grevesse, N., 1989, *Geochim. Cosmochim. Acta* 53, 197–214. **[Cam68]** Cameron, A. G. W., 1968, in *Origin and distribution of the elements* (Ahrens, L. H., ed.), Pergamon, Oxford, pp. 125–143. **[Cam73]** Cameron, A. G. W., 1973, *Space Sci. Rev.* 15, 121–146. **[SU56]** Suess, H. E., & Urey, H. C., 1956, *Revs. Modern Phys.* 28, 53–74.

2.2 Condensation Chemistry of the Elements in the Solar Nebula

The following table lists the condensation temperatures and condensates formed by equilibrium condensation from a solar composition gas. Condensation temperatures refer to the temperature where a compound starts forming in a cooling gas. Several elements are not abundant enough to form their own condensates but condense into solid solution with major element condensates. For these elements, the 50% condensation temperatures are listed; these refer to the temperature where 50% of the element is condensed. The column "major gases" identifies the most abundant gaseous species of an element at the condensation temperature.

Condensation temperatures typically increase as the total pressure increases; thus when referring to condensation temperatures, the total pressure should always be noted. Variations in the solar abundances assumed in the calculations may also introduce small changes in the condensation temperatures.

Table 2.3 Equilibrium Condensation Chemistry of the Elements in the Solar Nebula (at P = 10^{-4} bar)

	T_{cond} (K)	Initial Condensate	Major Gases	Cosmochem. Classification	Sources
H	180	H_2O (s)	H_2	atmop.	Lew72
He[*]	<5	He (s)	He	atmop.	Lew72
Li	1225	Li_2SiO_3 in $MgSiO_3$	LiCl, LiF	mod. vol.	WW77
Be	1490, 50% (10^{-3} bar)	$BeAl_2O_4$ in melilite & spinel	Be, $Be(OH)_2$, BeOH	refr. litho.	LL97
B	964, 50% (10^{-3} bar)	$CaB_2Si_2O_8$ in feldspar	HBO, HBO_2, $NaBO_2$	mod. vol.	LL97
C[†]	78	$CH_4 \cdot 6H_2O$ (s)	CO, CH_4	atmop.	Lew72
N[‡]	120	$NH_3 \cdot H_2O$ (s)	N_2, NH_3	atmop.	Lew72
O[§]	—	—	CO, H_2O	atmop.	—
F	736	$Ca_5(PO_4)_3F$	HF	mod. vol.	FL80
Ne[§]	~5	Ne (s)	Ne	atmop.	Lew72
Na	970, 50%	$NaAlSi_3O_8$ in feldspar	Na, NaCl	mod. vol.	FL80
Mg	1340, 50%	Mg_2SiO_4 (s)	Mg	major elem.	GL74, Was85

continued

Table 2.3 *(continued)*

	T_{cond} (K)	Initial Condensate	Major Gases	Cosmochem. Classification	Sources
Al	1670	Al_2O_3 (s)	Al, AlOH, Al_2O, AlS, AlH, AlO	refr. litho.	KF84
Si[#]	1529	$Ca_2Al_2SiO_7$(s)	SiO, SiS	major elem.	KF84
P	1151, 50%	Fe_3P (s)	PO, P, PN, PS	mod. vol.	FL80, Sea78
S	674	FeS (s)	H_2S, HS	mod. vol.	LKF96
Cl	863, 50%	$Na_4[AlSiO_4]_3Cl$ (s)	HCl, NaCl, KCl	mod. vol.	FL80
Ar	50	$Ar·6H_2O$ (s)	Ar	atmop.	SW78
K	1000, 50%	$KAlSi_3O_8$ (s) in feldspar	K, KCl, KOH	mod. vol.	FL80
Ca	1634	$CaAl_{12}O_{19}$(s)	Ca	refr. litho.	KF84
Sc	1652, 50%	Sc_2O_3 (s)	ScO	refr. litho.	KF86
Ti	1600	$CaTiO_3$ (s)	TiO, TiO_2	refr. litho.	KF84
V	1455, 50%	diss. in $CaTiO_3$	VO_2, VO	refr. litho.	KF86
Cr	1301, 50%	diss. in Fe alloy	Cr	mod. vol.	FP85
Mn	1190, 50%	Mn_2SiO_4 in olivine	Mn	mod. vol.	WW77
Fe	1337, 50%	Fe alloy	Fe	major elem.	FP85,Sea78
Co	1356, 50%	diss. in Fe alloy	Co	refr. sid.	FP85
Ni	1354, 50%	diss. in Fe alloy	Ni	refr. sid	FP85
Cu	1170, 50%	diss. in Fe alloy	Cu	mod. vol.	WW77
Zn	684, 50%	ZnS diss. in FeS	Zn	mod vol.	WW77
Ga	918, 50%	diss. in Fe alloy	GaOH, GaCl	mod vol.	WW79
Ge	825, 50%	diss. in Fe alloy	GeS, GeSe	mod vol.	WW79
As	1012, 50%	diss. in Fe alloy	As	mod vol.	WW79
Se	684, 50%	FeSe diss. in FeS	H_2Se, GeSe	mod vol.	WW77
Br[$]	~350	$Ca_5(PO_4)_3Br$ (s)	HBr, NaBr	highly vol.	FL80
Kr	54	$Kr·6H_2O$ (s)	Kr	atmop.	SW78
Rb[$]	~1080	diss. in feldspar	Rb, RbCl	mod. vol.	GL74,Was85
Sr	1217, 50%	diss. in $CaTiO_3$	Sr, $SrCl_2$, $Sr(OH)_2$, SrOH	refr. litho.	KF86
Y	1622, 50%	Y_2O_3 (s)	YO	refr. litho.	KF86
Zr	1717, 50%	ZrO_2 (s)	ZrO_2, ZrO	refr. litho.	KF86
Nb	1517, 50%	diss. in $CaTiO_3$	NbO_2, NbO	refr. litho.	KF86

continued

Table 2.3 *(continued)*

	T_{cond} (K)	Initial Condensate	Major Gases	Cosmochem. Classification	Sources
Mo	1595, 50%	refractory metal alloy	MoO, Mo, MoO_2	refr. sid.	FP85
Ru	1565, 50%	refractory metal alloy	Ru	refr. sid.	FP85
Rh	1392, 50%	refractory metal alloy	Rh	refr. sid.	FP85
Pd	1320, 50%	diss. in Fe alloy	Pd	mod. vol.	FP85
Ag	993, 50%	diss. in Fe alloy	Ag	mod. vol.	WW77
Cd[$]	430 (10^{-5}bar)	CdS in FeS	Cd	highly vol.	Lar73
In[$]	470, 50%	InS in FeS	In, InCl, InOH	highly vol.	Lar73
Sn	720, 50%	diss. in Fe alloy	SnS, SnSe	mod. vol.	WW77
Sb	912, 50%	diss. in Fe alloy	SbS, Sb	mod. vol.	WW79
Te	680, 50%	FeTe diss. in FeS	Te, H_2Te	mod. vol.	WW77
I	?	?	I, HI	highly/mod. vol.?	
Xe	74	$Xe \cdot 6H_2O$ (s)	Xe	atmop.	SW78
Cs	?	?	CsCl, Cs, CsOH	highly/mod. vol.?	
Ba	1162, 50%	diss. in $CaTiO_3$	$Ba(OH)_2$, BaOH, BaS, BaO	refr. litho.	KF86
La	1544, 50%	diss. in $CaTiO_3$	LaO	refr. litho.	KF86
Ce	1440, 50%	diss. in $CaTiO_3$	CeO_2, CeO	refr. litho.	KF86
Pr	1557, 50%	diss. in $CaTiO_3$	PrO	refr. litho.	KF86
Nd	1563, 50%	diss. in $CaTiO_3$	NdO	refr. litho.	KF86
Sm	1560, 50%	diss. in $CaTiO_3$	SmO, Sm	refr. litho.	KF86
Eu	1338, 50%	diss. in $CaTiO_3$	Eu	refr. litho.	KF86
Gd	1597, 50%	diss. in $CaTiO_3$	GdO	refr. litho.	KF86
Tb	1598, 50%	diss. in $CaTiO_3$	TbO	refr. litho.	KF86
Dy	1598, 50%	diss. in $CaTiO_3$	DyO, Dy	refr. litho.	KF86
Ho	1598, 50%	diss. in $CaTiO_3$	HoO, Ho	refr. litho.	KF86
Er	1598, 50%	diss. in $CaTiO_3$	ErO, Er	refr. litho.	KF86
Tm	1598, 50%	diss. in $CaTiO_3$	Tm, TmO	refr. litho.	KF86
Yb	1493, 50%	diss. in $CaTiO_3$	Yb	refr. litho.	KF86
Lu	1598, 50%	diss. in $CaTiO_3$	LuO	refr. litho.	KF86
Hf	1690, 50%	HfO_2(s)	HfO	refr. litho.	KF86
Ta	1543, 50%	diss. in $CaTiO_3$	TaO_2, TaO	refr. litho.	KF86
W	1794, 50%	refractory metal alloy	WO, WO_2, WO_3	refr. sid.	FP85

continued

Table 2.3 *(continued)*

	T_{cond} (K)	Initial Condensate	Major Gases	Cosmochem. Classification	Sources
Re	1818, 50%	refractory metal alloy	Re	refr. sid.	FP85
Os	1812, 50%	refractory metal alloy	Os	refr. sid.	FP85
Ir	1603, 50%	refractory metal alloy	Ir	refr. sid.	FP85
Pt	1411, 50%	refractory metal alloy	Pt	refr. sid.	FP85
Au	1284, 50%	Fe alloy	Au	mod. vol.	WW77
Hg	?	?	Hg	highly/mod. vol.?	
Tl$^\$$	448, 50%	diss. in Fe alloy	Tl	highly vol.	Lar73
Pb$^\$$	520, 50%	diss. in Fe alloy	Pb, PbS	highly vol.	Lar73
Bi	472, 50%	diss. in Fe alloy	Bi	highly vol.	Lar73
Th	1598, 50%	diss. in $CaTiO_3$	ThO_2	refr. litho.	KF86
U	1580, 50%	diss. in $CaTiO_3$	UO_2	refr. litho.	KF86

* This temperature is below cosmic background and condensation will not occur.

† Kinetic inhibition of the CO to CH_4 conversion yields either $CO \cdot 6H_2O(s)$ or CO(s) as the initial condensate.

‡ Kinetic inhibition of the N_2 to NH_3 conversion yields either $N_2 \cdot 6H_2O(s)$ or $N_2(s)$ as the initial condensate.

§ O is the most abundant element in rock, therefore a separate condensation temperature is meaningless. Most O condenses as H_2O ice; the remainder is CO or in rock.

Most Si condenses into $MgSiO_3$ and Mg_2SiO_4 (e.g., 1340 K at 10^{-4} bar).

\$ The condensation chemistry is uncertain and must be reevaluated.

Key to the cosmochemical classification of the elements: atmop. = atmophile, highly vol. = highly volatile, major elem. = major element, mod. vol. = moderately volatile, refr. litho. = refractory lithophile, refr. sid. = refractory siderophile.

Sources: **[FL80]** Fegley, Jr., B., & Lewis, J. S., 1980, *Icarus* 41, 439–455. **[FP85]** Fegley, Jr., B., & Palme, H., 1985, *Earth Planet. Sci. Lett.* 72, 311–326. **[GL74]** Grossman, L., & Larimer, J. W., 1974, *Rev. Geophys. Space Phys.* 12, 71–101. **[KF84]** Kornacki, A. S., & Fegley, Jr., B., 1984, *Proc. Lunar Planet. Sci. Conf. 14th, J. Geophys. Res.* 89, B588–B596. **[KF86]** Kornacki, A. S., & Fegley, Jr., B., 1986, *Earth Planet. Sci. Lett.* 75, 297–310. **[Lar73]** Larimer, J. W., 1973, *Geochim. Cosmochim. Acta* 37, 1603–1623. **[Lew72]** Lewis, J. S., 1972, *Icarus* 16, 241–252. **[LL97]** Lauretta, D. S., & Lodders, K., 1997, *Earth Planet. Sci. Lett.* 146, 315–327. **[LKF96]** Lauretta, D. S., Kremser, D. K., & Fegley, B., 1996, *Icarus* 122, 288–315. **[Sea78]** Sears, D. W., 1978, *Earth Planet Sci. Lett.* 41, 128–138. **[SW78]** Sill, G. T., & Wilkening, L. L., 1978, *Icarus* 33, 13–22. **[Was85]** Wasson, J. T., 1985, *Meteorites*, Springer Verlag, Berlin, pp. 267. **[WW77]** Wai, C. M., & Wasson, J. T., 1977, *Earth Planet. Sci. Lett.* 36, 1–13. **[WW79]** Wai, C. M., & Wasson, J. T., 1979, *Nature* 282, 790–793.

2.3 The Sun, the Planets, and Planetary Satellites

Table 2.4 The Sun, the Planets, and Planetary Satellites: Comparison of Some Orbital and Physical Data

Celestial Body	a (AU)	a (10^6 km)	e	i (deg.)	$P_{Orbital}$ (days)	$P_{Rotation}$ (days)	Radius (km)	Mass M (10^{24} kg)	ρ obs. (g cm^{-3})	GM (m^{3}s^{-2})	Gravity (m s^{-2})	v_{esc} (km s^{-1})
Sun	—	—	—	—	—	24.66225	695950.0	1989100	1.408	1.3272E20	274.03	617.92
Mercury	0.3871	57.91	0.2056	7.005 ec.	87.9694	58.6462	2437.6	0.3302	5.43	2.2033E13	3.701	4.250
Venus	0.7233	108.2	0.0068	3.395 ec.	224.695	R243.0187	6051.84	4.8685	5.243	3.2486E14	8.870	10.361
Earth	1.0000	149.598	0.0167	0.000 ec.	365.256	0.9972697	6371.01	5.9736	5.515	3.9859E14	9.820	11.186
Moon	2.570 E-3	0.38440	0.05490	5.15	27.32166	S	1737.1	0.07349	3.344	4.9037E12	1.624	2.376
Mars	1.5236	227.93	0.0934	1.850 ec.	686.980	1.02596	3389.92	0.64185	3.934	4.2828E13	3.727	5.026
1 Phobos	6.269E-5	9.378E-3	0.015	1.02	0.3189	S	13.5×10.8×9.4	0.96E-8	1.90	7.206E5	5.849E-3	0.0114
2 Deimos	1.568E-4	0.023459	0.0005	1.82	1.2624	S	7.5×6.1×5.5	1.9E-9	1.76	1.201E5	3.026E-3	6.18E-3
Jupiter	5.2026	778.30	0.0485	1.305 ec.	4330.595	0.41354	71492 (1 bar)	1898.6	1.326	1.267E17	25.376	60.236
1 Io	2.821E-3	0.4216	0.0041	0.04	1.769	S	1821.3	0.08918	3.53	5.960E12	1.797	2.558
2 Europa	4.488E-3	0.6709	0.0101	0.470	3.551	S	1560	0.04791	3.02	3.203E12	1.31	2.026
3 Ganymede	7.161E-3	1.070	0.0015	0.195	7.155	S	2634	0.14817	1.94	9.887E12	1.425	2.740
4 Callisto	0.012589	1.883	0.007	0.281	16.689	S	2400	0.10766	1.85	7.180E12	1.24	2.446
5 Amalthea	1.213E-3	0.1813	0.003	0.40	0.4981	S	~(135×83×75)	~7.2E-6	...	...	...	...
6 Himalia	0.0768	11.480	0.158	27.63	250.57	0.4	~93	~9.5E-6	...	...	...	...
7 Elara	0.0785	11.737	0.207	24.77	259.65	0.5	~38	~7.6E-7	...	...	...	...
8 Pasiphae	0.1571	23.500	0.378	145	R735	...	~25	~1.9E-7	...	...	...	...
9 Sinope	0.1585	23.700	0.275	153	R758	...	~18	~7.6E-8	...	...	...	...
10 Lysithea	0.0784	11.720	0.107	29.02	259.22	...	~18	~7.6E-8	...	...	...	...
11 Carme	0.1511	22.600	0.207	164	R 692	...	~20	~9.5E-8	...	...	...	...

continued

Table 2.4 *(continued)*

Celestial Body	a (AU)	a (10⁶ km)	e	i (deg.)	P_Orbital (days)	P_Rotation (days)	Radius (km)	Mass M (10²⁴ kg)	ρ obs. (g cm⁻³)	GM (m³s⁻²)	Gravity (m s⁻²)	v_esc (km s⁻¹)
12 Ananke	0.1417	21.200	0.169	147	R 631	...	~15	~3.8E-8	...	...	...	...
13 Leda	0.0742	11.094	0.148	26.07	238.72	...	~8	~5.7E-9	...	...	...	...
14 Thebe	1.483E-3	0.22190	0.015	0.8	0.6745	S	~(55×45)	~7.6E-7	...	...	...	...
15 Adrastea	8.623E-4	0.12898	~0	~0	0.2983	S	~(12.5×10×7.5)	~1.9E-8	...	...	...	...
16 Metis	8.555E-4	0.12796	<0.004	~0	0.2948	S	~20	~9.5E-8	...	...	...	...
Saturn	9.5719	1431.94	0.0532	2.485 ec.	10727.160	0.44401	60268 (1 bar)	568.46	0.6873	3.793E16	10.443	35.478
1 Mimas	1.240E-3	0.1855	0.0202	1.53	0.942	S	196	3.8E-5	1.14	2.502E9	0.0633	0.159
2 Enceladus	1.591E-3	0.2380	0.0045	0.02	1.370	S	250	8.0E-5	1.12	4.871E9	0.0785	0.198
3 Tethys	1.970E-3	0.2947	0.00	1.86	1.888	S	530	7.6E-4	1.00	4.150E10	0.1478	0.396
4 Dione	2.523E-3	0.3774	0.0022	0.02	2.737	S	560	1.05E-3	1.44	7.020E10	0.2238	0.501
5 Rhea	3.524E-3	0.5270	0.001	0.35	4.518	S	764	2.49E-3	1.24	1.541E11	0.2641	0.635
6 Titan	8.169E-3	1.2218	0.0292	0.33	15.945	...	2575	0.13455	1.881	8.978E12	1.354	2.641
7 Hyperion	9.944E-3	1.4811	0.1042	0.43	21.277	chaotic	175×120×100	~1.7E-5	...	...	...	...
8 Iapetus	0.02381	3.5613	0.0283	14.72	79.331	S	718	1.88E-3	1.02	1.061E11	0.2058	0.544
9 Phoebe	0.08660	12.952	0.163	177 ec.	R 550.48	0.4	115×110×105	4.0E-7	...	...	...	...
10 Janus	1.013E-3	0.1515	0.007	0.14	0.695	S	110×95×80	1.98E-6	0.65	1.321E8	0.0163	0.054
11 Epimetheus	1.012E-3	0.1514	0.009	0.34	0.694	S	70×58×50	5.5E-7	0.63	3.670E7	0.0102	0.035
12 Helene	2.523E-3	0.3774	0.005	0.2	2.737	...	18×16×15	...	...	...	...	...
13 Telesto	1.970E-3	0.2947	~0	~0	1.888	...	17×14×13	...	...	...	...	...
14 Calypso	1.970E-3	0.2947	~0	~0	1.888	...	17×11×11	...	...	...	...	...

continued

Table 2.4 *(continued)*

Celestial Body	a (AU)	a (10^6 km)	e	i (deg.)	$P_{Orbital}$ (days)	$P_{Rotation}$ (days)	Radius (km)	Mass M (10^{24} kg)	ρ obs. (g cm^{-3})	GM (m^3 s^{-2})	Gravity (m s^{-2})	v_{esc} (km s^{-1})
15 Atlas	9.204E-4	0.1377	0.002	0.3	0.602	...	19×?×14	...	...	...	...	...
16 Prometheus	9.317E-4	0.1394	0.0024	0.0	0.613	...	70×50×37	~1.4E-7	0.27	9.34E6	0.0026	0.019
17 Pandora	9.317E-4	0.1417	0.0042	0.1	0.629	...	55×43×33	~1.3E-7	0.42	8.67E6	0.0043	0.020
18 Pan	8.931E-4	0.1336	...	...	0.575	...	10	...	...	...	...	...
Uranus	19.194	2877.38	0.0429	0.773 ec.	30717.682	R 0.71833	24973 (1 bar)	86.625	1.318	5.794E15	8.85	21.267
1 Ariel	1.282E-3	0.1910	0.0034	0.31	R 2.520	S	579	1.353E-3	1.67	9.028E10	0.2684	0.558
2 Umbriel	1.786E-3	0.2663	0.0050	0.36	R 4.442	S	586	1.172E-3	1.40	7.820E10	0.221	0.513
3 Titania	2.932E-3	0.4359	0.0022	0.14	R 8.706	S	790	3.517E-3	1.71	2.347E11	0.378	0.772
4 Oberon	3.922E-3	0.5835	0.0008	0.10	R 13.463	S	762	3.01E-3	1.63	2.006E11	0.347	0.727
5 Miranda	8.651E-4	0.1294	0.0027	4.22	R 1.413	S	242	6.93E-5	1.20	4.624E9	0.0790	0.193
6 Cordelia	3.326E-4	0.04977	0.000	0.1	0.335	...	13	...	...	...	...	...
7 Ophelia	3.595E-4	0.05379	0.010	0.1	0.376	...	15	...	...	...	...	...
8 Bianca	3.956E-4	0.05917	<0.001	0.2	0.435	...	21	...	...	...	...	...
9 Cressida	4.130E-4	0.06178	<0.001	0.0	0.464	...	31	...	...	...	...	...
10 Desdemona	4.189E-4	0.06268	<0.001	0.2	0.474	...	27	...	...	...	...	...
11 Juliet	4.303E-4	0.06435	<0.001	0.1	0.493	...	42	...	...	...	...	...
12 Portia	4.419E-4	0.06609	<0.001	0.1	0.513	...	54	...	...	...	...	...
13 Rosalind	4.631E-4	0.06994	<0.001	0.3	0.558	...	27	...	...	...	...	...
14 Belinda	5.031E-4	0.07526	<0.001	0.0	0.624	...	33	...	...	...	...	...
15 Puck	5.750E-4	0.08601	<0.001	0.31	0.762	...	77	...	...	...	...	...

continued

Table 2.4 (*continued*)

Celestial Body	*a* (AU)	*a* (10⁶ km)	*e*	*i* (deg.)	$P_{Orbital}$ (days)	$P_{Rotation}$ (days)	Radius (km)	Mass M (10²⁴ kg)	ρ obs. (g cm⁻³)	GM (m³ s⁻²)	Gravity (m s⁻²)	v_{esc} (km s⁻¹)
16 S1997/U1	0.0521	7.795	0.2	146	654	...	30	...	...	...	...	...
17 S1997/U2	0.0432	6.466	0.4	153	495	...	60	...	...	...	...	...
Neptune	30.066	4497.81	0.010	1.768 ec.	60215.912	0.671252	24764 (1 bar)	102.43	1.638	6.835E15	11.14	23.492
1 Triton	2.372E-3	0.35476	1.6E-5	157.345	R 5.877	S	1353	0.02147	2.054	1.433E12	0.783	1.455
2 Nereid	0.03686	5.5134	0.7512	27.6	360.14	...	170	...	...	...	...	...
3 Naiad	3.224E-4	0.04823	<0.001	4.74	0.294	...	29	...	...	...	...	...
4 Thalassa	3.393E-4	0.05007	<0.001	0.21	0.311	...	40	...	...	...	...	...
5 Despina	3.512E-4	0.05253	<0.001	0.07	0.335	...	74	...	...	...	...	...
6 Galatea	4.142E-4	0.06195	<0.001	0.05	0.429	...	79	...	...	...	...	...
7 Larissa	4.917E-4	0.07355	0.0014	0.20	0.555	...	104×89	...	...	...	...	...
8 Proteus	7.866E-4	0.11765	<0.001	0.55	1.122	...	218×208×201	...	...	...	...	...
Pluto	39.537	5914.65	0.2501	17.121 ec.	90803.66	R 6.3872	1152	0.0131	2.05	8.340E11	0.645	1.211
1 Charon	1.297E-4	0.0196	<0.001	99	6.387	...	593	0.0019	2.02	1.268E11	0.369	0.658

Notes: a: Semimajor axis of revolution. *e*: Orbital eccentricity. *i*: Orbital inclination. Orbital and rotational periods in sidereal days. ec.: Orbital inclination to ecliptic; otherwise inclination to planetary equator. R: Retrograde motion. S: Synchronous rotation. Gravity: $= GM/R^2$. v_{esc}: Surface escape velocity $= (2GM/R)^{0.5}$. For more information, see individual planet tables.

Sources: Astronomical Almanac 1997, U.S. Printing Office, Washington D. C. Burns, J. A., & Matthews, M. S. (eds.), 1986, *Satellites*, Univ. of Arizona Press, Tucson, pp. 1021. Gladman, B. J., Nicholson, P. D., Burns, J. A., Kavelaars, J. J., Marsden, B. G., Williams, G. V., & Offutt, W. B., 1998, *Nature* 392, 897–899. Additional sources are given in the individual planet tables.

Table 2.5 Comparison of Some Planetary Properties

Property	Mercury	Venus	Earth	Moon	Mars	Jupiter	Saturn	Uranus	Neptune	Pluto
Mean distance to Sun (AU)	0.3871	0.7233	1.000	1.000	1.5236	5.2026	9.5719	19.194	30.066	39.533
Sidereal revolution period	87.9694 d	224.695 d	365.256 d	27.322 d	686.980 d	11.86 yrs	29.369 yrs	84.07 yrs	164.86 yrs	248.6 yrs
Synodic period	115.88 d	583.92 d	—	29.531 d	779.94 d	1.092 yrs	1.035 yrs	1.012 yrs	1.006 yrs	1.004 yrs
Sidereal rotational period	58.6462 d	R243.018 d	23.9345 h	27.3217 d	24.6230 h	9.925 h	10.65 h	R17.24 h	16.11 h	R6.3872 d
Obliquity to orbit	0.5°	177.4°	23.45°	6.68°	25.19°	3.12	26.73	97.86	29.56	122.5°
Mass (10^{24} kg)	0.33022	4.8685	5.9736	0.07349	0.6418	1898.6	568.46	86.825	102.43	0.0131
Mean radius (km)	2437.6	6051.84	6371.01	1737.1	3389.92	71492 *	60268 *	25559 *	24764 *	1152
Oblateness $(R_{eq} - R_{pol})/R_{eq}$	...	0.0	0.00335	0.00125	0.006476	0.064874	0.097962	0.022927	0.0182	...
Mean obs. density (g cm^{-3})	5.43	5.243	5.515	3.344	3.934	1.326	0.6873	1.318	1.638	1.79–2.06
Uncompressed dens. (g cm^{-3})	5.30	4.00	4.05	3.34	3.74	0.1	0.1	0.3	0.3	2.0
Albedo, geometric	0.059	0.76	0.3–0.5	0.07	0.16	0.51	0.50	0.66	0.62	0.5
Magnetic dipole moment (Tesla× R_{planet}^{3})	3.0×10^{-7}	$<3\times10^{-8}$	0.61×10^{-4}	...	$<6\times10^{-4}$	4.3×10^{-4}	0.21×10^{-4}	0.23×10^{-4}	0.133×10^{-4}	...
Tilt of magnetic dipole axis from spin axis	$<10°$	...	11.5°	...	...	9.6°	0.8°	58.6°	47°	...
$T_{surface}$ (K)	100–700	740	288–293	120–390	~140–300	165 *	134 *	76 *	71.5 *	40
$T_{blackbody}$ (K)	445	325	277	277	225	123	90	63	50	44
Solar constant (Wm^{-2})	9936.9	2613.9	1367.6	1367.6	589.0	50.5	15.04	3.71	1.47	...

* at 1 bar level. Equatorial radii are listed for the giant planets.

For more information and data sources, see individual planet chapters.

Table 2.6 Comparison of the Terrestrial Planets and the Moon

	Mercury	Venus	Earth	Moon	Mars
Bulk Planet					
Mass (10^{24} kg)	0.3302	4.8685	5.9736	0.07349	0.64185
Radius R (km)	2437.6	6051.84	6371.01	1737.1	3389.92
Volume (10^{10} km^3)	6.067	92.84	108.3	2.196	16.32
Obs. density (g cm^{-3})	5.43	5.243	5.515	3.344	3.934
Silicate Portion					
Mass (10^{24} kg)	0.104	3.14	4.03	0.0698	0.51
Mass (% of total)	31.5	64.5	67.5	95	79.4
R–R$_{core}$ (km)	540	2780	2890	1500	1630
Volume (10^{10} km^3)	3.20	78.2	90.6	2.19	14.0
Volume (% of planet)	53	84	84	99.7	86
P at 100 km (kbar)	12	29	32	5.4	13
P at core-mantle boundary (kbar)	80	1000	1390	80	210
Core					
Mass (10^{24} kg)	0.226	1.73	1.94	0.0037	0.132
Mass (% of total)	68.5	35.5	32.5	5	20.6
Radius (km)	1900	3270	3480	240	1760
Volume (10^{10} km^3)	2.87	14.6	17.7	0.0058	2.28

Table 2.7 Some Physical Properties of Planetary Atmospheres

	T$_{Surface}$ (K)	P$_{Surface}$ (bar)	Mean Surface Gravity, g$_P$ (m s^{-2})	Mean Molec. Weight, μ (g mol^{-1})	Pressure Scale Height, H (km)
Mercury	590–700 (sunward)	$<10^{-12}$	3.701	species dependent	species dependent
Venus	740	95.6	8.870	43.45	15.90
Earth	288	1.0	9.820	28.97	8.42
Mars	214	6.36×10^{-3}	3.727	43.34	11.07
Jupiter	165±5 (at 1 bar)	adiabat P>1 bar	25.376	2.22	24.35
Saturn	134±4 (at 1 bar)	adiabat P>1 bar	10.443	2.07	51.54
Titan	94	1.5	1.354	~28.6	~20.2
Uranus	76±2 (at 1 bar)	adiabat P>1 bar	8.85	2.64	27.05
Neptune	71.5±2 (at 1 bar)	adiabat P>1 bar	11.14	2.53–2.69	19.83–21.09
Triton	38±4	$(16±3) \times 10^{-6}$	0.783	~28 (?)	~15
Pluto	~40	$~3 \times 10^{-6}$ (?)	0.645	~16–25 (?)	~26

For more information and data sources see individual planet chapters.

3

The Sun

The sun is a young (population I) star in the G2V spectral class and orbits the center of the galaxy at a distance of ~8.5 kpc with a period of ~240 million years. The sun is in the top 5% of hotter stars in the solar neighborhood, which is dominantly populated by M dwarfs.

The interior of the sun is divided into three regions: (1) the core, where thermonuclear fusion occurs, extending out to 0.2 $R_\odot$, (2) the radiative zone, extending out to 0.7 $R_\odot$, and (3) the convective zone in the outermost 30% of the sun. The visible "surface" of the sun is the photosphere, from which visible sunlight is emitted. Outward from the photosphere are the chromosphere and the corona.

The variations in temperature, density, pressure, and other physical properties with depth inside the sun are calculated from standard solar models, which are computed using standard physics and the best available input data. The standard solar models are constrained by helioseismological observations of solar oscillations. Typical values for temperature, pressure, and density in the sun's core are 15.5×10^6 K (i.e., $T_6 = 15.5$), 2.5×10^{11} bar, and 148 g cm^{-3}, respectively.

Thermonuclear fusion of hydrogen to helium is the energy source for the sun. The mass deficit of ~0.0292 atomic mass units (AMU) between four ^{1}H nuclei and one ^{4}He nucleus corresponds to ~27.16 MeV energy release per ^{4}He nucleus formed. Two sets of nuclear reactions, the CNO cycle and the proton-proton (pp) chain, are responsible for fusion in the sun. The pp chain is the primary mechanism at temperatures in the sun's core and is also the primary fusion mechanism in stars below the sun on the main sequence (K, M stars). In contrast, the CNO cycle is more important at core temperatures $T_6 > 18$ and is more important than the pp chain in stars above the sun on the main sequence (F, A, B, O stars).

The first two reactions in the pp chain are

$$p + p \rightarrow {}^2H + e^+ + \nu_e \tag{1}$$

$$^2H + p \rightarrow {}^3He + \gamma \tag{2}$$

where p is a proton (^{1}H), ^{2}H is a deuteron (also written as D), e^+ is a positron, ν_e is an electron neutrino, and γ is a gamma ray photon. The pp chain splits into several branches. About 69% of the time, reactions (1) and (2) are followed by the reaction

$$^3He + {}^3He \rightarrow {}^4He + p + p \tag{3}$$

whereas approximately 31% of the time, ^7Be is formed via the reaction

$$^3He + {}^4He \rightarrow {}^7Be + \gamma \tag{4}$$

Approximately 99.7% of the time, reaction (4) is followed by the reactions

$$^7Be + e^- \rightarrow {}^7Li + \nu_e \tag{5}$$

$$^7Li + p \rightarrow {}^4He + {}^4He \tag{6}$$

whereas ~0.3% of the time, reaction (4) is followed by the reactions

$$^7Be + p \rightarrow {}^8B + \gamma \tag{7}$$

$$^8B \rightarrow {}^8Be + e^+ + \nu_e \tag{8}$$

$$^8Be \rightarrow {}^4He + {}^4He \tag{9}$$

Reactions (1)–(3) are the pp I chain, reactions (1), (2), and (4)–(6) are the pp II chain, and reactions (1), (2), (4), and (7)–(9) are the pp III chain.

The neutrinos produced in the different branches of the pp chain have been detected by several different experiments on Earth. These experiments (e.g., the *Homestake Mine, GALLEX, SAGE* experiments) measure the neutrino flux in solar neutrino units (1 SNU = 10^{-36} reactions per target atom per second) by measuring the radioactive elements produced by chemical reactions such as

$$^{37}Cl + \nu_e = {}^{37}Ar + e^- \tag{10}$$

$$^{71}Ga + \nu_e = {}^{71}Ge + e^- \tag{11}$$

or by measuring the Cerenkov light emission from neutrinos scattering electrons in a large water tank (*Kamiokande II* experiment). However, the observed neutrino flux of 2.55±0.25 SNU is significantly less than the predicted flux of $9.3^{+1.2}_{-1.4}$ SNU, a discrepancy known as the solar neutrino problem. Three possible explanations for the solar neutrino problem are: (1) standard solar models are incorrect in some respect; (2) hypothetical weakly interacting massive particles (WIMPS) transport energy inside the sun and cool the core, lowering the solar neutrino flux; and (3) the electron neutrinos produced inside the sun change into other types of neutrinos before they reach the earth (Mikheyev-Smirnov-Wolfenstein, or MSW, effect) and hence are not being detected by experiments sensitive to electron neutrinos of specific energies. These three explanations are being tested by current and planned solar neutrino experiments.

Sources: Bahcall, J. N., 1996, *ApJ.* 467, 475–484. Cox, A. N., Livingston, W. C., & Matthews, M. S. (eds.), 1991, *Solar interior and atmosphere,* Univ. of Arizona Press, Tucson, pp. 1403.

Table 3.1 Physical Parameters of the Sun

Property	Value	Property	Value
Mass (kg)	1.9891×10^{30}	$T_{surface}$ (photosphere) (K)	4400 (top)
Radius (km)	695950		6600 (bottom)
Surface area (km^2)	6.0865×10^{12}	$T_{central}$ (K)	1.55×10^7
Photospheric depth (km)	~400		$(1.49-1.57) \times 10^7$
Chromospheric depth (km)	~2500	$T_{blackbody} = T_{eff} = (L/\sigma)^{1/4}$ (K)	5778
GM ($m^3 s^{-2}$)	1.327124×10^{20}	Solar flux at surface (W m^{-2})	6.3167×10^7
Mean density (g cm^{-3})	1.408	Solar constant, flux at 1 AU (W m^{-2})	1367.6
Central density (g cm^{-3})	148 (140–180)	Absolute luminosity, L (W)	3.86×10^{26}
Surf. gravity, GM/R^2 (m s^{-2})	274.03	Mass-energy conversion rate (g s^{-1})	4.3×10^{12}
Moment of inertia, $I/(MR^2)$	0.059	Sun spot cycle (years)	11.4
$v_{esc.}$ at surface (km s^{-1})	617.592	Apparent visual magnitude, V, m_V	−26.78
Sidereal period (days)	25.38	Apparent bolometric magnitude, m_{bol}	−26.85
Obliquity to ecliptic	7°15'	Absolute visual magnitude, M_V	4.82
Rotation rate, mean	~27 days	Absolute bolometric magnitude, M_{bol}	4.75
Rotation rate, equatorial	~25 days	Spectral type	G2V
Rotation rate, polar	~34 days	Measured flux of solar neutrinos	2.55±0.25 SNU
Characteristic magnetic field strengths (Tesla):			
polar field		10^{-4}	
ephemeral (unipolar) active regions		20×10^{-4}	
bright, chromospheric network		25×10^{-4}	
chromospheric plages		200×10^{-4}	
prominences		$(10-100) \times 10^{-4}$	
sunspots		3000×10^{-4}	

1 Tesla = 10^4 Gauss

Sources: Bahcall, J. N., 1996, *ApJ*. 467, 475–484. Cox, A. N., Livingston, W. C., & Matthews, M. S. (eds.), 1991, *Solar interior and atmosphere*, Univ. of Arizona Press, Tucson, pp. 1403. Taylor, R. J., 1989, *Quart. J. Roy. Astron. Soc.* 30, 125–161. Turck-Chièze, S., Cahen, S., Cassé, M., & Doom, C., 1988, *ApJ.* 335, 415–424.

Table 3.2 Solar Interior Structure Standard Model

Radius ($R_\odot$)	Mass ($M_\odot$)	Density (g cm^{-3})	Luminosity ($L_\odot$)	Temperature (10^6 K)
0.000	0.0000	147.74	0.0000	15.513
0.010	0.0001	146.66	0.00089	15.48
0.022	0.001	142.73	0.009	15.36
0.061	0.020	116.10	0.154	14.404
0.090	0.057	93.35	0.365	13.37
0.120	0.115	72.73	0.594	12.25
0.166	0.235	48.19	0.845	10.53
0.202	0.341	34.28	0.940	9.30
0.246	0.470	21.958	0.985	8.035
0.281	0.562	15.157	0.997	7.214
0.317	0.647	10.157	0.992	6.461
0.370	0.748	5.566	0.9996	5.531
0.453	0.854	2.259	1.000	4.426
0.611	0.951	0.4483	1.000	2.981
0.7304	0.9809	0.1528	1.0000	2.035
0.862	0.9964	0.042	1.0000	0.884
0.965	0.9999	3.61×10^{-3}	1.0000	0.1818
1.000	1.0000	1.99×10^{-7}	1.0000	5570×10^{-3}

For solar composition X = 0.7046, Y = 0.2757, Z = 0.0197.

Source: Turck-Chièze, S., Cahen, S., Cassé, M., & Doom, C., 1988, *ApJ.* 335, 415–424.

Table 3.3   Solar Model Atmosphere

log τ_R	log τ_{5000}	T (K)	log P_g (dyn cm^{-2})	log P_e (dyn cm^{-2})	Depth (km)
-4.2	-4.182	4335	2.876	-1.155	0.0
-4.0	-3.983	4369	2.985	-1.048	26.3
-3.8	-3.785	4409	3.095	-0.940	52.9
-3.6	-3.589	4450	3.205	-0.831	79.8
-3.4	-3.392	4491	3.316	-0.722	107.1
-3.2	-3.195	4532	3.426	-0.613	134.5
-3.0	-2.998	4571	3.536	-0.505	162.3
-2.8	-2.800	4609	3.647	-0.398	190.2
-2.6	-2.602	4647	3.757	-0.291	218.4
-2.4	-2.403	4684	3.867	-0.185	246.7
-2.2	-2.204	4721	3.977	-0.079	275.3
-2.0	-2.005	4761	4.087	0.028	304.1
-1.8	-1.805	4806	4.197	0.136	333.2
-1.6	-1.606	4858	4.307	0.246	362.5
-1.4	-1.406	4922	4.417	0.360	392.5
-1.2	-1.207	5005	4.528	0.480	422.5
-1.0	-1.007	5112	4.638	0.609	453.3
-0.8	-0.806	5251	4.748	0.751	484.6
-0.6	-0.606	5439	4.856	0.917	516.5
-0.5	-0.506	5551	4.908	1.014	531.9
-0.4	-0.405	5683	4.958	1.127	547.0
-0.3	-0.305	5833	5.004	1.259	561.4
-0.2	-0.205	6004	5.046	1.411	574.8
-0.1	-0.105	6197	5.083	1.585	587.0
0.0	-0.006	6415	5.114	1.778	597.7
0.1	0.092	6660	5.140	1.987	606.9
0.2	0.190	6935	5.162	2.210	614.8
0.3	0.286	7267	5.178	2.459	621.3
0.4	0.381	7599	5.191	2.691	626.6
0.5	0.474	7887	5.202	2.877	631.1
0.6	0.567	8130	5.212	3.027	635.3
0.7	0.661	8340	5.221	3.149	639.4
0.8	0.751	8526	5.230	3.253	643.4
0.9	0.839	8695	5.239	3.345	647.7
1.0	0.928	8851	5.248	3.428	652.1

Model parameters (in cgs units): T_{eff} = 5780 K, log g = 4.44, [Metals/H] = 0.00, microturbulent velocity ξ = 1.15 km s^{-1}, τ_R: Rosseland mean optical depth.

Source: Edvardsson, B., Andersen, J., Gustafsson, B., Lambert, D. L., Nissen, P. E., & Tomkin, J., 1993, *Astron. & Astrophys.* 275, 101–152.

Table 3.4 Solar Luminosity Through Time Standard Model

Time since beginning (10^9 years)	Radius ($R_\odot$)	Luminosity ($L_\odot$)	Central Temperature (10^6 K)
0	0.872	0.7688	14.35
0.143	0.885	0.7248	13.46
0.856	0.902	0.7621	13.68
1.863	0.924	0.8156	14.08
2.193	0.932	0.8352	14.22
3.020	0.953	0.8855	14.60
3.977	0.981	0.9522	15.12
4.587 present	1.000	1.000	15.51
5.506	1.035	1.079	16.18
6.074	1.059	1.133	16.65
6.577	1.082	1.186	17.13
7.027	1.105	1.238	17.62
7.728	1.143	1.318	18.42
8.258	1.180	1.399	18.74
8.757	1.224	1.494	18.81
9.805	1.361	1.760	19.25

For solar composition X = 0.7046, Y = 0.2757, Z = 0.0197.
The sun's luminosity for the past is approximated by:

$$\frac{L}{L_o} \cong \frac{1}{\left(1+0.4\left(1-\frac{1}{t_p}\right)\right)} \quad \text{where } t_p = 4.6 \times 10^9 \text{ years}$$

Source: Turck-Chièze, S., Cahen, S., Cassé, M., & Doom, C., 1988, *ApJ*. 335, 415–424.

Table 3.5 Elemental Abundances in the Sun's Photosphere [a]

Z	Value	Z	Value	Z	Value	Z	Value
1 H	12.0	22 Ti	5.04	44 Ru	1.84	66 Dy	1.1
2 He	11.0	23 V	4.00	45 Rh	1.12	67 Ho	(0.26)
3 Li	1.16	24 Cr	5.67	46 Pd	1.69	68 Er	0.93
4 Be	1.15	25 Mn	5.39	47 Ag	(0.94)	69 Tm	(0.00)
5 B	(2.6)	26 Fe	7.51	48 Cd	1.86	70 Yb	1.08
6 C	8.55	27 Co	4.92	49 In	(1.66)	71 Lu	(0.76)
7 N	7.97	28 Ni	6.25	50 Sn	2.0	72 Hf	0.88
8 O	8.87	29 Cu	4.21	51 Sb	1.0	73 Ta	...
9 F	4.56	30 Zn	4.60	52 Te	...	74 W	(1.11)
10 Ne	8.07	31 Ga	2.88	53 I	...	75 Re	...
11 Na	6.33	32 Ge	3.41	54 Xe	...	76 Os	1.45
12 Mg	7.58	33 As	...	55 Cs	...	77 Ir	1.35
13 Al	6.47	34 Se	...	56 Ba	2.13	78 Pt	1.8
14 Si	7.55	35 Br	...	57 La	1.22	79 Au	(1.01)
15 P	5.45	36 Kr	...	58 Ce	1.55	80 Hg	...
16 S	7.21	37 Rb	2.60	59 Pr	0.71	81 Tl	(0.9)
17 Cl	5.5	38 Sr	2.90	60 Nd	1.50	82 Pb	1.95
18 Ar	6.60	39 Y	2.24	62 Sm	1.00	83 Bi	...
19 K	5.12	40 Zr	2.60	63 Eu	0.51	90 Th	0.27
20 Ca	6.36	41 Nb	1.42	64 Gd	1.12	92 U	(<−0.47)
21 Sc	3.20	42 Mo	1.92	65 Tb	(−0.1)		

[a] On the atomic scale where $\log N_H = [H] = 12$. Values in parentheses are uncertain.

Table 3.6 Elemental Abundances in the Sun's Corona [a]

Z	Value	Z	Value	Z	Value	Z	Value
1 H	...	11 Na	6.38	18 Ar	5.89	25 Mn	5.38
2 He	10.14	12 Mg	7.59	19 K	5.14	26 Fe	7.65
6 C	7.90	13 Al	6.47	20 Ca	6.46	27 Co	...
7 N	7.40	14 Si	7.55	21 Sc	(4.04)	28 Ni	6.22
8 O	8.30	15 P	5.24	22 Ti	5.24	29 Cu	4.31
9 F	(4.0)	16 S	6.93	23 V	(4.23)	30 Zn	4.76
10 Ne	7.46	17 Cl	4.93	24 Cr	5.81		

[a] on the atomic scale where $\log N_{Si} = [Si] = 7.55$. Values in parentheses are uncertain.

Sources: Anders, E., & Grevesse, N., 1989, Geochim. Cosmochim. Acta 53, 197–214. Grevesse, N., & Anders, E., 1991, in Solar interior and atmosphere (Cox, A. N., Livingston, W. C., & Matthews, M. S., eds.), Univ. of Arizona Press, Tucson, pp. 1227–1234. Grevesse, N., & Noels, A., 1993, in Origin and evolution of the elements (Prantzos, N., Vangioni-Flam, E., & Cassé, M., eds.), Cambridge Univ. Press, pp. 14–25.

4

MERCURY

Mercury has been known to mankind for approximately 5000 years. It is one of the brightest planets in the sky, but it is difficult to observe from Earth because it is always within 27°45' of the sun. Mercury can be observed before sunrise and at sunset and was known to the Greeks as Apollo (the morning star) and Hermes (the evening star). Three flybys by *Mariner 10* in 1974 and 1975 (29 March 1974, 21 September 1974, and 16 March 1975) provided a large amount of information about Mercury. *Mariner 10* and Earth-based observations of Mercury are summarized in [VCM88]. An atlas of *Mariner 10* images is given by [DDG78].

The difficulties in visual and photographic observations of Mercury are exemplified by the long history of erroneous measurements of Mercury's rotation rate, which was first thought to be ~24 hours, later revised to ~88 days, synchronous with its orbital period. However, radar observations by [PD65] showed that Mercury's rotation rate is ~59 days, in a 3:2 resonance with its orbital period. Subsequently, *Mariner 10* photography refined the sidereal rotation period to 58.646 days [Kl76]. Mercury's solar day is ~176 terrestrial days (twice the orbital period and three times the sidereal rotation period) because Mercury moves 2/3 of the way through its orbit during one rotation. It thus takes three rotations for the same spot on Mercury to face the sun.

Mercury's perihelion precesses by ~10' of arc per century. There is a 43" discrepancy between the observed precession and that predicted by Newtonian mechanics, which led dynamicists to postulate either an asteroid belt (Vulcanoids) or another planet (Vulcan) inside Mercury's orbit. Despite an erroneous "detection" in the mid-19th century, there is no evidence for a planet inside Mercury's orbit. Observational, dynamical, and thermodynamic constraints restrict any Vulcanoids to 2–100 km diameter objects at 0.1–0.25 AU [CDW96]. In 1915, Einstein successfully explained the precession of Mercury's perihelion, including the 43" discrepancy, with the General Theory of Relativity.

Mariner 10 data indicate that Mercury's spin axis is offset by about 2° from the perpendicular to its orbital plane, and dynamical considerations indicate that the "true" offset is probably 0.5° [Kl76, Pe88]. Mercury does not have seasonal changes like those on Earth. However, the orbital eccentricity leads to approximately twice as much sunlight at the subsolar point at perihelion (~0.31 AU) than at aphelion (~0.47 AU). At perihelion, Mercury's (constant) rotational velocity is slightly slower than its (variable) orbital velocity. As a result, at equatorial longitudes 90° east and west of the subsolar point, the sun rises, sets, and rises again before traversing the sky and setting twice.

Mercury is the second smallest planet (after Pluto) and is smaller in diameter but more massive than Ganymede or Titan. Mercury's bulk density of ~5.43 g cm^{-3} is the second highest (after Earth) for any planet or satellite. The high density is due to a large metal/silicate mass ratio (70:30 to 66:34), about twice that of any other terrestrial planet or satellite. The origin of Mercury's high density is unknown, and the explanations advanced fall into three categories: (1) physical fractionation of metal and silicate in the solar nebula because of density, magnetism, or mechanical strength, (2) chemical fractionation based on the different volatilities of metal and silicate in the solar nebula, and (3) a large collision to shatter and blow off the majority of proto-Mercury's silicate mantle [CFB88].

Mercury's internal structure and its moment of inertia C/MR^2 are unknown. *Mariner 10* detected a weak dipolar magnetic field aligned with the spin axis within 10° and with a strength and field moment of ~1% and 6×10^{-4} of the earth's magnetic field, respectively. The origin of the magnetic field is unknown and is variously explained by (1) an active dynamo in a (partially?) molten core, (2) a fossil magnetic field, and (3) a solar wind induced field. If the interior is differentiated, the bulk density implies a core ~66–70% of Mercury's radius. Recent reanalysis of radar ranging data suggests a crustal thickness of 100–300 km [AJL96]. Current interior structure models for Mercury are nonunique.

Mariner 10 mapped ~45% of Mercury's surface and observed impact craters (the most abundant feature), lobate scarps (compressional thrust faults), ridges and valleys, plains, and mountains that occur in four main geological provinces: (1) heavily cratered terrain, (2) intercrater and smooth plains, (3) the ~1300 km diameter Caloris Basin, and (4) "weird" terrain antipodal to Caloris. Formation of the lobate scarps implies contraction of Mercury's radius by ~0.1% (1–2 km). The "weird" terrain apparently formed by seismic energy from the Caloris impact traveling through Mercury's interior. The spatial density of impact craters gives an average age of 3–4 Ga for Mercury's surface.

Temperatures on Mercury's surface range from ~100 K before sunrise to ~700 K at "noontime." There are two hot "poles", which face the sun during alternate perihelion passages, at opposite ends of the equatorial bulge raised by the sun at perihelion [SU67]. However, some regions near the north and south poles are permanently shadowed and cold. Thermal modeling by [PWV92] indicates that flat areas near the poles can be at 167 K; shadowed regions inside craters may be even colder.

Radar observations by [SBM92, BMS93, HSV94] indicate the presence of highly scattering material, interpreted as water ice and/or other ices in the north and south polar regions. Many of the highly scattering regions are found in craters near the north and south poles. The radar data do not uniquely constrain either the chemical composition or the thickness of the scattering material. [Le95] points out that water ice or other ices could be derived from short period comets, extinct periodic comets, or C-type Mercury crossing asteroids. [SHL95] proposed that elemental sulfur was responsible for the radar scattering in the polar regions. Further observations are necessary to resolve the nature of the radar scattering regions.

The surface composition has also been studied spectroscopically, and there is a marginal detection of Fe^{2+} silicate absorptions in the 0.9 μm region [Vi88]. Basalt and anorthosite were recently reported from IR spectra in the 7.3- to13.5-μm region [SKW94].

Mercury has a rarefied atmosphere with a pressure $<10^{-12}$ bar. The observed constituents are H, He, and O discovered by the ultraviolet spectrometer on *Mariner 10*, and Na and K discovered by Earth-based observations. However, the total pressure of the observed constituents is less than the 10^{-12} bar upper limit, and other species may be present. The solar wind, surface sputtering, outgassing, and vaporization of impactors are sources for H, He, O, Na, and K. Atmospheric abundances of Na and K are spatially and temporally variable. Photoionization, followed by reaction with the surface or transport to space, is the dominant sink for the atmospheric constituents [HMS88].

Sources and further reading: [AJL96] Anderson, J. D., Jurgens, R. F., Lau, E. L., Slade, M. A., & Schubert, G., 1996, *Icarus* 124, 690–697. [BMS93] Butler, B. J., Muhleman, D. O., & Slade, M. A., 1993, *J. Geophys. Res.* 98, 15003–15023. [CFB88] Cameron, A. G. W., Fegley, B., Jr., Benz, W., & Slattery, W. L., 1988, in *Mercury* (Vilas, F., Chapman, C. R., & Matthews, M. S., eds.), Univ. of Arizona Press, Tucson, pp. 692–708. [CDW96] Campins, H., Davis, D. R., Weidenschilling, S. J., & Magee, M., 1996, *Astron. Soc. Pac. Conf. Ser.* 107, pp. 85–96.

[DDG78] Davies, M. E., Dwornik, S. E., Gault, D. E., & Strom, R. G., 1978, *Atlas of Mercury*, NASA, SP-423, Washington, D.C. [HMS88] Hunten, D. M., Morgan, T. H., & Shemansky, D. E., 1988, in *Mercury* (Vilas, F., Chapman, C. R., & Matthews, M. S., eds.), Univ. of Arizona Press, Tucson, pp. 561–612. [HSV94] Harmon, J. K., Slade, M. A., Velez, R. A., Crespo, A., Dryer, M. J., & Johnson, J. M., 1994, *Nature* 369, 213–215. [Kl76] Klaasen, K. P., 1976, *Icarus* 28, 469–478. [Le95] Lewis, J. S., 1995, *Physics and chemistry of the solar system*, Academic Press, New York, pp. 556. [PD65] Pettengill, G. H., & Dyce, R. B., 1965, *Nature* 206, 1240. [Pe88] Peale, S. J., 1988, in *Mercury* (Vilas, F., Chapman, C. R., & Matthews, M. S., eds.), Univ. of Arizona Press, Tucson, pp. 461–493. [PWV92] Paige, D. A., Wood, S. E., & Vasava, A. R., 1992, *Science* 258, 643–646. [SBM92] Slade, M. A., Butler, B. J., & Muhleman, D. O., 1992, *Science* 258, 635–640. [SHL95] Sprague, A. L., Hunten, D. M., & Lodders, K., 1995, *Icarus* 118, 211–215. [SKW94] Sprague, A. L., Kozlowski, R. W. H., Witteborn, F. C., Cruikshank, D. P., & Wooden, D. H., 1994, *Icarus* 109, 156–167. [SU67] Soter, S. L., & Ulrichs, J., 1967, *Nature* 214, 1315–1316. [VCM88] Vilas, F., Chapman, C. R., & Matthews, M. S. (eds.), 1988, *Mercury*, Univ. of Arizona Press, Tucson, pp. 794. [Vi88] Vilas, F., 1988, in *Mercury* (Vilas, F., Chapman, C. R., & Matthews, M. S., eds.), Univ. of Arizona Press, Tucson, pp. 59–76.

Table 4.1 Some Physical Properties of Mercury

Property	Value	Property	Value
Mean radius (km) *	2437.6±2.9	Sidereal revolution period (⊕ days)	87.9694
Equatorial radius (km) *	2440.0±1.2	Mean synodic period (⊕ days)	115.88
Equatorial ellipticity †	$(54±5.4)×10^{-5}$	Eccentricity of orbit	0.2056
Ellipsoidal flattening †	$(2.89±3.68)×10^{-3}$	Inclination of orbit to ecliptic	7.004°
Mass (kg)	$0.33022×10^{24}$	Mean orbital velocity (km s^{-1})	47.89
Mean density (g cm^{-3})	5.43±0.01	Inclination of equator to orbit	0.5°
GM (m^{3}s^{-2})	$2.2033×10^{13}$	Sidereal rotation period (⊕ days)	58.6462
Equatorial gravity (m s^{-2})	3.701	(2/3 of orbital period)	±0.005
Polar gravity (m s^{-2})	3.701	Surface temperature (K)	100–700
Escape velocity (km s^{-1})	4.25	Temperature sunward (K)	590–700
$J_2 ×10^5$	6±2	Surface pressure (bars)	$<10^{-12}$
C/MR^2	...	Magnetic dipole moment (Tesla R$_{Merc}^3$)	$3.0×10^{-7}$
Solar constant (Wm^{-2})	9936.9	Magnetic axis offset	<10°

* The mean radius is (a + b + c)/3 and the equatorial radius is (a + b)/2 where the three ellipsoidal radii (in km) are a = 2440.6±0.1, b = 2439.3±0.1, c = 2432.9±8.8.
† The equatorial ellipticity is (a – b)/a and the ellipsoidal flattening is [(ab)^{½} – c]/(ab)^{½}.

Sources: Anderson, J. D., Colombo, G., Esposito, P. B., Lau, E. I., & Trager, G. B., 1987, *Icarus* 71, 337–349. Anderson, J. D., Jurgens, R. F., Lau, E. L., Slade, M. A., & Schubert, G., 1996, *Icarus* 124, 690–697. Klaasen, K. P., 1976, *Icarus* 28, 469–478. Connerney, J. E. P., & Ness, N. F., 1988, in *Mercury,* (Vilas, F., Chapman, C. R., & Matthews, M. S., eds.), Univ. of Arizona Press, Tucson, pp. 494–513.

Table 4.2 Composition of Mercury's Atmosphere

Element	Number density (cm^{-3})		Element	Number density (cm^{-3})	
	[SH95]	[HMS88]		[SH95]	[HMS88]
H	200	23 (hot) – 230 (cold)	Na *	20000	17000–38000
He	6000	6000	Ar	$<3×10^7$	$<6.6×10^6$
Li	<2	...	K *	500	500
O	<40000	44000	Ca	<247	...

* Abundances are spatially and temporally variable.

Sources: **[HMS88]** Hunten, D. M., Morgan, T. H., & Shemansky, D. E., 1988, in *Mercury,* (Vilas, F., Chapman, C. R., & Matthews, M. S., eds.), Univ. of Arizona Press, Tucson, pp. 562–612. **[SH95]** Sprague, A. L., & Hunten, D. M., 1995, in *Volatiles in the earth and the solar system*, (Farley, K. A., ed.), AIP Conf. Proc. 341, pp. 200–208.

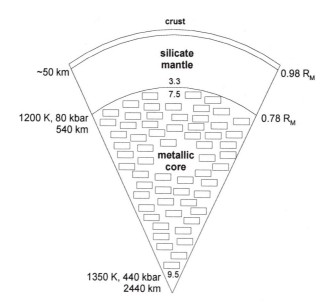

Figure 4.1 Mercury's interior structure. Numbers at the phase boundaries are densities in g cm^{-3}. Internal structure models of Mercury are poorly constrained and nonunique.

Table 4.3 Model Mass Distribution Within Mercury

Planet Portion	Mass (kg)	% Mass of Mercury	Thickness (km)	Density (STP, g cm^{-3})
Crust	$(1.16–1.12)\times10^{22}$	3.5–3.4	~50	3.0
Mantle	$(9.44–8.98)\times10^{22}$	28.6–27.2	~490	3.2–3.3
Core	$(2.24–2.29)\times10^{23}$	67.9–69.4	~1900	7.94–7.93
Bulk Mercury	3.302×10^{23}	100	2440	5.43 (observed)

Sources: Basaltic Volcanism Study Project: *Basaltic volcanism on the terrestrial planets*, 1981, Pergamon Press, pp. 1286. Anderson, J. D., Jurgens, R. F., Lau, E. L., Slade, M. A., & Schubert, G., 1996, *Icarus* 124, 690–697, crustal thickness 100–300 km. Lewis, J. S., 1995, *Physics and chemistry of the solar system*, Academic Press, New York, pp. 556.

Table 4.4 Model Compositions of Mercury

Model	I [MA80]	II [BVSP81]	III [BVSP81]	IV [BVSP81]	V [BVSP81]	VI [FC87]	VII [Goe88]
Mantle & Crust							
MgO	33.7	40.5	40.8	47.7	54.6	34.5	32–38
Al_2O_3	6.4	9.6	7.2	4.7	0	18.1	3.5–7
CaO	5.2	8.6	6.6	4.1	1.8	14.6	3.5–7
SiO_2	47.2	40.8	45.0	43.5	43.6	32.1	38–48
TiO_2	0.33	0.49	0.37	...	...	0.7	0.15–0.3
Na_2O	0.08	0	0	0	0	...	0.2–1
K_2O	0.0083	0	0	0	0	0	...
FeO	3.7	0.05	0.04	0	0	0	0.5–5
Cr_2O_3	3.3	...	...	...	...	...	...
MnO	0.06	...	...	...	...	...	...
H_2O	0.016	0	0	...	...	...	some
Th ppm	0.122	0.19	0.14	0.12	0	0.401	...
U ppm	0.034	0.053	0.040	0.026	0	0	...
Core							
Fe	93.5	94.1	92.4	94.5	94.5	...	88–91
Co	0.25	...	...	...	...	...	...
Ni	5.4	5.9	7.6	5.5	5.5	...	6.5–7.5
S	0.35	0	0	0	0	...	0.5–5
O	...	...	...	0	0	...	...
P	0.57	...	...	...	...	...	...
Relative masses							
Mantle & crust	32.0	31.6	32.1	30.9	30.6	36	...
Core	68.0	68.4	67.9	69.1	69.4	64	...
$C/(MR^2)$	0.335	0.3350	0.3346	0.334	0.334	0.337	...

mass% if not noted otherwise.
I: Four component meteorite model [MA80]. II: Equilibrium condensation, [BVSP81] model Me1. III: Equilibrium condensation model including feeding zones, [BVSP81] model Me2. IV: Extreme dynamically mixed model to satisfy mean density, [BVSP81] model Me3. V: Extreme collisionally differentiated model to satisfy mean density, [BVSP81] model Me4. VI: Vaporization model [FC87]. VII: Preferred model by [Goe88].

Sources: **[BVSP81]** Basaltic Volcanism Study Project: *Basaltic volcanism on the terrestrial planets*, 1981, Pergamon Press, pp. 1286. **[FC87]** Fegley, B., & Cameron, A. G. W., 1987, *Earth Planet. Sci. Lett.* 82, 207–222. **[Goe88]** Goettel, K. A., 1988, in *Mercury* (Vilas, F., Chapman, C. R., & Matthews, M. S., eds.) Univ. of Arizona Press, Tucson, pp. 613–621. **[MA80]** Morgan, J. W., & Anders, E., 1980, *Proc. Natl. Acad. Sci.* 77, 6973–6977.

Table 4.5 Locations and Sizes of Craters on Mercury (diameter 150 km and larger)

Name	Latitude	Longitude	Diameter (km)
Beethoven	20.8°S	123.6°W	643
Dostoevskij	45.1°S	176.4°W	411
Tolstoj	16.3°S	163.5°W	390
Goethe	78.5°N	44.5°W	383
Shakespeare	49.7°N	150.9°W	370
Raphael	19.9°S	75.9°W	343
Homer	1.2°S	36.2°W	314
Monet	44.4°N	10.3°W	303
Vyāsa	48.3°N	81.1°W	290
Van Eyck	43.2°N	158.8°W	282
Mozart	8.0°N	190.5°W	270
Haydn	27.3°S	71.6°W	270
Renoir	18.6°S	51.5°W	246
Pushkin	66.3°S	22.4°W	231
Rodin	21.1°N	18.2°W	229
Vālmiki	23.5°S	141.0°W	221
Wren	24.3°N	35.2°W	221
Michelangelo	45.0°S	109.1°W	216
Mendes Pinto	61.3°S	17.8°W	214
Bach	68.5°S	103.4°W	214
Vivaldi	13.7°N	85.0°W	213
Sholem Aleichem	50.4°N	87.7°W	200
Chekhov	36.2°S	61.5°W	199
Hugo	38.9°N	47.0°W	198
Stravinsky	50.5°N	73.5°W	190
Smetana	48.5°S	70.2°W	190
Strindberg	53.7°N	135.3°W	190
Al-Hamadhani	38.8°N	89.7°W	186
Milton	26.2°S	174.8°W	186
Matisse	24.0°S	89.8°W	186
Schubert	43.4°S	54.3°W	185
Praxiteles	27.3°N	59.2°W	182

continued

Table 4.5 *(continued)*

Name	Latitude	Longitude	Diameter (km)
Cervantes	74.6°S	122.0°W	181
Dürer	21.9°N	119.0°W	180
Ma Chih-Yuan	60.4°S	78.0°W	179
Rubens	59.8°N	74.1°W	175
Petrarch	30.6°S	26.2°W	171
Chao Meng-Fu	87.3°S	134.2°W	167
Händel	3.4°N	33.8°W	166
Chaikovskij	7.4°N	50.4°W	165
Sōtatsu	49.1°S	18.1°W	165
Wang Meng	8.8°N	103.8°W	165
Shelley	47.8°S	127.8°W	164
Verdi	64.7°N	168.6°W	˙163
Chŏng Chŏl	46.4°N	116.2°W	162
Phidias	8.7°N	149.3°W	160
Bramante	47.5°S	61.8°W	159
Derzhavin	44.9°N	35.3°W	159
Kurosawa	53.4°S	21.8°W	159
Imhotep	18.1°S	37.3°W	159
Ibsen	24.1°S	35.6°W	159
Sayat-Nova	28.4°S	122.1°W	158
Proust	19.7°N	46.7°W	157
Melville	21.5°N	10.1°W	154
Pigalle	38.5°S	9.5°W	154
Lermontov	15.2°N	48.1°W	152
Kuan Han-Ch'ing	29.4°N	52.4°W	151
Darío	26.5°S	10.0°W	151
Giotto	12.0°N	55.8°W	150
Sophocles	7.0°S	145.7°W	150

Sources: IAU recommendations published by U.S. Geological Survey, Branch of Astrogeology, Flagstaff, AZ.

5

VENUS

Venus is usually the brightest object in the sky after the sun and moon, but until the space age our knowledge of our "twin planet" was very limited because its surface is hidden by a global cloud layer. Observed by the ancient Babylonians 5,000 years ago, Venus has played an important role in myth, literature, and science throughout recorded history. In 1610, Galileo's observations of the phases of Venus were important evidence supporting the Copernican theory of the solar system. In 1761 and 1769, observations of the solar transits of Venus (which occur in pairs 8 years apart at >100 year intervals) were used to measure the distance of the earth from the sun. Lomonosov's observations during the 1761 transit also provided the first evidence of an atmosphere on Venus, but the atmospheric composition was unknown until Adams and Dunham discovered CO_2 in 1932 [AD32]. Starting in the 1960s, the use of high-altitude telescopes (e.g., on balloons or airplanes) and of Fourier transform IR spectrometers led to the discovery of H_2O, CO, HCl, HF, and other trace gases. In the late 1980s, the discovery of IR spectral windows allowed Earth-based IR observations of the subcloud atmosphere on Venus' nightside.

Venus appears yellow-white in visible light, but the first UV images of Venus in the 1920s showed dark Y- or V-shaped cloud features. The UV absorber responsible for the features is unknown, but it may be elemental sulfur, Cl_2, a S-Cl gas, Cl compounds dissolved in cloud particles, or another sulfur gas. Tracking of these features, first by Earth-based observations and later by spacecraft observations (e.g., by *Pioneer Venus* and *Mariner 10*) shows ~100 m s^{-1} retrograde zonal winds known as the 4 day super rotation. In situ measurements by and Doppler tracking of the *Pioneer Venus*, *Venera*, and *Vega* entry probes show that the zonal winds decrease with decreasing altitude and are ~1 m s^{-1} or less at the surface. The origin of the 4 day super rotation is still incompletely understood.

Starting in the early 1960s, Venus was the target of numerous flyby, probe, lander, and orbiter spacecraft missions by the United States (*Mariner*, *Pioneer Venus*, *Magellan, Galileo*) and the former Soviet Union (*Venera* and *Vega*), and of increasingly sophisticated Earth-based and Earth-orbital observations. Results of early spacecraft missions and Earth-based observations are summarized by [HCD83, JGR80, LP84, LTP92], whereas [BB92, Kr86, JGR92, PSS93, BHP97] summarize results from the *Venera 13/14* entry probes, the *Venera 15/16* orbiters, the *Magellan* orbiter, the *Galileo* flyby, and recent Earth-based observations.

Venus' orbit and rotation are notable in several respects. First, its orbit has the lowest eccentricity (0.0068 or 1/147) of any major planet. Second, the sidereal orbital period is 224.695 days, but the sidereal rotation period is 243.018 days (retrograde). Thus, a Venusian "day" is 116.75 Earth days long (1/day = 1/243.02 + 1/224.70), with the sun rising in the west and setting in the east after 58.375 Earth days of daylight and rising again after another 58.375 Earth days of night. Venus' rotation period is close to, but not exactly equal to, 243.16 days, which would be in 3:2 resonance with the earth's orbital period. The orientation of Venus' spin axis is almost normal to the ecliptic with a tilt of ~177°. The relationship between the sidereal orbital periods of Venus and Earth is such that inferior conjunction (i.e., closest approach of Venus to Earth) occurs once every 583.92 Earth days (1/t =1/224.70–1/365.25), at ~19 month intervals. Venus presents almost exactly the same face to Earth at each inferior conjunction because the 583.92 day synodic period is 5.001 Venusian days.

Although the incident solar flux on Venus is approximately 1.9 times larger (=1/0.723^2) than that on Earth, Venus absorbs only ~62% as much solar energy because of the high albedo of the global cloud cover. The high temperature and pressure at Venus' surface (740 K and 95.6 bar, respectively, at the modal radius of 6051.4 km) are due to a super greenhouse effect maintained by the high IR opacity of CO_2, SO_2, and H_2O in its atmosphere. The origin, duration, and present stability of the Venusian super greenhouse are not well understood. Temperature and pressure decrease adiabatically with altitude (dT/dz ~ – 7.7 K/km and d(lnP)/d(lnT) ~ 6.0) throughout the lower troposphere. Temperature is 660 K and pressure is 48.0 bar on top of Maxwell Montes, which is ~10.4 km above the modal radius and the highest point on the planet. The high temperatures lead to high atmospheric abundances of CO_2, SO_2, OCS, HCl, and HF, which are present at much lower levels in the earth's atmosphere because most of the C, S, Cl, and F at the surface of the earth is in the crust and oceans.

Venus' atmosphere is dominantly CO_2 (96.5%) and N_2 (3.5%), with smaller amounts of SO_2, H_2O, CO, OCS, HCl, HF, the noble gases, and other reactive species. The observed noble gas abundances and isotopic ratios are different from those on Earth; e.g., the $^{36}Ar/^{40}Ar$ ratio of ~0.9 is ~270 times larger than on Earth. Abundances of CO_2, N_2, the noble gases, and HCl and HF are constant throughout most of Venus' atmosphere, but other gases such as SO_2, H_2O, CO, and OCS have spatially and temporally variable abundances throughout much of Venus' atmosphere. However, the atmospheric composition below ~22 km, comprising ~80% of Venus' atmosphere, is poorly constrained because in situ and Earth-based observations of this region are extremely difficult.

Seven *Venera* and *Vega* landers have made elemental analyses of the surface of Venus. The *Venera 8, 9, 10,* and *Vega 1/2* spacecraft analyzed K, U, and Th by γ-ray spectroscopy and the *Venera 13, 14,* and *Vega 2* spacecraft analyzed Si, Ti, Al, Fe, Mn, Mg, Ca, K, S, and Cl by X-ray fluorescence (XRF) spectroscopy. Elements lighter than Mg could not be detected, and the Na content was estimated using geochemical methods. The γ-ray and XRF analyses show that different rock types were sampled at several of the landing sites. Comparisons with elemental analyses of terrestrial rocks, normative calculations of mineralogy, geochemical correlations, and geological interpretations of *Magellan* radar images of the landing sites suggest basaltic rocks, similar to terrestrial midocean ridge basalts (MORB), and related basalts at the *Venera 9, 10, 14, Vega 1* and *2* landing sties, and alkaline rocks or related rock types at the *Venera 8* and *13* landing sites [BB92, KKB93, FKL97].

Chemistry at the base of Venus' atmosphere is driven by the high temperatures and pressures, because short wavelength UV sunlight is absorbed higher in the atmosphere and only longer wavelength yellow to red sunlight reaches the surface. Theoretical modeling of high temperature equilibria between atmospheric gases and minerals expected on Venus' surface predicts that the observed abundances of several gases, including CO_2, SO_2, OCS, HCl, and HF are controlled by reactions with reactive minerals on Venus' surface. For example, the CO_2 pressure on Venus is plausibly regulated by the "Urey reaction",

$$CaCO_3 \text{ (calcite)} + SiO_2 \text{ (silica)} = CaSiO_3 \text{ (wollastonite)} + CO_2 \qquad (1)$$
$$\log_{10} P_{CO_2} = 7.97 - 4456/T \qquad (2)$$

because the observed CO_2 pressure of ~92 bar at 740 K is virtually identical to the equilibrium CO_2 pressure from reaction (1) at that temperature. Theoretical modeling also predicts that the atmospheric abundances of the reactive hydrogen halides are probably regulated by equilibria involving Cl- and F-bearing minerals, which are common in alkaline rocks on Earth and by analogy also on Venus. The chemistry of atmosphere-surface reactions on Venus is reviewed by [LP84, FKL97].

Radar observations from the *Pioneer Venus* and *Magellan* spacecraft show global variations in surface radar emissivity that are probably related to some types of atmosphere-surface reactions postulated to explain the observed abundances of atmospheric gases on Venus. For example, low radar emissivity regions in the Venusian highlands may be due to the presence of high dielectric minerals, such as perovskites and pyrochlores. These minerals are commonly found in alkaline and carbonatite rocks that have the necessary mineralogy for regulating the abundances of CO_2, HCl, and HF in Venus' atmosphere [KKF94]. Alternatively, some of the low emissivity regions may be due to high dielectric minerals condensed out of volcanic gases [BFA95, PFS96].

The most prominent feature of Venus' middle atmosphere is the global cloud layer that begins at ~45 km altitude and extends to ~70 km altitude, with thinner hazes up to 20 km above and several km below these altitudes. About 70% of all solar energy absorbed by Venus is absorbed in the clouds and atmosphere at altitudes of 50 km and above. In contrast, about two-thirds of all solar energy absorbed by Earth is absorbed at the surface. About half of the absorbed sunlight in Venus' cloud region is absorbed by an unknown UV absorber (responsible for the dark V- or Y-shaped cloud features) at wavelengths ≤ 500 nm.

The clouds are low-density hazes because the visibility inside the densest region of the clouds is a few km. The clouds are composed of three different types of particles: aerosols ~0.3 μm diameter (mode 1 particles) with peak densities of ~0.1 mg m^{-3} in the upper and middle clouds, spherical droplets ~2μm diameter (mode 2 particles) composed of 75% sulfuric acid ($H_2SO_4 \cdot 2H_2O$) with peak densities of ~1 mg m^{-3} throughout the clouds, and mode 3 particles of ~7 μm diameter with unknown composition and peak densities of ~10 mg m^{-3} in the middle to lower clouds. The aqueous sulfuric acid droplets are visible from Earth. The mode 1 and mode 3 particles may also be sulfuric acid particles. Some data suggest that the mode 3 particles may be crystalline and could be composed of Fe or Al chlorides, solid perchloric acid hydrates, or phosphorus oxides.

The aqueous sulfuric acid droplets in the clouds are formed by UV sunlight photooxidation of SO_2. Photooxidation reduces the SO_2 abundance from ~150 ppmv below the clouds to ~0.01–0.1 ppmv at the cloud tops. Photooxidation of SO_2 is closely tied to CO_2 photolysis because the O_2 produced is used to convert SO_2 to SO_3, which then forms sulfuric acid. Spectroscopic observations show temporal trends in the SO_2 abundance at the cloud tops. The observed variations are probably due to atmospheric dynamics. Sulfuric acid is a powerful desiccant. The atmospheric H_2O content decreases from ~30 ppmv below the clouds to only a few ppmv above the clouds, because of formation of the concentrated aqueous sulfuric acid cloud droplets.

Venus' CO_2-rich atmosphere, like that of Mars, is continually converted by UV sunlight to O and CO:

$$CO_2 + hv \rightarrow CO + O(^3P) \qquad (\lambda < 227.5 \text{ nm}) \qquad (3)$$
$$\rightarrow CO + O(^1D) \qquad (\lambda < 167.0 \text{ nm}) \qquad (4)$$

The electronically excited ^1D oxygen atoms formed in reaction (4) are rapidly converted to the ground state (^3P) by collisions with other molecules. The direct recombination of O atoms and CO

$$CO + O(^3P) \rightarrow CO_2 \qquad (5)$$

is spin forbidden by quantum mechanics and is much slower than O atom recombination to form O_2:

$$O(^3P) + O(^3P) \rightarrow O_2 \qquad (6)$$

Photolysis would completely destroy the CO_2 above the clouds in ~14,000 years, all the CO_2 in Venus' atmosphere in ~5 million years, and would produce observable amounts of O_2 (which is not seen and has an upper limit of <0.3 ppmv) in ~5 years unless CO_2 is reformed by another route. Gas phase catalytic reformation of CO_2 by H, Cl, or N gases has been proposed to solve this problem [YD82, Pri85]. The relative importance of the catalytic schemes depends on the H_2 abundance in Venus' stratosphere, which is unknown, because H_2 is involved in chemistry forming OH radicals. For example, the reaction

$$CO + OH \rightarrow CO_2 + H \qquad (7)$$

is important at H_2 levels of tens of ppmv. At very low H_2 levels of ~0.1 ppbv, reaction (7) is no longer important and the reactions

$$COCl + O_2 + M \rightarrow ClCO_3 + M \qquad (8)$$
$$ClCO_3 + Cl \rightarrow CO_2 + ClO + Cl \qquad (9)$$
$$ClCO_3 + O \rightarrow CO_2 + ClO + O \qquad (10)$$

where M is any third body, recycle CO to CO_2. At intermediate H_2 levels of ~0.1 ppmv, the reaction

$$NO + HO_2 \rightarrow NO_2 + OH \qquad (11)$$

precedes reaction (7), which then recycles CO to CO_2.

Venus is often regarded as Earth's "twin planet" because of its similar size (~95%), mass (~82%), and gravity (~90%) compared to Earth. However, radar images from Earth-based observatories and from the *Pioneer*

Venus, *Venera 15/16*, and *Magellan* spacecraft reveal both important similarities and differences to the earth. The three types of terrain on Venus are (1) lowlands, or plains, comprising ~40% of Venus' surface that lies below the mean radius, (2) rolling plains, also comprising ~40% of the surface and lying at an elevation of 0–2 km, and (3) highlands that are ~20% of the surface and are >2 km above the mean radius. The total range of elevations on Venus is ~13 km from the lowest valleys to the top of Maxwell Montes, but about 80% of the surface is within ±1 km of the mean radius. The unimodal topographic distribution is in contrast to Earth, which has a bimodal hypsometric curve.

The two major highlands regions (terrae) on Venus are Ishtar Terra in the high northern latitudes and Aphrodite Terra in the equatorial regions. Both regions are continental size, with Ishtar roughly the size of Australia and Aphrodite roughly the size of South America. The western part of Ishtar is dominated by the Lakshmi Planum plateau that resembles, but is larger than, the Tibetan plateau on Earth. Maxwell Montes, which is higher than Mount Everest on Earth, is in eastern Ishtar. Aphrodite is rougher and more complex than Ishtar and is characterized by several deep narrow valleys, such as Diana Chasma, and by several distinct mountain ranges that reach up to 6 km high. Three smaller highlands regions are Alpha Regio, Beta Regio, and Phoebe Regio.

Venus' surface shows extensive evidence of widespread volcanism: (1) large shield volcanoes (e.g., Sif Mons) similar to the shield volcanoes of the Hawaiian islands, (2) volcanic plains, (3) volcanic calderas, (4) smaller volcanic landforms such as cones and pancake domes, and (5) long sinuous channels that can meander for several thousand km across the surface. In some cases, the different landforms indicate different types of magmas, for example, the pancake domes were apparently formed by viscous SiO_2-rich magmas, whereas the long sinuous channels were apparently formed by fluid magmas, such as carbonatites.

Tectonic features are also present on the surface. Tesserae, which are tectonically deformed regions formed by piling up blocks of crust, are common in the highlands. The lowlands and rolling plains contain wrinkle ridges formed by buckling of the crust. Other features formed by volcanism and tectonism are coronae, circular- or oval-shaped features a few hundred km in diameter that may have raised outer rims and arachnoids, which are caldera-like collapse features surrounded by fractures.

Venus' surface has ~900 impact craters ranging in diameter from ~3 km to a few hundred km. The small size cutoff is due to atmospheric disruption of small impactors. The smaller craters are more irregularly shaped, indicating the impact of several fragments instead of one object. The crater ejecta patterns are unlike those on other bodies and probably have been affected by the dense atmosphere and prevailing winds. Venus' interior structure is unknown but spacecraft data allow several inferences. No intrinsic magnetic field has been detected, and any dipole field is $<10^{-4}$ that of Earth. The ~243 day rotation rate may be too slow to generate a field by dynamo action in a core. Radar imaging does not show evidence of plate tectonics, which may be due to a lack of water and/or to difficulty in subducting the lithosphere, which is hotter and perhaps more buoyant than on Earth because of Venus' high surface temperature. Unlike Earth, gravity is strongly correlated with topography on Venus, suggesting that higher regions are above regions of mantle upwelling.

Sources: [AD32] Adams, J., B., & Dunham, T., 1932, *Publ. Astron. Soc. Pac.* 44, 243–247. [BB92] Barsukov, V. L., Basilevsky, A. T., Volkov, V. P., & Zharkov, V. N. (eds.), 1992, *Venus geology, geochemistry, and geophysics*, Univ. of Arizona Press, Tucson, pp. 421. [BFA95] Brackett, R. A., Fegley, B., Jr., & Arvidson, R. E., 1995, *J. Geophys. Res.* 100, 1553–1563. [BHP97] Bougher, S. W., Hunten, D. M., & Phillips, R. (eds.), 1997, *Venus II*, Univ. of Arizona Press, Tucson, pp. 1362. [FKL97] Fegley, B., Klingelhöfer, G., Lodders, K., & Widemann, T., 1997, in *Venus II* (Bougher, S. W., Hunten, D. M., & Phillips, R., eds.), pp. 591–636. [HCD83] Hunten, D. M., Colin, L., Donahue, T. M., & Moroz, V. I. (eds.), 1983, *Venus*, Univ. of Arizona Press, Tucson, pp. 1143. [JGR80] Pioneer Venus papers in *J. Geophys. Res.* 85 No. A13 (30 Dec. 1980). [JGR92] Magellan papers in *J. Geophys. Res.* 97 No. E8 & E10 (1992). [KKB93] Kargel, J. S., Komatsu, G., Baker, V. R., & Strom, R. G., 1993, *Icarus* 103, 253–275. [KKF94] Kargel, J. S., Kirk, R. L., Fegley, B., Jr., & Treiman, A. H., 1994, *Icarus* 112, 219–252. [Kr86] Krasnopolsky, V. A., 1986, *Photochemistry of the atmospheres of Mars and Venus,* Springer Verlag, Berlin, pp. 334. [LP84] Lewis, J. S., & Prinn, R. G., 1984, *Planets and their atmospheres*, Academic Press, NY, pp. 470. [LTP92] Luhmann, J. G., Tatrallyay, M., & Pepin, R. O. (eds.), *Venus and Mars: Atmospheres, ionospheres, and solar wind interactions*, AGU Geophysical Monograph 66, pp. 430 [PFS96] Pettengill, G. H., Ford, P. G., & Simpson, R. A., 1996, *Science* 272, 1628–1631. [Pri85] Prinn, R. G., 1985 in *The Photochemistry of Atmospheres*, (Levine, J. S., ed.), Academic Press, NY, pp. 281–336. [PSS93] *Galileo* papers in *Planetary Space Sci.* 41, No. 7, (1993). [YD82] Yung, Y. L., & DeMore, W. B., 1982, *Icarus* 51, 199–247.

Table 5.1 Spacecraft Missions to Venus

Mission	Launch Date	Type	Remarks
Sputnik 7; USSR	4 Feb. 1961	flyby	failed to depart low Earth orbit
Venera 1; USSR	12 Feb. 1961	flyby	communications failed, now in solar orbit
Mariner 1; USA	22 July 1962	flyby	launch failure
Sputnik 23; USSR	25 Aug. 1962	flyby	failed to depart low Earth orbit
Mariner 2; USA	27 Aug. 1962	flyby	flyby (36,000 km), confirmed high surface temp. [BCJ64], 1st USA success
Sputnik 24; USSR	1 Sept. 1962	flyby	failed to depart low Earth orbit
Sputnik 25; USSR	12 Sept. 1962	flyby	failed to depart low Earth orbit
Venera 1964A & B; USSR	19 Feb. 1964 1 March 1964	flyby	launch failure in both cases
Cosmos 27; USSR	27 March 1964	flyby	communications failure
Zond 1; USSR	2 April 1964	flyby	communications failure
Venera 2; USSR	12 Nov. 1965	flyby	communications failure before arrival
Venera 3; USSR	16 Nov. 1965	atm. probe	communications failed before atmospheric entry; probe crashed on Venus
Cosmos 96; USSR	23 Nov. 1965	probe	failed to depart low Earth orbit
Venera 1965A; USSR	23 Nov. 1965	flyby	launch failure
Venera 4; USSR	12 June 1967	atm. probe	measured % CO_2, P & T, atmos. compos. expts. [AMR68, LP84], crash landed
Mariner 5; USA	14 June 1967	flyby	flyby (3,900 km), atmospheric structure & composition expts. [KLC67, LP84]
Cosmos 167; USSR	17 June 1967	probe	failed to depart low Earth orbit
Venera 5; USSR	5 Jan. 1969	probe	atm. composition expts., V5 failed at 26 km, V6 failed at 11 km [AMR70, LP84]
Venera 6; USSR	10 Jan. 1969	probe	
Venera 7; USSR	17 Aug. 1970	lander	first soft landing on Venus, atmos. composition & structure [AMR71, LP84]
Cosmos 359; USSR	22 Aug. 1970	lander	failed to depart low Earth orbit
Venera 8; USSR	27 March 1972	lander	atm. comp. & structure, photometry, K, U, Th γ-ray analysis on surface, survived for 50 min. [AMM73, LP84, VSK73]
Cosmos 482; USSR	31 March 1972	lander	failed to depart low Earth orbit
Mariner 10; USA	3 Nov. 1973	flyby	5700 km flyby en route to Mercury, IR, UV spectra, imaging of clouds, particles & fields experiments [JAS75, Sci74]
Venera 9; USSR	9 June 1975	orbiters & landers	atm. comp. & structure, photometry, TV images of surface, γ-ray analysis of K, U, Th on surface [Ke77, VoZ83]
Venera 10; USSR	14 June 1975		

continued

Table 5.1 *(continued)*

Mission	Launch Date	Type	Remarks
Pioneer Venus 1; USA	20 May 1978	orbiter	first radar mapping of another planetary
Pioneer Venus 2; USA	8 Aug. 1978	bus & probes	surface, atm. science from bus & 4 probes [BHP97, HCD83, Ica82, JGR80]
Venera 11; USSR	9 Sept. 1978	flybys & probes	atmospheric science from 2 probes, no TV or surface analyses [HCD83, Kr86]
Venera 12; USSR	14 Sept. 1978		
Venera 13; USSR	30 Oct. 1981	flybys & probes	atmospheric science, XRF analyses & color images of surface [BB92, Kr86]
Venera 14; USSR	4 Nov. 1981		
Venera 15; USSR	2 June 1983	two orbiters	radar imaging from N. pole to 30° N and atm. spectroscopy expts. [BB92, BHP97]
Venera 16; USSR	7 June 1983		
Vega 1; USSR	15 Dec. 1984	landers & balloons	atmospheric science, balloons floated for 48 hours at ~54 km, XRF & γ-ray analyses of the surface [BHP97, Kr86]
Vega 2; USSR	21 Dec. 1984		
Magellan; USA	4 May 1989	orbiter	radar mapping, altimetry, emissivity data for surface, radio occultation expts. atm. science [BHP97, JGR92]
Galileo; USA	18 Oct. 1989	flyby	imaging & spectroscopy of atmosphere [BHP97, PSS93]

Sources: **[AMR68]** Avduevskii, V. S., Marov, M Ya., & Rozhdestvenskii, M. K., 1968, *Cosmic Res.* 7, 209–219. **[AMR70]** Avduevskii, V. S., Marov, M Ya., & Rozhdestvenskii, M. K., 1970, *Cosmic Res.* 8, 800–808. **[AMR71]** Avduevskii, V. S., Marov, M Ya., & Rozhdestvenskii, M. K., 1971, *J. Atmos. Sci.* 28, 263–269. **[AMM73]** Avduevskii, V. S., Marov, M. Ya., Moshkin, B. E., & Ekonomov, A. P., 1973, *J. Atmos. Sci.* 30, 1215–1218. **[BB92]** Barsukov, V. L., Basilevsky, A. T., Volkov, V. P., & Zharkov, V. N. (eds.), 1992, *Venus geology, geochemistry, and geophysics*, Univ. of Arizona Press, Tucson, pp. 421. **[BCJ64]** Barath, F. T., Barrett, A. H., Copeland, J., Jones, D. E., & Lilley, A. E., 1964, *Astron. J.* 69, 49–58. **[BHP97]** Bougher, S. W., Hunten, D. M., & Phillips, R. J. (eds.), 1997, *Venus II*, Univ. of Arizona Press, Tucson, pp. 1362. **[HCD83]** Hunten, D. M., Colin, L., Donahue, T. M., & Moroz, V. I. (eds.), 1983, *Venus* Univ. of Arizona Press, Tucson, pp. 1143. **[Ica82]** Two special issues on Venus in *Icarus* 51 No. 2 (Aug. 1982) & 52 No. 2 (Nov. 1982). **[JAS75]** *Mariner 10* papers in *J. Atmos. Sci.* 32 No. 6 (June 1975). **[JGR80]** *Pioneer Venus* papers in *J. Geophys. Res.* 85 No. A13 (30 Dec. 1980). **[JGR92]** Magellan papers in *J. Geophys. Res.* 97 No. E8 & E10 (1992). **[Ke77]** Keldysh, M. V., 1977 *Icarus* 30, 605–625. **[KLC67]** Kliore, A. J., Levy, G. L., Cain, D. L., Fjeldbo, G., & Rasool, S. I. 1967, *Science* 158, 1683–1688. **[Kr86]** Krasnopolsky, V. A., 1986, *Photochemistry of the atmospheres of Mars and Venus,* Springer Verlag, Berlin, pp. 334. **[LP84]** Lewis, J. S., & Prinn, R. G., 1984, *Planets and their atmospheres*, Academic Press, NY, pp. 470. **[NSSDC]** National space science data center, Greenbelt, MD. **[PSS93]** *Galileo* papers in *Planetary Space Sci.* 41 No. 7 (1993). **[Sci74]** *Mariner 10* papers in *Science* 183 No. 4131 (29 March 1974). **[VoZ83]** Von Zahn, U., Kumar, S., Niemann, H., & Prinn, R., 1983, in *Venus* (Hunten, D. M., Colin, L., Donahue, T. M., & Moroz, V. I., eds.), Univ. of Arizona Press, Tucson, pp. 299–430. **[VSK73]** Vinogradov, A. P., Surkov, Yu. A., & Kirmazov, F. F., 1973, *Icarus* 20, 253–259.

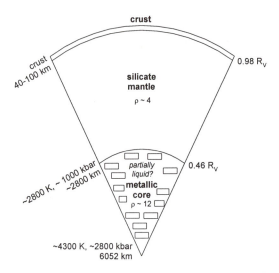

Figure 5.1 Interior of Venus based on geochemical models

Table 5.2 Some Physical Properties of Venus

Property	Value	Property	Value
Modal radius (km)	6051.37	Sidereal revolution period ($\oplus$ days)	224.695
Median radius (km)	6051.64	Mean synodic period ($\oplus$ days)	583.92
Mean radius (km)	6051.84	Eccentricity of orbit	0.0068
Oblateness ($R_{eq} - R_{pol})/R_{eq}$	0.0	Inclination of orbit to ecliptic	3.39°
Mass (kg)	4.8685×10^{24}	Mean orbital velocity (km s^{-1})	35.03
Mean density (g cm^{-3})	5.243	Inclination of equator to orbit	177.4°
GM m^{3}s^{-2}	3.2486×10^{14}	Sidereal rotation period ($\oplus$ days), retrograde	243.0187
Gravity (m s^{-2})	8.870	Temperature at modal radius (K)	740
Escape velocity (km s^{-1})	10.361	Pressure at modal radius (bars)	95.6
$J_2 \times 10^6$	6±3	Mean visible cloud temperature (K)	230±10
C/MR^2	...	Magnetic dipole moment (Tesla R$_{Ven}^3$)	$<3 \times 10^{-8}$
Solar constant (Wm^{-2})	2613.9		

Sources: Ananda, M. P., Sjorgren, W. L., Phillips, R. J., Wimberly, R. N., & Bills, B. G., 1980, *J. Geophys. Res.* 85, 8303–8318. Davies, M. E., Colvin, T. R., Rogers, P. G., Chodas, P. W., Sjogren, W. L., Akim, E. L., Stepanyantz, V. A., Vlasova, Z. P., & Zakharov, A. I., 1992, *J. Geophys. Res.* 97, 13141–14151. Ford, P. G. & Pettengill, G. H., 1992, *J. Geophys. Res.* 97, 13103–13114. McNamee, J. B., Borderies, N. J., & Sjogren, W. L., 1993, *J. Geophys. Res.* 98, 9113–9128. Russell, C. T., Elphic, R. C. & Slavin, J. A., 1980, *J. Geophys. Res.* 85, 8319–8322.

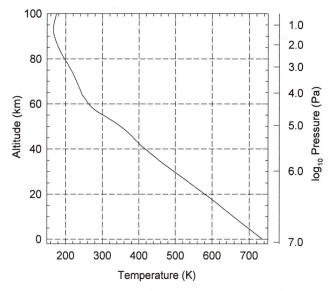

Figure 5.2 Temperature and pressure in Venus' atmosphere

Table 5.3 Temperature, Pressure, and Density in Venus' Atmosphere

Altitude (km)	Temperature (K)	Pressure (Pa)	Density (kg m⁻³)
0	735.3	9,210,000	64.79
10	658.2	4,739,000	37.72
20	580.7	2,252,000	20.39
30	496.9	958,100	10.15
40	417.6	350,100	4.404
50	350.5	106,600	1.594
60	262.8	23,570	0.469
70	229.8	3,690	8.39(−2)
80	197.1	447.6	1.19(−2)
90	169.4	37.36	1.15(−3)
100	175.4	2.660	7.89(−5)

Exponents in parentheses.

Source: Seiff, A., Schofield, J. T., Kliore, A. J., Taylor, F. W., Limaye, S. S., Revercomb, H. E., Sromovsky, L. A., Kerzhanovich, V. V., Moroz, V. I., & Marov, M. Ya., 1986, in *The Venus International Reference Atmosphere,* Advances in Space Research, Vol. 5. (Kliore, A. J., Moroz, V. I., & Keating, G. M., eds.), Pergamon Press, pp. 3–32.

Table 5.4 Chemical Composition of the Atmosphere of Venus

Gas	Abundance	Gas Source(s)	Gas Sink(s)	Sources
CO_2	96.5±0.8%	outgassing	carbonate formation	VoZ83
N_2	3.5±0.8%	outgassing	—	VoZ83
SO_2*	150±30 ppm (22–42 km)	outgassing & reduc-	H_2SO_4 formation &	VoZ83
	25–150 ppm (12–22 km)	tion of OCS, H_2S	$CaSO_4$ formation	FKL97
H_2O*	30±15 ppm (0–45 km)	outgassing	H escape &	TCB97
	30–70 ppm (0–5 km)	—	Fe^{2+} oxidation	IMM97
Ar	70±25 ppm	outgassing, primordial	—	VoZ83
CO*	45±10 ppm (cloud top)	CO_2 photolysis	photooxidation	CCK68
	30±18 ppm (42 km)			OCW80
	28±7 ppm (36–42 km)			GZL79
	20±3 ppm (22 km)			VoZ83
	17±1 ppm (12 km)			MVS89
He	12^{+24}_{-8} ppm	outgassing (U, Th)	—	VoZ83
Ne	7±3 ppm	outgassing, primordial	—	VoZ83
OCS	4.4±1 ppm (33 km)	outgassing	conversion to SO_2	PDG93
H_2S *	3±2 ppm (<20 km)	outgassing	conversion to SO_2	HHD80
HDO*	1.3±0.2 ppm (sub-cloud)	outgassing	H escape	DBO91
HCl	0.6±0.12 ppm (cloud top)	outgassing	Cl-mineral	CCB67
	0.5 ppm (35–45 km)		formation	BDC90
Kr	~25 ppb	outgassing, primordial	—	DHH81
SO*	20±10 ppb (cloud top)	photochemistry	photochemistry	NES90
S_{1-8}*	20 ppb (<50 km)	sulfide weathering	conversion to SO_2	MGE80
HF	$5^{+5}_{-2.5}$ ppb (cloud top)	outgassing	F-mineral formation	CCB67
	4.5 ppb (35–45 km)			BDC90
Xe	~1.9 ppb	outgassing, primordial	—	Don86,Pe91

* Abundances of these species are altitude dependent [see VoZ83, FT92, FLK97].

Sources: **[BDC90]** Bézard, B., DeBergh, C., Crisp, D., & Maillard, J. P., 1990, *Nature* 345, 508–511. **[CCB67]** Connes, P., Connes, J., Benedict, W. S., & Kaplan, L. D., 1967, *ApJ.* 147, 1230–1237. **[CCK68]** Connes, P., Connes, J., Kaplan, L. D., & Benedict, W. S., 1968, *ApJ.* 152, 731–743. **[DBO91]** DeBergh, C., Bézard, B., Owen, T., Crisp, D., Maillard, J. P., & Lutz, B. L., 1991, *Science* 251, 547–549. **[DHH81]** Donahue, T. M., Hoffman, J. H., & Hodges, R. R., 1981, *Geophys. Res. Lett.* 8, 513–516. **[Don86]** Donahue, T. M., 1986, *Icarus* 66, 195–210. **[FKL97]** Fegley, B., Klingelhöfer, G., Lodders, K., & Widemann, T., 1997, in *Venus II* (Boucher, S. W., Hunten, D. M., & Phillips, R., eds.), Univ. of Arizona Press, Tucson, pp. 591–636. **[FT92]** Fegley, A. & Treiman, A. H., 1992, in *Venus and Mars: Atmospheres, ionospheres, and solar wind interactions* (Luhmann, J. G., Tatrallyay, M., & Pepin, R. O., eds.), AGU Geophysical Monograph 66, pp. 7–71. **[GZL79]** Gel'man, B. G., Zolotukhin, V. G., Lamonov, B. V., Levchuk, B. V., Lipatov, A. N., Mulkhin, L. M., Nenarokov, D. F., Rotin, V. A., & Okhotnikov, B. P., 1979, *Cosmic Res.* 17, 585–589.

[HHD80] Hoffman, J. H., Hodges, R. R., Donahue, T. M., & McElroy, M. B., 1980, *J. Geophys. Res.* 85, 7882–7890. [IMM97] Ignatiev, N. I., Moroz, V. I., Moshkin, B. E., Ekonomov, A. P., Gnedykh, V. I., Grigoriev, A. V., & Khatuntsev, I. V., 1997, *Planet. Space Sci.* 45, 427–438. [MGE80] Moroz, V. I., Golovin, Yu. M., Ekonomov, A. P., Moshkin, B. E., Parfent'ev, N. A., & San'ko, N. F., 1980, *Nature* 284, 243–244. [MVS89] Marov, M. Ya., Volkov, V. P., Surkov, Yu. A., & Ryvkin, M. L., 1989, in *The planet Venus: atmosphere, surface, interior structure* (Barsukov, V. L., & Volkov, V. P., eds.), Nauka, Moscow, USSR, pp. 25–67. [NES90] Na, C. Y., Esposito, L. W., & Skinner, T. E.,1990, *J. Geophys. Res.* 95, 7485–7491. [OCW80] Oyama, V. I., Carle, G. C., Woeller, F., Pollack, J. B., Reynolds, R. T., & Craig, R. A., 1980, *J. Geophys. Res.* 85, 7891–7902. [Pe91] Pepin, R. O., 1991, *Icarus* 92, 2–79. [PDG93] Pollack, J. B., Dalton, J. B., Grinspoon, D., Wattson, R. B., Freedman, R., Crisp, D., Allen, D. A., Bézard, B., DeBergh, C., Giver, L. P., Ma, Q., & Tipping, R., 1993, *Icarus* 103, 1–42. [TCB97] Taylor, F. W., Crisp, D., & Bézard, B., 1997, in *Venus II* (Boucher, S. W., Hunten, D. M., & Phillips, R., eds.), Univ. of Arizona Press, Tucson, pp.325–351. [VoZ83] Von Zahn, U., Kumar, S., Niemann, H., & Prinn, R., 1983, in *Venus* (Hunten, D. M., Colin, L., Donahue, T. M., & Moroz, V. I., eds.), Univ. of Arizona Press, Tucson, pp. 299–430.

Table 5.5 Isotopic Composition of the Atmosphere of Venus

Isotopic Ratio	Observed Value	Notes	Sources
D/H	0.016 ± 0.002	Pioneer Venus MS *	DHH82
	0.019 ± 0.006	IR spectroscopy	DBO91
$^{12}C/^{13}C$	86 ± 12	IR spectroscopy	BBM87
	88.3 ± 1.6	Venera 11/12 MS	IGK80
$^{14}N/^{15}N$	273 ± 56	Pioneer Venus MS	HHD79
$^{16}O/^{18}O$	500 ± 25	Pioneer Venus MS	HHD80
	500 ± 80	IR spectroscopy	BBM87
$^{20}Ne/^{22}Ne$	11.8 ± 0.7	Pioneer Venus MS	Don86
$^{35}Cl/^{37}Cl$	2.9 ± 0.3	IR spectroscopy	CCB67, You72
$^{36}Ar/^{38}Ar$	5.56 ± 0.62	Pioneer Venus MS	Don86
	5.08 ± 0.05	Venera 11/12 MS	IGK80
$^{40}Ar/^{36}Ar$	1.03 ± 0.04	Pioneer Venus MS	HHD80
	1.19 ± 0.07	Venera 11/12 MS	IGK80

Note: No isotopic compositions are available for Kr and Xe on Venus.
* MS = Mass Spectrometer

Sources: [BBM87] Bézard, B., Baluteau, J. P., Marten, A., & Coron, N., 1987, *Icarus* 72, 623–634.[CCB67] Connes, P., Connes, J., Benedict, W. S., & Kaplan, L. D., 1967, *ApJ.* 147, 1230–1237. [DBO91] DeBergh, C., Bézard, B., Owen, T., Crisp, D., Maillard, J. P., & Lutz, B. L., 1991, *Science* 251, 547–549. [DHH82] Donahue, T. M., Hoffman, J. H., Hodges, R. R., & Watson, A. J., 1982, *Science* 216, 630–633. [Don86] Donahue, T. M., 1986, *Icarus* 66, 195–210. [HHD79] Hoffman, J. H., Hodges, R. R., Donahue, T. M., McElroy, M. B., & Koplin, M., 1979, *Science* 205, 49–52. [HHD80] Hoffman, J. H., Hodges, R. R., Donahue, T. M., & McElroy, M. B., 1980, *J. Geophys. Res.* 85, 7882–7890. [IGK80] Istomin, V. G., Grechnev, K. V., & Kochnev, V. A., 1980, 23rd COSPAR Meeting, Budapest, Hungary. [You72] Young, L. D. G., 1972, *Icarus* 17, 632–658.

Table 5.6 XRF Elemental Analyses of Venus' Surface

Oxide	Mass Percent (±1σ)		
	Venera 13 [SBM84]	Venera 14 [SBM84]	Vega 2 ‡ [SMK86]
SiO_2	45.1±3.0	48.7±3.6	45.6±3.2
TiO_2	1.59±0.45	1.25±0.41	0.2±0.1
Al_2O_3	15.8±3.0	17.9±2.6	16±1.8
FeO *	9.3±2.2	8.8±1.8	7.7±1.1
MnO	0.2±0.1	0.16±0.08	0.14±0.12
MgO	11.4±6.2	8.1±3.3	11.5±3.7
CaO	7.1±0.96	10.3±1.2	7.5±0.7
Na_2O †	2.0±0.5	2.4±0.4	2
K_2O	4.0±0.63	0.2±0.07	0.1±0.08
SO_3	1.62±1.0	0.88±0.77	4.7±1.5
Cl	<0.3	<0.4	<0.3
Total	98.1	98.7	95.4

* All Fe reported as FeO for all analyses.
† Calculated by [SBM84, SMK86].
‡ In addition to Cl, [SMK86] report the following upper limits (in mass%): Cu, Pb <0.3; Zn <0.2; Sr, Y, Zr, Nb, Mo <0.1; As, Se, Br <0.08.

Sources: **[SBM84]** Surkov, Yu. A., Barsukov, V. L., Moskalyeva, L. P., Kharyukova, V. P., & Kemurdzhian, A. L., 1984, *J. Geophys. Res. (Proc. 14th LPSC)* 89, B393–B402.
[SMK86] Surkov, Yu. A., Moskalyova, L. P., Kharyukova, V. P., Dudin, A. D., Smirnow, G. G., & Zaitseva, S. Ye., 1986, *J. Geophys. Res. (Proc. 17th LPSC)* 91, E215–E218.

Table 5.7 Gamma Ray Analyses of Venus' Surface

Space Probe	K (mass%)	U (ppm)	Th (ppm)
Venera 8	4.0±1.2	2.2±0.7	6.5±2.2
Venera 9	0.47±0.08	0.60±0.16	3.65±0.42
Venera 10	0.30±0.16	0.46±0.26	0.70±0.34
Vega 1	0.45±0.22	0.64±0.47	1.5±1.2
Vega 2	0.40±0.20	0.68±0.38	2.0±1.0

Source: Surkov, Yu. A., Kirnozov, F. F., Glazov, V. N., Dunchenko, A. G., Tatsy, L. P., & Sobornov, O. P., 1987, *J. Geophys. Res., Proc. 17th LPSC,* 92, E537–E540.

Table 5.8 Model Elemental Abundances in Venus (silicates plus core)

Element	Unit	Value	Element	Unit	Value	Element	Unit	Value
H	ppm	35	Zn	ppm	82	Pr	ppm	0.135
Li	ppm	1.94	Ga	ppm	3.4	Nd	ppm	0.723
Be	ppm	0.047	Ge	ppm	8.4	Sm	ppm	0.218
B	ppb	10	As	ppm	3.1	Eu	ppm	0.083
C	ppm	468	Se	ppm	5.4	Gd	ppm	0.30
N	ppm	4.3	Br	ppm	0.111	Tb	ppm	0.056
O	mass%	30.90	Rb	ppm	0.509	Dy	ppm	0.382
F	ppm	15	Sr	ppm	15.2	Ho	ppm	0.084
Na	ppm	1390	Y	ppm	2.74	Er	ppm	0.242
Mg	mass%	14.54	Zr	ppm	7.5	Tm	ppm	0.037
Al	mass%	1.48	Nb	ppm	0.84	Yb	ppm	0.240
Si	mass%	15.82	Mo	ppm	2.47	Lu	ppm	0.405
P	ppm	1860	Ru	ppm	1.23	Hf	ppm	0.241
S	mass%	1.62	Rh	ppb	265	Ta	ppb	24.4
Cl	ppm	20.9	Pd	ppb	870	W	ppb	189
K	ppm	150	Ag	ppb	49	Re	ppb	64
Ca	mass%	1.61	Cd	ppb	17.2	Os	ppb	920
Sc	ppm	10.1	In	ppb	2.24	Ir	ppb	890
Ti	ppm	850	Sn	ppb	430	Pt	ppm	1.76
V	ppm	86	Sb	ppb	39	Au	ppb	250
Cr	ppm	4060	Te	ppb	830	Hg	ppb	8.3
Mn	ppm	460	I	ppb	14.3	Tl	ppb	4.05
Fe	mass%	31.17	Cs	ppb	17	^{204}Pb	ppb	1.66
Co	ppm	820	Ba	ppm	4.2	Bi	ppb	3.08
Ni	mass%	1.77	La	ppm	0.397	Th	ppb	53.7
Cu	ppm	35	Ce	ppm	1.06	U	ppb	15.0

Note: Four-component meteorite model

Source: Morgan, J. W., & Anders, E., 1980, *Proc. Natl. Acad. Sci.* 77, 6973–6977.

Table 5.9   Model Compositions of Venus

Model:	I [MA80]	II [BVSP81]	III [BVSP81]	IV [BVSP81]	V [BVSP81]
Mantle & Crust					
MgO	35.5	37.6	38.3	33.3	38.0
Al_2O_3	4.1	3.8	3.9	3.4	3.9
CaO	3.3	3.6	3.6	3.4	3.2
SiO_2	49.8	52.9	53.9	40.4	45.9
TiO_2	0.21	0.20	0.20	0.24	0.3
Na_2O	0.28	1.6	1.5	0.15	0.1
K_2O	0.027	0.174	0.159	0.018	0.015
FeO	5.4	0.24	2.1	18.7	8.1
Cr_2O_3	0.87	...	...	0.3	0.3
MnO	0.09	...	...	0.2	0.1
P_2O_5	0.022	...	...	...	...
H_2O	0.22	0	0	...	...
Th ppm	0.079	0.073	0.075	0.066	0.075
U ppm	0.022	0.021	0.021	0.019	0.021
Core					
Fe	88.6	94.4	84.7	78.7	86.2
Co	0.26	...	...	...	...
Ni	5.5	5.6	5.2	6.6	4.8
S	5.1	0	10	4.9	1.0
O	0	...	...	9.8	8.0
P	0.58	...	...	...	...
Relative Masses (mass%)					
Mantle & Crust	68.0	69.8	69.1	76.4	71.8
Core	32.0	30.2	30.9	23.6	28.2

mass% if not noted otherwise
I: Four component meteorite model [MA80]
II: Equilibrium condensation, [BVSP81] model Ve1.
III: Equilibrium condensation model including feeding zones, [BVSP81] model Ve2
IV: Pyrolite model, [BVSP81] model Ve4
V: Iron-deficient model, [BVSP81] model Ve5

Sources: **[BVSP81]** Basaltic Volcanism Study Project: *Basaltic volcanism on the terrestrial planets*, 1981, Pergamon Press, pp. 1286. **[MA80]** Morgan, J. W., & Anders, E., 1980, *Proc. Natl. Acad. Sci.* 77, 6973–6977.

6

THE EARTH AND THE MOON

6.1 Earth

The Solid Earth

The Earth is composed of the atmosphere, hydrosphere, crust, mantle, and core. Each of these major regions is subdivided further. First, we discuss the solid Earth. The atmosphere and hydrosphere are discussed subsequently.

The average thickness of the (granitic) continental crust is about 35 km; most of the continental crust is 20–50 km thick. A small percentage may be as thin as 10 km or as thick as 70 km. The average density of the continental crust is about 2.8 g cm^{-3}. Continental crust is enriched in incompatible elements and contains most of the Cs, Rb, Ba and a large fraction of the U, Pb, and K in the silicate earth. The oldest regions of the continental crust are ancient cratons in North America, Africa, Australia, and Russia. The oldest known rocks are ~4.1 Ga old.

The (basaltic) oceanic crust is about 5–10 km thick and has an average density of about 3.0 g cm^{-3}. The oceanic crust is fairly young, with an average age of 60 Ma; the oldest oceanic crust is ~200 Ma old. To a first approximation, the oceanic crust is composed of sediments, mid-ocean ridge basalt (MORB), and ocean island basalts (OIB). The oceanic crust is not as enriched in incompatible elements as the continental crust.

The Mohorovičić discontinuity (the Moho) separates the crust and mantle and was recognized from the sharp increase in P wave velocities. The Moho is not a sharp boundary, and seismic velocity changes occur over a depth of ~5 km.

Although the Moho divides the compositionally different crust and mantle, the upper 100–200 km of the earth is made up of about 12 plates, which are rigid lithospheric blocks that ride on the weaker part of the upper mantle called the asthenosphere. Oceanic plates are ~60 km thick and continental plates are ~100–200 km thick. Plate interactions fall into three categories: (1) Divergent boundaries, such as the Mid-Atlantic Ridge, occur where plates spread apart at rift zones. (2) Convergent boundaries are where plates move toward each other. The higher density oceanic plates

Earth & Moon

are subducted under the lower density continental plates at convergent boundaries, whereas mountain ranges such as the Himalayas are formed by collision of two continental plates at convergent boundaries. (3) Transform or strike slip boundaries occur where two plates slide parallel in opposite directions. Transform boundaries are often associated with stresses and high seismicity, such as along the San Andreas fault in California. Most volcanic and tectonic activity occurs near plate margins, such as the volcanic Ring of Fire along the Pacific Rim.

The mantle constitutes most of the bulk silicate earth. Seismic data, analyses of ultramafic rocks and nodules from the upper mantle (e.g., xenoliths in kimberlite pipes), analyses of basaltic magmas probably generated by partial melting of the upper mantle (e.g., MORB and OIB), experimental studies of high temperature and pressure phase equilibria, and cosmochemical constraints (e.g., elemental and isotopic analyses of meteorites) provide our knowledge of mantle composition and mineralogy. Seismic data show that the mantle is further subdivided into the upper mantle and the lower mantle by seismic discontinuities at depths of about 410 and 670 km. The two discontinuities are probably due to crystal structure changes and disproportionation of major mantle minerals: the transformation of olivine to β-phase at 410 km and the transformation of silicate with γ-spinel structure to perovskite-structure silicate plus magnesiowüstite $(Mg,Fe)O$ at 670 km. The lower 200 to 300 km of the lower mantle, near the core mantle boundary (CMB), comprise the D" layer. The nature of the D" layer is currently under debate; it may be a thermal boundary layer, a region where subducted oceanic slabs pile up at the base of the mantle, a higher pressure phase change in mantle mineralogy, the residue left over from core formation, or a chemical reaction layer between the core and mantle. Recent seismic studies indicate that the lowest 20 km or so in D" may be molten.

Seismic data also show that the core is divided into a molten outer core, slightly less dense than pure molten Fe, and a solid inner core. Geochemical arguments and laboratory studies indicate that a light element, such as S, O, or C, makes up ~10% of the outer core. (It is also possible that smaller amounts of many light elements are present depending on their solubilities in molten Fe at high pressures and temperatures.) Studies of enstatite chondrites and achondrites, which have the same oxygen isotopic composition as the Earth, also suggest that ^{40}K may be present in the outer core. The ^{40}K would provide a heat source to drive the convective motions in the fluid outer core that are responsible for the generation of Earth's magnetic field. However, differential precession of the outer core and mantle, the escape of primordial heat left from planetary accretion, or compression may also power the dynamo action in the outer core.

The inner core is solid with anisotropic seismic properties. It is slightly more dense than pure solid Fe and is probably an Fe-Ni alloy. The inner core is formed by high pressure "freezing" of the molten outer core. The heat released may help to support the outer core-driven dynamo. The differential rotation of the inner and outer core provides yet another heat source.

Sources and further reading: Jeanloz, R., 1990, *Annu. Rev. Earth Planet. Sci.* 18, 357–386. Lodders, K., 1995, *Meteoritics* 30, 93–101. Newsom, H. E., & Jones, J. H. (eds.), 1990, *Origin of the earth*, Oxford Univ. Press, New York, pp. 378. Taylor, S. R., & McLennan, S. M., 1985, *The continental crust: its composition and evolution*, Blackwell Sci. Publ., Oxford, pp. 312. Turcotte, D. L., & Schubert G., 1982, *Geodynamics*, John Wiley, New York, pp. 450. Wilson, M., 1989, *Igneous petrogenesis*, Unwin Hyman, London, pp. 466.

Table 6.1 Some Physical Properties of the Earth

Property	Value	Property	Value
Equatorial radius R_{eq} (km)	6378.136	Sidereal revolution period ($\oplus$ days)	365.256
Polar radius R_p (km)	6356.753	Eccentricity of orbit	0.0167°
Mean radius R_E (km)	6371.01	Inclination of orbit to ecliptic	0.00°
Total surface area (km²)	5.10×10^8	Mean orbital velocity (km s^{-1})	29.79
Oceanic surface area (km²)	3.62×10^8	Inclination of equator to orbit	23.45°
Volume (km³)	1.0832×10^{12}	Sidereal rotation period ($\oplus$ hours)	23.9345
Oblateness $(R_{eq} - R_p)/R_{eq}$	3.3529×10^{-3}	Lengthening of day (msec. Cy^{-1})	2.0±0.2
f^{-1} = oblateness^{-1}	298.257	Mean surface temperature (K)	288 K
Mass (kg)	5.9736×10^{24}	Temperature extremes (°C)	~60, ~ −90
Mean density (g cm^{-3})	5.515	Surface pressure (bars)	1.0
GM (m³s^{-2})	3.9860×10^{14}	Magnetic dipole moment (Tesla $R_\oplus^3$)	0.61×10^{-4}
g_o=GM/R_E^2 (m s^{-2})	9.82022	Magnetic axis offset	11.5°
Equatorial gravity (m s^{-2})	9.78033	$J_2 \times 10^6$	1082.636
Polar gravity (m s^{-2})	9.83219	$J_3 \times 10^5$	−0.254
Eq. escape velocity (km s^{-1})	11.18	$J_4 \times 10^8$	−1.61
C/(MR$_E^2$)	0.3307	Heat flow (mW m^{-2}):	
Moments of inertia (kg m²):		global mean	87±2.0
equatorial (A)	8.0096×10^{37}	oceanic mean	101±2.2
equatorial (B)	8.0094×10^{37}	continental mean	65±1.6
polar (C)	8.0358×10^{37}	Total global heat loss (10^{13} W)	4.4±0.1

Sources: IERS Standards, 1992, *Technical Report*, Central Bureau of IERS, Observatoire de Paris. Lambeck, K., 1980, *The Earth's variable rotation—Geophysical causes and consequences*, Cambridge Univ. Press, NY, pp. 449. Pollack, H. N., Hurter, S. J., & Johnston, R., 1993, *Rev. Geophys.* 31, 267–280. Press, F., & Siever, R., 1978, *Earth*, Freeman, San Francisco, pp. 649

Table 6.2 Mass Distribution Within the Earth

Region	Mass (kg)	% of Whole Earth	% of Bulk Silicate Earth
Total atmosphere	5.137×10^{18}	8.65×10^{-5}	—
Troposphere	4.22×10^{18}	7.06×10^{-5}	—
Stratosphere	9.06×10^{17}	1.52×10^{-5}	—
Upper atmosphere	4×10^{15}	6.70×10^{-8}	—
Biosphere	1.148×10^{16}	1.92×10^{-7}	—
Hydrosphere	1.664×10^{21}	0.0279	0.0413
Crust	2.367×10^{22}	0.3951	0.585
Silicate mantle	4.007×10^{24}	67.077	99.37
Core	1.941×10^{24}	32.5	—
Whole Earth	5.9736×10^{24}	100	—

Sources for tables 6.2–6.5: Bloxham, J., & Jackson, A., 1991, *Rev. Geophys.* 29, 97–120. Gubbins, D., Masters, T. G., & Jacobs, J. A., 1979, *Geophys. J. Roy. Astr. Soc.* 59, 57–99. Jeanloz, R., 1990, *Annu. Rev. Earth Planet. Sci.* 18, 357–386. Labrosse, S., Poirier, J. P., & Le Mouël, J. L., 1997, *Phys. Earth Planet. Int.* 99, 1–17. Loper, D. E., & Roberts, P. H., 1981, *Phys. Earth Planet. Int.* 24, 302–307. Poirier, J. P., 1988, *Geophys. J. Roy. Astr. Soc.* 92, 99–105. Press, F., & Siever, R., 1978, *Earth*, Freeman, San Francisco, pp. 649. Warneck, P., 1988, *Chemistry of the natural atmosphere*, Academic Press, New York, pp. 757.

Table 6.3 Some Properties of the Earth's Crust

Property	Continental Crust	Oceanic Crust	Total Crust
Mass (kg)	1.522×10^{22}	8.450×10^{21}	2.367×10^{22}
Average thickness (km)	35	4.7	—
Area (km^2)	149×10^6	361×10^6	510×10^6
Area (% of Earth surface)	29.2 %	70.8	100
Mean density (g cm^{-3})	2.7–2.8	3.0	2.8
Mean height of continents above sea level (m)			874
Mean depth of oceans (m)			3794
Volume of oceans (km^3)			1.37×10^9

For sources, see Table 6.2.

Table 6.4 Some Properties of the Earth's Silicate Mantle

Property		Value
Mass (kg)		4.007×10^{24}
Mean density (g cm^{-3})		4.5
Thickness (km)	lithosphere	~100
	asthenosphere	~300
	Moho to core	2900

For sources, see Table 6.2.

Table 6.5 Some Properties of the Earth's Core

Property	Value
Mass (kg)	1.941×10^{24}
Radius of inner core (km)	1220
Radius of inner + outer core (km)	3485
Density of inner core (g cm^{-3})	12.8–13.1
Density of outer core (g cm^{-3})	9.9–12.2
Rotation rate (s^{-1})	7.3×10^{-5}
Typical core velocity (m s^{-1})	10^{-3}
Kinematic viscosity (m^2 s^{-1})	3.0×10^{-7}
Thermal diffusivity (m^2 s^{-1})	1.5×10^{-5}
Total heat flux at core-mantle boundary (W)	$(2.5–6) \times 10^{12}$
Heat of freezing at inner core boundary (W)	$(0.8–1) \times 10^{12}$
Heat from differentiation of core (W)	2.5×10^{12}

For sources, see Table 6.2.

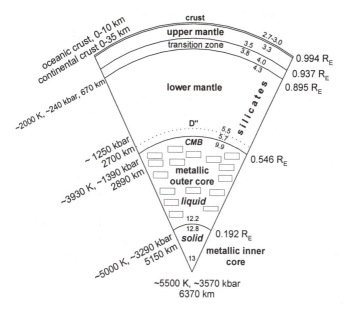

Figure 6.1 The interior structure of the earth. Numbers at phase boundaries are densities in g cm⁻³. See text for explanation and sources.

Table 6.6 Interior Structure of the Earth

Depth (km)	Layer or Boundary	Density (g cm⁻³)	T (K)	P (kbar)	Velocity (km s⁻¹) P-wave	S-wave
0–35	Continental crust	2.7–2.8	290–770	0–13	1.5–5.9	0–3.7
0–10	Oceanic crust	3.0	290–500	0–3.3	6.5	0–3.7
35 *	*Mohorovičić discontinuity*				↑↓	↑↓
35–400	Upper mantle	3.3–3.5	200–1770	2–140	8.0–8.9	4.5–4.8
400–670	Transition zone	3.8–4.0	1770–2000	140–240	9.1–10.3	4.9–5.6
670–2890	Lower mantle	4.3–5.7	2000–3930	240–1390	10.7–13.6	5.9–7.2
2700–2890	D"	5.5–5.7		1250–1390	13.7	7.3
2890	*Wiechert-Oldham-Gutenberg discontinuity; CMB*	3930	1390		↑↓	↑↓
2890–5150	Outer core	9.9–12.2	3930–5000	1390–3290	8.0–10.4	0.0
5150–6370	Inner core	12.8–13.1	5000–5500	3290–3570	11.0–11.3	3.5–3.7

* 7 km below oceans, 30–70 km below continents. CMB is core-mantle boundary.

Table 6.7　Geologic Time Scale

Eon	Era	Period	Epoch	Time Before Present (Myr)	Marking Event
Phanerozoic					
	Cenozoic				
		Quaternary	Holocene	0–10,000 years	
			Pleistocene	0.01–1.6	*Homo erectus*
		Tertiary Neogene	Pliocene	1.6–5.3	ape man fossils
			Miocene	5.3–23.7	origin of grass
		Tertiary Paleogene	Oligocene	23.7–36.6	cats, dogs, & pigs
			Eocene	36.6–57.8	hoofed mammals
			Paleocene	57.8–66.4	early primates
	Mesozoic				
		Cretaceous		66.4–138	extinction of dinosaurs
		Jurassic		138–208	first birds
		Triassic		208–245	dinosaurs appear
	Paleozoic				
		Permian		245–286	flowers, insect pollination
		Carboniferous			
		Pennsylvanian		286–320	conifers, reptiles
		Mississippian		320–360	amphibians
		Devonian		360–408	first vertebrates ashore
		Silurian		408–438	spore-bearing plants
		Ordovician		438–505	first animals ashore
		Cambrian		505–570	vertebrates appear
Proterozoic/Precambrian III					
	Late			570–900	first plants, jellyfish
	Middle			900–1600	
	Early			1600–2500	
Archean/Precambrian II					
	Late			2500–3000	photosynthetic bacteria
	Middle			3000–3400	
	Early			3400–3800	most ancient fossils 3.4 Gyrs
Hadean/Precambrian I				3800–4550	oldest rocks 4.1 Gyrs Earth forms 4.55 Gyrs ago

Sources: Calder, N., 1984, *1984 and beyond*, Viking Press, New York, pp. 207. Harland, W. B., Cox, A. V., Llewellyn, P. G., Pickton, C. A. G., Smith, A. G., & Walters, R, 1982, *A geologic timescale*, Cambridge Univ. Press, Cambridge, pp. 131. Palmer, A. R., 1983, *Geology* 11, 503–504.

Table 6.8 Elemental Abundances in the Whole Earth (silicates & core)

Element	Unit	[KL93]	[MA80]	Element	Unit	[KL93]	[MA80]
H	ppm	36.9	33	Ge	ppm	10.2	7.6
^{4}He	ppm	...	2.0×10^{-4}	As	ppm	1.73	3.2
Li	ppm	1.69	1.85	Se	ppm	3.16	9.6
Be	ppm	0.052	0.045	Br	ppm	0.13	0.106
B	ppm	0.292	0.0096	^{84}Kr	ppm	3.7×10^{-6}	8.8×10^{-7}
C	ppm	44	446	Rb	ppm	0.76	0.458
N	ppm	0.59	4.1	Sr	ppm	14.4	14.5
O	wt%	31.67	30.12	Y	ppm	2.88	2.62
F	ppm	15.8	13.5	Zr	ppm	7.74	7.2
^{20}Ne	ppm	2.7×10^{-5}	4.5×10^{-6}	Nb	ppm	0.517	0.8
Na	ppm	2450	1250	Mo	ppm	1.71	2.35
Mg	wt%	14.86	13.90	Ru	ppm	1.31	1.18
Al	wt%	1.433	1.41	Rh	ppb	227	252
Si	wt%	14.59	15.12	Pd	ppb	831	890
P	ppm	1180	1920	Ag	ppb	99	44
S	wt%	0.893	2.92	Cd	ppb	68	16.4
Cl	ppm	264	19.9	In	ppb	4.9	2.14
^{36}Ar	ppm	6.5×10^{-5}	3.5×10^{-5}	Sn	ppb	340	390
K	ppm	225	135	Sb	ppb	61	35
Ca	wt%	1.657	1.54	Te	ppb	390	1490
Sc	ppm	11.1	9.6	I	ppb	36	13.6
Ti	ppm	797	820	^{132}Xe	ppb	4.0×10^{-4}	9.9×10^{-4}
V	ppm	104	82	Cs	ppb	55	15.3
Cr	ppm	3423	4120	Ba	ppm	4.33	4.0
Mn	ppm	2046	750	La	ppm	0.434	0.379
Fe	wt%	32.04	32.07	Ce	ppm	1.114	1.01
Co	ppm	779	840	Pr	ppm	0.165	0.129
Ni	wt%	1.72	1.82	Nd	ppm	0.836	0.69
Cu	ppm	82.7	31	Sm	ppm	0.272	0.208
Zn	ppm	47.3	74	Eu	ppm	0.1035	0.079
Ga	ppm	4.42	3.1	Gd	ppm	0.363	0.286

continued

Table 6.8 *(continued)*

Element	Unit	[KL93]	[MA80]	Element	Unit	[KL93]	[MA80]
Tb	ppm	0.0671	0.054	Os	ppb	898	880
Dy	ppm	0.448	0.364	Ir	ppb	889	840
Ho	ppm	0.1027	0.08	Pt	ppm	1.77	1.67
Er	ppm	0.294	0.231	Au	ppb	157	257
Tm	ppm	0.0447	0.035	Hg	ppb	6.5	7.9
Yb	ppm	0.300	0.229	Tl	ppb	7.35	3.86
Lu	ppm	0.0449	0.386	^{204}Pb	ppb	172	1.58
Hf	ppm	0.203	0.23	Bi	ppb	5.72	2.94
Ta	ppb	28.1	23.3	Th	ppb	54.3	51.2
W	ppb	171	180	U	ppb	15.2	14.3
Re	ppb	67.4	60				

Noble gas data from [KL93] for silicate portion were taken as representative for the bulk Earth and are scaled accordingly.

Sources: **[KL93]** Kargel, J. S., & Lewis, J. S., 1993, *Icarus* 105, 1–25. **[MA80]** Morgan, J. W., & Anders, E., 1980, *Proc. Natl. Acad. Sci.* 77, 6973–6977.

Table 6.9 Elemental Abundances in the Bulk Silicate Earth and Present Depleted Mantle

Element	Unit	Bulk Silicates (Mantle, Crust, & Hydrosphere = Primitive Mantle)										Present Mantle	
		[MS95]	[KL93]	[Ri91]	[Ho88]	[ZH86]	[TM85]	[WDJ84]	[And83]	[Sun82]	[We81]	[WDJ84]	[We81]
H	ppm	...	54.7	...	...	...	...	...	...	...	...	...	...
Li	ppm	1.6	1.71	1.6	...	...	0.83	2.15	2.09	1.4	2.3–3.4	2.07	2
Be	ppb	68	77.5	80	...	...	60	...	...	...	...	100	...
B	ppb	300	433	500	...	...	600	...	...	...	...	800	...
C	ppm	120	65	250	...	...	...	46.2	...	...	108	24	100
N	ppm	2	0.88	...	...	...	...	...	...	...	...	...	...
O	wt%	...	44.42	...	...	...	...	...	...	...	...	...	...
F	ppm	25	20.7	26	...	...	...	19.4	28	26	28	16.3	16
^{20}Ne	ppm	...	4×10^{-5}	...	...	...	...	...	...	...	...	...	...
Na	ppm	2670	2932	2545	2460.	...	2500	2890	2040	2890	1830–3740	2745	1290
Mg	wt%	22.80	22.01	22.45	22.80	22.80	21.20	22.23	20.52	22.90	25.62	22.22	25.57
Al	wt%	2.35	2.123	2.36	2.15	2.15	1.93	2.22	2.02	2.28	1.146–1.787	2.17	0.963
Si	wt%	21.00	21.61	20.93	21.50	21.50	23.3	21.48	22.40	20.80	21.44	21.31	20.82
P	ppm	90	79.4	95	...	...	...	64.5	57	92	127–186	60	110
S	ppm	250	274	350	...	...	...	13.2	48	350–1000	420	8	400
Cl	ppm	17	36.4	30	35 [JDD95]	...	...	11.8	8	21–38	44	0.5	1.3
^{36}Ar	ppm	...	9.7×10^{-5}	...	...	...	...	...	...	...	...	...	...
K	ppm	240	232.4	240	258.2	...	180	231	151	230	440	127	50

continued

Table 6.9 *(continued)*

Element	Unit	Bulk Silicates (Mantle, Crust, & Hydrosphere = Primitive Mantle)										Present Mantle	
		[MS95]	[KL93]	[Ri91]	[Ho88]	[ZH86]	[TM85]	[WDJ84]	[And83]	[Sun82]	[We81]	[WDJ84]	[We81]
Ca	wt%	2.53	2.455	2.57	2.30	2.34	2.07	2.53	2.20	2.50	1.36–1.739	2.50	1.253
Sc	ppm	16.2	16.4	17.34	14.88	...	13	17	15	...	8.6–10	16.9	8.2
Ti	ppm	1205	1180	1280	1090	...	960	1350	1225	1300	550–960	1320	430
V	ppm	82	90	82	...	...	128	82.1	77	87	53–63	81.3	50
Cr	ppm	2625	2905	2935	...	...	3000	3010	2342	3000	2770	3010	2770
Mn	ppm	1045	1057	1080	...	...	1000	1020	1016	1100	1060	1016	1040
Fe	wt%	6.26	6.269	6.53	5.86	...	6.22	5.89	6.11	6.50	6.425–6.80	5.86	6.317
Co	ppm	105	104.6	105	104	...	100	105	101	110	110	105	110
Ni	ppm	1960	1948	1890	2080	...	2000	2110	1961	2000	2220	2108	2220
Cu	ppm	30	31.2	30	28.0	...	28	28.5	29	30	14–18	28.2	13
Zn	ppm	55	53.9	56	...	...	50	48.5	37	56	58–64	48	56
Ga	ppm	4.0	4.29	3.9	...	...	3	3.8	4	4.5–5.0	3–4.3	3.7	2.5
Ge	ppm	1.1	1.15	1.1	...	...	1.2	1.32	1.13	...	...	1.31	...
As	ppb	50	170	130	...	...	100	152	...	...	...	140	...
Se	ppb	75	44.3	50	...	...	41	13.5	20	...	20–30	12.6	20
Br	ppb	50	35.2	75	100 [JDD95]	...	...	45.6	...	60–90	...	4.6	...
^{84}Kr	ppm	...	5.5×10^{-6}	...	...	...	...	...	...	...	...	...	...
Rb	ppb	600	598	635	535.3	...	550	742	390	660	2200	276	500

continued

Table 6.9 *(continued)*

Element	Unit	Bulk Silicates (Mantle, Crust, & Hydrosphere = Primitive Mantle)										Present Mantle	
		[MS95]	[KL93]	[Ri91]	[Ho88]	[ZH86]	[TM85]	[WDJ84]	[And83]	[Sun82]	[We81]	[WDJ84]	[We81]
Sr	ppm	19.9	20.7	21.05	18.21	19.6	17.8	27.7	16.2	...	29–51	26	22
Y	ppm	4.30	3.91	4.55	3.940	...	3.4	...	3.26	...	3.6–5.9	2.9	2.9
Zr	ppm	10.5	11.47	11.22	9.714	...	8.3	...	13	...	21–32	7.8	18
Nb	ppb	658	765	713	617.5	...	560	...	970	...	1600–2700	600	1300
Mo	ppb	50	59.3	65	...	...	59	...	...	...	...	...	...
Ru	ppb	5	4.23	4.2	...	...	4.3	...	...	...	...	...	...
Rh	ppb	0.9	1.31	1	...	...	1.7	1.18	...	...	...	...	...
Pd	ppb	3.9	4.63	5	...	...	3.9	...	...	...	...	...	...
Ag	ppb	8	8.45	8	...	...	19	2.92	3	5–10	50	2.51	50
Cd	ppb	40	31.7	40	...	...	40	26.1	20	...	40–50	25.5	40
In	ppb	11	6.9	13	...	...	18	18.5	6	10–15	...	18.1	...
Sn	ppb	130	280	175	150	...	<1000	...	600	...	...	...	...
Sb	ppb	5.5	8	5	...	...	25	5.7	...	3–6	...	4.5	...
Te	ppb	12	16.1	13	...	...	22	19.9	...	...	...	19.9	...
I	ppb	10	10.7	11	10 [DDJ92]	...	...	13.3	...	...	...	4.2	...
^{132}Xe	ppm	...	4.4×10^{-7}	...	...	...	...	...	...	...	...	...	...
Cs	ppb	21	13.1	33	(26.8)	...	18	9.14	20	8–17	34	1.44	6
Ba	ppm	6.6	6.41	6.989	6.049	...	5.1	5.6	5.22	...	32	2.4	20

continued

Table 6.9 *(continued)*

Element	Unit	Bulk Silicates (Mantle, Crust, & Hydrosphere = Primitive Mantle)										Present Mantle	
		[MS95]	[KL93]	[Ri91]	[Ho88]	[ZH86]	[TM85]	[WDJ84]	[And83]	[Sun82]	[We81]	[WDJ84]	[We81]
La	ppb	648	622	708	613.9	...	551	520	570	...	1600	350	920
Ce	ppb	1675	1592	1833	1601	...	1436	1730	1400	...	3100	1410	1930
Pr	ppb	254	242	278	241.9	...	206	...	...	...	...	...	...
Nd	ppb	1250	1175	1366	1189	1170	1067	1430	1020	...	2000	1280	1440
Sm	ppb	406	360	444	386.5	380	347	520	320	...	520–960	490	400
Eu	ppb	154	145	168	145.6	...	131	188	130	...	190–300	180	160
Gd	ppb	544	529	595	512.8	...	459	740	...	...	...	690	...
Tb	ppb	99	95.5	108	94	...	87	126	90	...	140–220	120	120
Dy	ppb	674	656	737	637.8	...	572	766	...	...	...	730	...
Ho	ppb	149	146	163	142.3	...	128	181	...	...	...	170	...
Er	ppb	438	430	479	416.7	...	374	460	...	...	...	440	...
Tm	ppb	68	62.7	74	64.3	...	54	...	...	...	...	47	...
Yb	ppb	441	437	481	414.4	420	372	490	320	...	450–710	...	380
Lu	ppb	67.5	64.8	73.7	63.7	...	57	74	60	...	77–120	71	65
Hf	ppb	283	300	309	267.6	...	270	280	330	...	400–600	260	300
Ta	ppb	37	41.2	41	35.1	...	40	25.6	40	...	150–300	12.6	100
W	ppb	29	12	21	...	...	16	24.1	...	...	...	16.4	...
Re	ppb	0.28	0.293	0.28	...	...	0.25	0.236	0.21	...	...	0.23	...

continued

Table 6.9 (continued)

Element	Unit	Bulk Silicates (Mantle, Crust, & Hydrosphere = Primitive Mantle)										Present Mantle	
		[MS95]	[KL93]	[Ri91]	[Ho88]	[ZH86]	[TM85]	[WDJ84]	[And83]	[Sun82]	[We81]	[WDJ84]	[We81]
Os	ppb	3.4	4.05	3.4	...	...	3.8	3.106	2.90	...	...	3.1	...
Ir	ppb	3.2	3.25	3.3	...	...	3.2	2.81	2.97	...	...	2.8	...
Pt	ppb	7.1	11.2	6.8	...	...	8.7	...	...	...	...	...	...
Au	ppb	1	1.03	0.75	...	...	1.3	0.524	0.50	...	...	0.5	...
Hg	ppb	10	1.8	10	...	...	...	...	...	...	...	...	...
Tl	ppb	3.5	10.4	7	...	...	6	...	10	4–6	20	...	10
Pb	ppb	150	149	185	175	...	120	...	120	3.2*	460	...	200
Bi	ppb	2.5	4.2	2.5	...	...	10	...	3.3	1–4	10–17	...	10
Th	ppb	79.5	78.2	84.1	81.3	...	64	...	76.5	...	200	...	70
U	ppb	20.3	22	21	20.3	20.8	18	29.3	19.6	...	50	21	25

* for ^{204}Pb

Sources: [And83] Anderson, D. L., 1983, Proc. 14th. Lunar Planet. Sci. Conf., J. Geophys. Res. 88, B41–B52. [DDJ92] Deruelle, B., Dreibus, G., & Jambon, A., 1992, Earth Planet. Sci. Lett. 108, 217–227. [Ho88] Hofmann, A. W., 1988, Earth Planet. Sci. Lett. 90, 297–314. [JDD95] Jambon, A., Deruelle, B., Dreibus, G., & Pineau, F., 1995, Chem. Geol. 126, 101–117. [KL93] Kargel, J. S., & Lewis, J. S., 1993, Icarus 105, 1–25. [MS95] McDonough, W. F., & Sun, S. S., 1995, Chem. Geol. 120, 223–253. [Ri91] Ringwood, A. E., 1991, Geochim. Cosmochim. Acta 55, 2083–2110. [Sun82] Sun, S. S., 1982, Geochim. Cosmochim. Acta 46, 179–192. [TM85] Taylor, S. R., & McLennan, S. M., 1985, The continental crust: Its origin and evolution, Blackwell Sci. Publ. Oxford, pp 312. [WDJ84] Wänke, H., Dreibus, G., & Jagoutz, E., 1984, in Archean geochemistry, (Kröner, A., Hanson. G. N., & Goodwin, A. M., eds.) Springer Verlag, Berlin, pp. 1–24. [We81] Wedepohl, K. H., 1981, Fortschr. Mineral. 59, 203–205. [ZH86] Zindler, A., & Hart, S., 1986, Annu. Rev. Earth Planet. Sci. 14, 493–571.

Table 6.10 Elemental Abundances in the Earth's Crust

	Unit	Crust & Hydrosphere		Present Bulk Continental Crust				Oceanic Crust	
		calc.*	[We81]	[We95]	[TM85]	[WT84]	[MM82]	[Ho88]†	[TM85]
Li	ppm	17	13.8	18	13	...	20	...	10
Be	ppm	2.2	...	2.4	1.5	...	2.8	...	0.5
B	ppm	10.6	...	11	10	...	10	...	4
C	ppm	1865	365	1990	...	...	200	...	...
N	ppm	68	...	60	...	...	20	...	...
F	ppm	490	522	525	...	...	625	...	...
Na	wt%	2.276	2.446	2.36	2.30	3.2	2.83	1.987	2.08
Mg	wt%	2.064	2.388	2.20	3.20	1.63	2.09	4.569	4.64
Al	wt%	7.4355	8.235	7.96	8.41	8.36	8.13	8.08	8.47
Si	wt%	26.90	28.051	28.80	26.77	29.8	27.72	23.58	23.10
P	ppm	709	760	757	...	740	1050	...	...
S	ppm	710	877	697	...	...	260	...	...
Cl	ppm	1720	1910	472	...	...	130	174 [JDD95]	
K	wt%	2.00	1.755	2.14	0.91	1.7	2.59	0.0884	0.125
Ca	wt%	3.60	4.863	3.85	5.29	3.2	3.63	8.08	8.08
Sc	ppm	15	21	16	30	...	22	41.37	38
Ti	ppm	3750	5320	4010	5400	3000	4400	9680	9000
V	ppm	92	134	98	230	...	135	...	250
Cr	ppm	120	146	126	185	61	100	...	270
Mn	ppm	670	830	716	1400	540	950	...	1000
Fe	wt%	4.04	4.871	4.32	7.07	3.65	5.0	8.107	8.16
Co	ppm	22	25	24	29	...	25	47.07	47
Ni	ppm	52	69	56	105	39	75	149.5	135
Cu	ppm	23	47	25	75	...	55	74.4	86
Zn	ppm	61	77	65	80	...	70	...	85
Ga	ppm	14	18	15	18	...	15	...	17
Ge	ppm	1.3	...	1.4	1.6	...	1.5	...	1.5
As	ppm	1.6	...	1.7	1.0	...	1.8	...	1.0
Se	ppm	0.11	0.15	0.120	0.05	...	0.05	...	0.16
Br	ppm	5.3	...	1.0	...	...	2.5	0.4 [JDD95]	
Rb	ppm	73	79	78	32	55	90	1.262	2.2
Sr	ppm	312	293	333	260	498	375	113.2	130
Y	ppm	22	30	24	20	13	33	35.82	32

continued

Table 6.10 *(continued)*

	Unit	Crust & Hydrosphere		Present Bulk Continental Crust				Oceanic Crust	
		calc.*	[We81]	[We95]	[TM85]	[WT84]	[MM82]	[Ho88]†	[TM85]
Zr	ppm	190	140	203	100	213	165	104.24	80
Nb	ppm	18	14	19	11	12	20	3.507	2.2
Mo	ppm	1.0	...	1.1	1.0	...	1.5	...	1.0
Ru	ppb	0.1	...	0.1	...	...	10	...	1.0
Rh	ppb	0.06	...	0.06	...	...	5	...	0.2
Pd	ppb	0.4	...	0.4	1	...	10	...	<0.2
Ag	ppb	65	70	70	80	...	70	...	26
Cd	ppb	93	100	100	98	...	200	...	130
In	ppb	47	...	50	50	...	100	...	72
Sn	ppm	2.2	...	2.3	2.5	...	2	1.382	1.4
Sb	ppb	280	...	300	200	...	200	...	17
Te	ppb	4.7	...	(5)	...	...	10	...	3
I	ppb	1500	...	800	...	...	500	8 [DDJ92]	
Cs	ppm	3.2	1.3	3.4	1.0	...	3	0.0141	0.03
Ba	ppm	545	543	584	250	731	425	13.87	25
La	ppm	28	29	30	16	27	30	3.895	3.7
Ce	ppm	56	54	60	33	55	60	12.001	11.5
Pr	ppm	6.3	...	6.7	3.9	...	8.2	2.074	1.8
Nd	ppm	25	25.6	27	16	23	28	11.179	10.0
Sm	ppm	5.0	5.6	5.3	3.5	3.9	6.0	3.752	3.3
Eu	ppm	1.2	1.4	1.3	1.1	1.07	1.2	1.335	1.3
Gd	ppm	3.7	...	4.0	3.3	...	5.4	5.077	4.6
Tb	ppm	0.61	1.0	0.650	0.60	0.50	0.9	0.885	0.87
Dy	ppm	3.6	...	3.8	3.7	...	3.0	6.304	5.7
Ho	ppm	0.75	...	0.800	0.78	...	1.2	1.342	1.3
Er	ppm	2.0	...	2.1	2.2	...	2.8	4.143	3.7
Tm	ppm	0.28	...	0.300	0.32	0.23	0.5	0.621	0.54
Yb	ppm	1.9	3.3	2.0	2.2	1.46	3.4	3.900	5.1
Lu	ppm	0.33	0.56	0.350	0.30	...	0.5	0.589	0.56
Hf	ppm	4.6	3.4	4.9	3.0	4.4	3	2.974	2.5
Ta	ppm	1.0	2.2	1.1	1.0	...	2	0.192	0.3
W	ppm	0.93	...	1.0	1.0	...	1.5	...	0.5
Re	ppb	0.4	...	0.4	0.5	...	1	...	0.9

continued

Table 6.10 (continued)

	Unit	Crust & Hydrosphere		Present Bulk Continental Crust				Oceanic Crust	
		calc.*	[We81]	[We95]	[TM85]	[WT84]	[MM82]	[Ho88]†	[TM85]
Os	ppb	0.05	...	0.05	...	...	5	...	<0.004
Ir	ppb	0.05	...	0.05	0.1	...	1	...	0.02
Pt	ppb	0.4	...	0.4	...	...	10	...	2.3
Au	ppb	2.3	...	2.5	3.0	...	4	...	0.23
Hg	ppb	37	...	40	...	...	80	...	20
Tl	ppb	490	400	520	360	...	500	...	12
Pb	ppm	13.8	12	14.8	8.0	15	13	0.489	0.8
Bi	ppb	79	70	85	60	...	200	...	7
Th	ppm	7.9	7	8.5	3.5	5.1	7.2	0.187	0.22
U	ppm	1.6	1.2	1.7	0.91	1.3	1.8	0.071	0.10

* Crust and hydrosphere calculated from continental crust [We95] and ocean water abundances.

† Taking N-MORB (mid-ocean ridge basalt) as representative for the oceanic crust.

Sources: **[DDJ92]** Deruelle, B., Dreibus, G., & Jambon, A., 1992, *Earth Planet. Sci. Lett.* 108, 217–227. **[Ho88]** Hofmann, A. W., 1988, *Earth Planet. Sci. Lett.* 90, 297–314. **[JDD95]** Jambon, A., Deruelle, B., Dreibus, G., & Pineau, F., 1995, *Chem. Geol.* 126, 101–117. **[MM82]** Mason, B., & Moore, C. B., 1982, *Principles of geochemistry*, 4th ed., J. Wiley & Sons, New York, pp. 344. **[TM85]** Taylor, S. R., & McLennan, S. M., 1985, *The continental crust: Its origin and evolution*, Blackwell Sci. Publ., Oxford, pp. 312. **[We81]** Wedepohl, K. H., 1981, *Fortschr. Mineral.* 59, 203–205. **[We95]** Wedepohl, K. H., 1995, *Geochim. Cosmochim. Acta* 59, 1217–1232. **[WT84]** Weaver, B. L., & Tarney, J., 1984, in *Physics and chemistry of the earth* (Pollack, H. N., & Murthy, V. R., eds.), Vol. 15, Pergamon, Oxford, pp. 39–68.

Table 6.11 Elemental Abundances in the Earth's Present Continental Crust

Element	Unit	Lower Continental Crust		Upper Continental Crust		
		[We95]	[TM85]	[We95]	[TM85]	[SDK76]
Li	ppm	13	11	22	20	...
Be	ppm	1.7	1.0	3.1	3	...
B	ppm	5	8.3	17	15	...
C	ppm	588	...	3240	...	...
N	ppm	34	...	83	...	...
F	ppm	429	...	611	...	...
Na	wt%	2.120	2.08	2.567	2.89	...
Mg	wt%	3.155	3.80	1.351	1.33	...
Al	wt%	8.212	8.52	7.744	8.04	...
Si	wt%	27.133	25.42	30.348	30.8	...
P	ppm	872	...	665	...	...
S	ppm	408	...	953	...	...
Cl	ppm	640	...	278	...	...
K	wt%	1.314	0.28	2.865	2.80	...
Ca	wt%	4.860	6.07	2.945	3.00	...
Sc	ppm	25.3	36	7	11	7
Ti	ppm	5010	6000	3117	3000	...
V	ppm	149	285	53	60	...
Cr	ppm	228	235	35	35	35
Mn	ppm	929	1670	527	600	...
Fe	wt%	5.706	8.24	3.089	3.50	...
Co	ppm	38	35	11.6	10	12
Ni	ppm	99	135	18.6	20	19
Cu	ppm	37.4	90	14.3	25	14
Zn	ppm	79	83	52	71	52
Ga	ppm	17	18	14	17	...
Ge	ppm	(1.4)	1.6	1.4	1.6	...
As	ppm	2.0	0.8	1.3	1.5	...
Se	ppm	0.170	0.05	0.083	0.05	...
Br	ppm	0.28	...	1.6	...	...
Rb	ppm	41	5.3	110	112	110
Sr	ppm	352	230	316	350	316
Y	ppm	27.2	19	20.7	22	21
Zr	ppm	165	70	237	190	240
Nb	ppm	11.3	6	26	25	26
Mo	ppm	0.6	0.8	1.4	1.5	...
Pd	ppb	...	1.0	...	0.5	...
Ag	ppb	80	90	55	50	...
Cd	ppb	101	98	102	98	...
In	ppb	52	50	61	50	...

continued

Table 6.11 *(continued)*

Element	Unit	Lower Continental Crust		Upper Continental Crust		
		[We95]	[TM85]	[We95]	[TM85]	[SDK76]
Sn	ppm	2.1	1.5	2.5	5.5	...
Sb	ppb	300	200	310	200	...
Te	ppb	...	...	...	...	...
I	ppb	140	...	1400	...	...
Cs	ppm	0.8	0.1	5.8	3.7	...
Ba	ppm	568	150	668	550	1070
La	ppm	26.8	11	32.3	30	32
Ce	ppm	53.1	23	65.7	64	66
Pr	ppm	7.4	2.8	6.3	7.1	...
Nd	ppm	28.1	12.7	25.9	26	26
Sm	ppm	6.0	3.17	4.7	4.5	4.5
Eu	ppm	1.6	1.17	0.95	0.88	0.94
Gd	ppm	5.4	3.13	2.8	3.8	2.8
Tb	ppm	0.81	0.59	0.50	0.64	0.48
Dy	ppm	4.7	3.6	2.9	3.5	...
Ho	ppm	0.99	0.77	0.62	0.80	0.62
Er	ppm	...	2.2	...	2.3	...
Tm	ppm	...	0.32	...	0.33	...
Yb	ppm	2.5	2.2	1.5	2.2	1.5
Lu	ppm	0.43	0.29	0.27	0.32	0.23
Hf	ppm	4.0	2.1	5.8	5.8	5.8
Ta	ppm	0.84	0.6	1.5	2.2	...
W	ppm	0.6	0.7	1.4	2.0	...
Re	ppb	...	0.5	...	0.5	...
Os	ppb	...	...	...	...	...
Ir	ppb	...	0.13	...	0.02	0.02
Pt	ppb	...	...	...	...	...
Au	ppb	...	3.4	...	1.8	1.8
Hg	ppb	21	...	56	...	...
Tl	ppb	260	230	750	750	520
Pb	ppm	12.5	4.0	17	20	17
Bi	ppb	37	38	123	127	...
Th	ppm	6.6	1.06	10.3	10.7	10
U	ppm	0.93	0.28	2.5	2.8	2.5

Note: [TM85] assume lower continental crust = 75% of total crust; [We95] assumes lower continental crust = 50% of total crust.

Sources: **[SDK76]** Shaw, D. M., Dostal, J., & Keays, R. R., 1976, *Geochim. Cosmochim. Acta* 40, 73–83. **[TM85]** Taylor, S. R., & McLennan, S. M., 1985, *The continental crust: Its origin and evolution*, Blackwell Sci. Publ., Oxford, pp. 312. **[We95]** Wedepohl, K. H., 1995, *Geochim. Cosmochim. Acta* 59, 1217–1232.

Table 6.12 Elemental Abundances in Some Terrestrial Rocks

	Unit	Ultrama- fic Rocks [WeM79]	Kimber- lites [WeM79]	Nephe- linites [WeM79]	Alk.-Ol. Basalt [a] [WeM79]	Diabase (basalt) [MM82]	Tholeiitic Basalt [WeM79]	Shale [MM82]	Granite [MM82]
H	ppm	...	...	...	...	600	...	6300	400
Li	ppm	2	25	16	12	15	7	66	22
Be	ppm	~0.4	~1	...	...	0.8	0.7	3	3
B	ppm	7	...	...	...	15	5.8	100	1.7
C	wt%	0.01	1.62	0.0205	...	0.01	...	0.10–1.2	0.2
N	ppm	14	...	...	...	52	...	60–1000	59
O	wt%	42.90	...	...	...	44.90	...	49.50	45.5
F	ppm	97	1900	...	...	250	...	740	700
Na	wt%	0.223	0.203	2.53	2.30	1.60	1.758	0.96	2.46
Mg	wt%	24.75	16.00	7.12	4.52	3.99	3.691	1.50	0.24
Al	wt%	1.43	1.89	6.16	7.939	7.94	8.04	8.0	7.43
Si	wt%	20.33	14.70	18.84	22.62	24.61	23.92	27.3–28	33.96
P	ppm	220	3880	3800	2090	610	960	700	390
S	ppm	~4000	2000	620	...	123	...	2400	260
Cl	ppm	110	300	518	...	200	...	180	70
K	wt%	0.039	1.04	1.22	1.328	0.53	0.697	2.66	4.51
Ca	wt%	2.72	7.04	9.00	6.504	7.83	7.226	1.6–2.21	0.99
Sc	ppm	15	15	21	20	35	30	13	2.9
Ti	ppm	780	11800	16800	14390	6400	9710	4600	1500
V	ppm	50	120	221	213	264	251	130	17
Cr	ppm	3090	1100	344	202	114	168	90	20
Mn	ppm	1040	1160	1500	1472	1280	1356	850	195
Fe	wt%	6.483	7.16	9.108	9.078	7.76	8.554	4.72	1.37
Co	ppm	110	77	52	43	47	48	19	2.4
Ni	ppm	1450	1050	291	145	76	134	68	1
Cu	ppm	47	80	63	85	110	90	45	13
Zn	ppm	56	80	102	108	86	100	95	45
Ga	ppm	2.5	~10	15	15	16	17	19	20
Ge	ppm	1	~0.5	1.6	...	1.4	1.4	1.6	1.1

continued

Table 6.12 *(continued)*

	Unit	Ultrama-fic Rocks [WeM79]	Kimber-lites [WeM79]	Nephe-linites [WeM79]	Alk.-Ol. Basalt [a] [WeM79]	Diabase (basalt) [MM82]	Tholeiitic Basalt [WeM79]	Shale [MM82]	Granite [MM82]
As	ppm	1	...	...	...	1.9	1.5	13	0.5
Se	ppb	20	~150	...	...	300	...	600	7
Br	ppm	0.24	...	...	...	0.4	...	4–20	0.4
Rb	ppm	1.2	65	39	32	21	22	140	220
Sr	ppm	22	740	1350	530	190	328	300	250
Y	ppm	2.88	22	36	33	25	28	26	13
Zr	ppm	16	250	205	189	105	137	160	165
Nb	ppm	1.3	110	103	69	9.5	13	11	24
Mo	ppm	0.2	~0.5	...	...	0.6	1	2.6	6.5
Ru	ppb	...	7.0	...	...	...	...	...	...
Rh	ppb	...	7.0	...	...	...	...	...	...
Pd	ppb	10	53	...	...	25	...	4	2
Ag	ppb	50	...	...	...	80	11	70	50
Cd	ppb	60	70	52	82	150	170	300	30
In	ppb	20	100	34	...	70	70	100	20
Sn	ppm	0.52	15	...	...	3.2	1.5	6.0	3.5
Sb	ppb	300	...	...	...	...	300	1500	310
Te	ppb	~1	...	...	...	...	...	80	...
I	ppm	0.13	...	...	...	<0.030	...	2.2–19	<0.030
Cs	ppb	6.0	2300	...	...	900	1100	5000	1500
Ba	ppm	20	1000	1046	528	160	246	580	1220
La	ppm	0.92	150	89	54	10	15	24–32	101
Ce	ppm	1.93	200	171	105	23	32.9	50–70	170
Pr	ppm	0.32	22	18	28	3.4	4.7	6.1–7.9	19
Nd	ppm	1.44	85	66	49	15	18.9	24–31	55
Sm	ppm	0.4	13	14.5	9.1	3.6	4.9	5.7–5.8	8.3
Eu	ppm	0.16	3.0	4.0	3.5	1.1	1.5	1.1–1.2	1.3
Gd	ppm	0.74	8.0	12.1	8.1	4	5.5	5.2	5
Tb	ppm	0.12	1.0	1.7	1.8	0.65	1.2	0.85	0.54

continued

Table 6.12 *(continued)*

	Unit	Ultrama- fic Rocks [WeM79]	Kimber- lites [WeM79]	Nephe- linites [WeM79]	Alk.-Ol. Basalt [a] [WeM79]	Diabase (basalt) [MM82]	Tholeiitic Basalt [WeM79]	Shale [MM82]	Granite [MM82]
Dy	ppm	0.57	...	7.3	4.7	4	4.9	4.0–4.3	2.4
Ho	ppm	0.16	0.55	1.7	1.9	0.69	1.3	1.04–1.2	0.35
Er	ppm	0.40	1.45	3.3	2.4	2.4	2.8	2.7–3.4	1.2
Tm	ppm	0.067	0.23	0.88	0.7	0.30	0.46	0.5	0.15
Yb	ppm	0.38	1.2	2.3	1.9	2.1	2.6	2.2–3.1	1.1
Lu	ppm	0.065	0.16	0.39	0.5	0.35	0.46	0.48–0.6	0.19
Hf	ppm	0.6	7	5	...	2.7	2.5	2.8	5.2
Ta	ppb	≤100	9000	19000	...	500	500	800	1500
W	ppb	0.300 ?	...	11000	...	500	700	1800	400
Re	ppb	0.23	1.0	...	...	...	...	...	...
Os	ppb	3.1	5.0	...	...	...	...	0.05	...
Ir	ppb	3.2	7	...	...	...	...	0.08	...
Pt	ppb	60	190	...	...	1.2	0.02	...	1.9
Au	ppb	7.0	4	...	1.6	...	4.0	2.5	...
Hg	ppb	30	10	20	17	200	10	400	100
Tl	ppm	0.01	0.22	0.009	0.05	0.3	0.1	1–1.2	0.15
Pb	ppm	0.2	10	7.8	4.3	7.8	3.7	20	48
Bi	ppb	6	30	14	30	50	30	400–430	70
Th	ppm	0.07	16	11	3	2.4	1.8	12	50
U	ppm	0.025	3.1	3.2	0.7	0.6	0.5	3.7	3.4

[a] Alk.-Ol. basalt = Alkali-olivine basalt
? Data uncertain

Sources: **[MM82]** Mason, B., & Moore, C. B., 1982, *Principles of geochemistry*, 4th ed., J. Wiley & Sons, New York, pp. 344, including data collected by Li, Y.-H., 1991, *Geochim. Cosmochim. Acta*, 55, 3223–3240. **[WeM79]** Wedepohl, K. H., & Muramatsu , Y., 1979, in *Proc. 2nd Intern. Kimberlite Conf.* 11, (Boyd, F. R., & Meyer, H. O. A., eds.), AGU, Washington, pp. 300–312.

Table 6.13 Selected Volcanic Gas Analyses*

No.	St. Augustine		Momotombo		St. Helens		Mt. Etna	
	1	**2**	**3**	**4**	**5**	**6**	**7**	**8**
T (°C)	648	775	658	820	663	802	1075	1075
H_2O	97.23	96.73	97.90	97.11	98.52	91.58	53.69	27.71
H_2	0.381	0.56	0.17	0.70	0.269	0.8542	0.57	0.30
CO_2	1.90	1.50	1.47	1.44	0.913	6.942	20.00	22.76
CO	0.0035	0.0072	0.0015	0.0096	0.0013	0.06	0.42	0.48
SO_2	0.006	0.33	0.30	0.50	0.073	0.2089	24.85	47.70
H_2S	0.057	0.29	0.16	0.23	0.137	0.3553	0.22	0.22
OCS	...	...	...	...	0.00002	0.0008	...	...
S_2	...	...	0.0009	0.0003	0.0003	0.0039	0.21	0.76
SO	...	...	...	...	...	...	0.03	0.06
HCl	0.365	0.55	0.12	0.20	0.089	...	...	...
HF	0.056	0.030	0.011	0.018	...	...	...	...
H_2O/H_2	255	173	576	139	366	107	94	92
CO_2/CO	543	208	980	150	702	116	48	47
SO_2/H_2S	0.1	1.1	2	2	0.5	0.6	113	217
HCl/HF	6.5	18	11	11	...	...	...	...
$\Sigma C/\Sigma S$	30	2.4	3	2	6	12	0.8	0.5
$\log_{10}fO_2$	−17.54	−14.46	−16.52	−13.55	−16.76	−14.25	−9.47	−9.47

continued

Table 6.13 *(continued)*

No.	Erta' Ale		Nyiragongo		Surtsey		Kilauea (E. rift)	
	9	10	11	12	13	14	15	16
T (°C)	1130	1075	960	1020	1125	1125	935	1032
H_2O	77.24	69.41	55.62	45.90	81.13	91.11	78.7	80.4
H_2	1.39	1.57	2.18	1.59	2.80	1.42	1.065	0.9289
CO_2	11.26	17.16	36.35	45.44	9.29	3.31	3.17	3.52
CO	0.44	0.75	2.13	2.72	0.69	0.11	0.0584	0.0784
SO_2	8.34	9.46	0.81	2.30	4.12	2.81	11.5	14.0
H_2S	0.68	1.02	2.45	1.41	0.89	0.06	3.21	0.511
OCS	...	0.02	0.08	0.08	...	...	0.0054	0.0014
S_2	0.21	0.59	0.38	0.55	0.25	0.01	1.89	0.197
SO	...	...	...	...	...	...	...	...
HCl	0.42	...	...	...	...	...	0.167	0.174
HF	...	...	...	...	...	...	0.20	0.19
H_2O/H_2	56	44	26	29	29	64	74	87
CO_2/CO	26	23	17	17	14	30	54	45
SO_2/H_2S	12	9	0.3	2	5	47	4	27
HCl/HF	...	...	...	...	...	...	0.8	0.9
$\Sigma C/\Sigma S$	1.2	1.5	9	10	2	2	0.2	0.2
$\log_{10}fO_2$	−9.16	−10.12	−12.4	−11.3	−9.80	−9.11	−11.91	−10.19

*In volume% (= mol%)

The selected analyses are chosen to show typical volcanic gas compositions from convergent plate, divergent plate, and hot spot volcanoes, temporal variations in gas chemistry, variations with magma chemistry, and the maximum and minimum oxygen fugacity (fO_2 in log bar units), or in the case of Mt. Etna, the maximum and minimum steam contents in volcanic gases at each volcano. The gas analyses are from Symonds, R. B., Rose, W. I., Bluth, G. J. S., & Gerlach, T. M., 1994, in *Volatiles in magmas*, (Carroll, M. R., & Holloway, J. R., eds.), Mineral. Soc. of America, Washington, D. C., pp. 1–66. The key to the analyses and references to papers for the gas collections and computations follow:

continued

Table 6.13   *(continued)*

Convergent plate volcanism

Mount St. Augustine, Alaska, USA, andesite

1. 79A3G, July 1979	Kodosky, L. G., Motyka, R. J., & Symonds, R. B., 1991,
2. Spine-1D, 6 July 1989	*Bull. Volcanol.* 53, 381–394. Symonds, R. B., Rose, W.
	I., Bluth, G. J. S., & Gerlach, T. M., 1994, in *Volatiles in*
	magmas, (Carroll, M. R., & Holloway, J. R., eds.), Min-
	eral. Soc. of America, Washington, D. C., pp. 1–66.

Momotombo, Nicaragua, tholeiite

3. MoMo-2, Dec. 1980	Bernard, A., 1985, Ph.D. dissertation, Univ. of Brussels,
4. MoMo-1, Dec. 1980	Belgium. Symonds, R. B., pers. communication, 1998.

Mount St. Helens, Washington, USA, dacite

5. CNRS, 16 Sept. 1981	Gerlach, T. M., & Casadevall, T. J., 1986, *J. Volcanol.*
6. 800925-710, 25 Sept. 1980	*Geotherm. Res.* 28, 107–140.

Mount Etna, Italy, hawaiite

7. #14A, 12 July 1970	Huntingdon, A. T., 1973, *Phil. Trans. Roy. Soc. Lond.*
8. horn. 1 #1, 12 July 1970	274A, 119–128. Gerlach, T. M., 1979, *J. Volcanol. Geo-*
	therm. Res. 6, 165–178.

Divergent plate volcanism

Erta' Ale, Ethiopia, tholeiite

9. #910, 23 Jan. 1974	Tazieff, H., LeGuern, F., Carbonnelle, J., & Zettwoos, P.,
10. #1032, 3 Dec. 1971	1972, *Compt. Rend. Acad. Sci. Paris, Ser. D* 274,
	1003–1006. Giggenbach, W. F., & LeGuern, F., 1976,
	Geochim. Cosmochim. Acta 40, 25–30. Gerlach, T. M.,
	1980, *J. Volcanol. Geotherm. Res.* 7, 415–441.

Nyiragongo, Zaire, nephelinite

11. #13, 1959	Chaigneau, M., Tazieff, H., & Febre, R., 1960, *Compt.*
12. #12, 1959	*Rend. Acad. Sci. Paris, Ser. D,* 250, 2482–2485. Gerlach,
	T. M., 1980, *J. Volcanol. Geotherm. Res.* 8, 177–189.

Surtsey, N. Atlantic near Iceland, alkaline basalt

13. #12, 15 Oct. 1964	Sigvaldason, G. E., & Elisson, G., 1968, *Geochim. Cos-*
14. #29, 31 March 1967	*mochim. Acta* 32, 797–805. Gerlach, T. M., 1980, *J. Vol-*
	canol. Geotherm. Res. 8, 191–198.

Intraplate hot spot volcanism

Kilauea, east rift zone, Hawaii, USA, tholeiite

15. Pele 12, 15 Jan. 1983.	Gerlach, T. M., 1993, *Geochim. Cosmochim. Acta* 57,
16. Pele 7, 15 Jan. 1983.	795–814.

Table 6.14 Terrestrial Impact Craters

fame & Location	Latitude	Longitude	Diameter (km)	Age (Ma)
Acraman, Australia	32°01'S	135°27'E	160	>570
Amguid, Algeria	26°05'N	4°23'E	0.45	<0.1
Aouelloul, Mauritania	20°15'N	12°41'W	0.39	3.1±0.3
Araguainha Dome, Brazil	16°46'S	52°59'W	40	<249±1.9
Azuara, Spain	41°10'N	0°55'W	30	<130
B.P. Structure, Libya	25°19'N	24°20'E	2.8	<120
Barringer, Arizona, USA	35°02'N	111°01'W	1.2	0.049±0.003
Beaverhead, Montana, USA	45°00'N	113°00'W	15	~600
Bee Bluff, Texas, USA	29°02'N	99°51'W	2.4	<40
Beyenchime-Salaatin, Russia	71°50'N	123°30'E	7.5	<65
Bigach, Kazakhstan	48°30'N	82°00'E	7	6±3
Boltysh, Ukraine	48°45'N	21°10'E	25	88±3
Bosumtwi, Ghana	6°32'N	1°25'W	10.5	1.03±0.2
Boxhole, Northern Terr., Australia	22°37'S	135°12'E	0.18	...
Brent, Ontario, Canada	46°05'N	78°29'W	3.8	450±30
Campo del Cielo, Argentina	27°38'S	61°42'W	0.05	<0.004
Carswell, Saskatchewan, Canada	58°27'N	109°30'W	39	115±10
Charlevoix, Quebec, Canada	47°32'N	70°18'W	54	357±15
Clearwater Lake East, Quebec, Canada	56°05'N	74°07'W	22	290±20
Clearwater Lake West, Quebec, Canada	56°13'N	74°30'W	32	290±20
Connolly Basin, Western Australia	23°32'S	124°45'E	9	<60
Crooked Creek, Missouri, USA	37°50'N	91°23'W	7	320±80
Dalgaranga, Western Austr., Australia	27°43'S	117°05'E	0.021	0.025
Decaturville, Missouri, USA	37°54'N	92°43'W	6	<300
Deep Bay, Saskatchewan, Canada	56°24'N	102°59'W	13	100±50
Dellen, Sweden	61°55'N	16°32'E	15	109.6±1.0
Des Plaines, Illinois, USA	42°03'N	78°52'W	8	<280

continued

Table 6.14 *(continued)*

Name & Location	Latitude	Longitude	Diameter (km)	Age (Ma)
Dobele, Latvia	56°35'N	23°15'E	4.5	300±35
Eagle Butte, Alberta, Canada	49°42'N	110°30'W	19	<65
El'gygytgyn, Siberia, Russia	67°30'N	172°05'E	18	3.75
Flynn Creek, Tennessee, USA	36°17'N	85°40'W	3.55	360±20
Glasford, Illinois, USA	40°36'N	89°47'W	4	<430
Glover Bluff, Wisconsin, USA	43°58'N	89°32'W	10	<500
Goat Paddock, Western Australia	18°20'S	126°40'E	5.1	<50
Gosses Bluff, Northern Terr., Australia	23°50'S	132°19'E	22	142.5±0.5
Gow Lake, Saskatchewan, Canada	56°27'N	104°29'W	4	<250
Gusev, Russia	48°20'N	40°15'E	3.5	65
Haughton, NW Territories, Canada	75°22'N	89°40'W	20.5	21.5±1.2
Haviland, Kansas, USA	37°35'N	99°10'W	0.011	0
Henbury, Northern Terr., Australia	24°34'S	133°10'E	0.157	<0.005
Holleford, Ontario, Canada	44°28'N	76°38'W	2.35	550±100
Ile Rouleau, Quebec, Canada	50°41'N	73°53'W	4	<300
Ilumetsa, Estonia	57°58'N	25°25'E	0.08	>0.002
Ilyinets, Ukraine	49°06'N	29°12'E	4.5	395±5
Janisjärvi, Russia	61°58'N	30°55'E	14	698±22
Kaalijärvi, Estonia	58°24'N	22°40'E	0.11	0.004±0.001
Kaluga, Russia	54°30'N	36°15'E	15	380±10
Kamensk, Russia	48°20'N	40°15'E	25	65±2
Kara, Russia	69°10'N	65°00'E	65	73±5
Kara-Kul, Tajikistan	38°57'N	73°24'E	52	<225
Karla, Russia	57°54'N	48°00'E	12	10±5
Kelly West, Northern Terr., Australia	19°30'S	132°50'E	10	>550
Kentland, Indiana, USA	40°45'N	87°24'W	13	<300
Kjardla, Estonia	57°00'N	22°42'E	4	455

continued

Table 6.14 *(continued)*

Name & Location	Latitude	Longitude	Diameter (km)	Age (Ma)
Kursk, Russia	51°40'N	36°00'E	5.5	250±80
Lac Couture, Quebec, Canada	60°08'N	75°20'W	8	430±25
Lac la Moinerie, Quebec, Canada	57°26'N	66°36'W	8	400±50
Lappajärvi, Finland	63°09'N	23°42'E	14	77.3±0.4
Lawn Hill, Queensland, Australia	12°24'S	134.03'E	18	>515
Liverpool, Northern Terr., Australia	12°24'S	134°03'E	1.6	150±70
Logancha, Russia	65°30'N	95°50'E	20	50±20
Logoisk, Byelorussia	54°12'N	27°48'E	17	40±5
Lonar, India	19°58'N	76°31'E	1.83	0.052±0.006
Macha, Russia	57°30'N	116°00'E	0.3	<0.007
Manicouagan, Quebec, Canada	51°23'N	68°42'W	100	212±2
Manson, Iowa, USA	42°35'N	94°31'W	35	65.7±1.0
Marquez Dome, Texas, USA	31°17'N	96°18'W	15	58±2
Middlesboro, Kentucky, USA	36°37'N	83°44'W	6	<300
Mien, Sweden	56°25'N	14°52'E	9	121.0±2.3
Misarai, Lithuania	53°00'N	23°54'E	5	395±145
Mishina Gora, Russia	58°40'N	28°00'E	4	<360
Mistastin, Newfoundland/Labrador, Can.	55°53'N	63°18'W	28	38±4
Montagnais, Nova Scotia, Canada	42°53'N	64°13'W	45	50.5±0.76
Monturaqui, Chile	23°56'S	68°17'W	0.46	1
Morasko, Poland	52°29'N	16°54'E	0.1	0.01
New Quebec, Quebec, Canada	61°18'N	73°40'W	3.44	1.4±0.1
Nicholson Lake, NW Territories, Canada	62°40'N	102°41'W	12.5	<400
Oasis, Libya	24°35'N	24°24'E	11.5	<120
Obolon', Ukraine	49°30'N	32°55'E	17	215±25
Odessa, Texas, USA	31°45'N	102°29'W	0.168	<0.05
Quarkziz, Algeria	29°00'N	7°33'W	3.5	<70

continued

Table 6.14 *(continued)*

Name & Location	Latitude	Longitude	Diameter (km)	Age (Ma)
Piccaninny, Western Austr., Australia	17°32'S	128°25'E	7	<360
Pilot Lake, NW Territories, Canada	60°17'N	111°01'W	5.8	445±2
Popigai, Russia	71°30'N	111°00'E	100	35±5
Presqu'ile, Quebec, Canada	49°43'N	78°48'W	12	<500
Pretoria Salt Pan, South Africa	25°24'S	28°05'E	1.13	0.2
Puchezh-Katunki, Russia	57°06'N	43°35'E	80	220±10
Rogozinskaja, Russia	58°18'N	62°00'E	8	55±5
Red Wing Creek, North Dakota, USA	47°36'N	103°33'W	9	200±5
Riachao Ring, Brazil	7°43'S	46°39'W	4.5	<200
Ries, Germany	48°53'N	10°37'E	24	14.8±0.7
Rochechouart, France	45°30'N	0°56'E	23	160±5
Roter Kamm, Namibia	27°46'S	16°18'E	2.5	3.7±0.3
Rotmistrova, Ukraine	49°00'N	32°00'E	2.7	140±20
Sääksjärvi, Finland	61°23'N	22°25'E	5	514±12
Saint Martin, Manitoba, Canada	51°47'N	98°32'W	40	220.5±18
Serpent Mound, Ohio, USA	39°02'N	83°24'W	6.4	<320
Serra da Cangalha, Brazil	8°05'S	46°52'W	12	<300
Shunak, Kazakhstan	42°42'N	72°42'E	3.1	12
Sierra Madera, Texas, USA	30°36'N	102°55'W	12	<100
Sikote Alin, Russia	46°07'N	134°40'E	0.027	0
Siljan, Sweden	61°02'N	14°52'E	55	368±1
Slate Islands, Ontario, Canada	48°40'N	87°00'W	30	<350
Sobolev, Russia	46°18'N	138°52'E	0.053	0
Söderfjärden, Finland	63°02'N	21°35'E	6	550
Spider, Western Austr., Australia	16°30'S	126°00'E	13	>570
Steen River, Alberta, Canada	59°31'N	117°38'W	25	95±7
Steinheim, Germany	48°41'N	10°04'E	3.8	14.8±0.7

continued

Table 6.14 *(continued)*

Name & Location	Latitude	Longitude	Diameter (km)	Age (Ma)
Strangways, Northern Terr., Australia	15°12'S	133°35'E	25	<470
Sudbury, Ontario, Canada	46°36'N	81°11'W	200	1850±3
Tabun-Khara-Obo, Mongolia	44°06'N	109°36'E	1.3	>120
Talemzane, Algeria	33°19'N	4°02'E	1.75	<3
Teague, Western Austr., Australia	25°50'S	120°55'E	28	1685±5
Tenoumer, Mauritania	22°55'N	10°24'W	1.9	2.5±0.5
Ternovka, Ukraine	48°01'N	33°05'E	12	330±30
Tin Bider, Algeria	27°36'N	5°07'E	6	<70
Upheaval Dome, Utah, USA	38°26'N	109°54'W	5	<65
Ust-Kara, Russia	69°18'N	65°18'E	25	73±3
Vargeao Dome, Brazil	26°50'S	52°07'W	12	<70
Veevers, Western Austr., Australia	22°58'S	125°22'E	8	160±30
Vepriaj, Lithuania	55°06'N	24°36'E	8	160±30
Vredefort, South Africa	27°00'S	27°30'E	140	1970±100
Wabar, Saudi Arabia	21°30'N	50°28'E	0.097	0.006±0.002
Wanapitei Lake, Ontario, Canada	46°44'N	80°33'W	7.5	36±2
Wells Creek, Tennessee, USA	36°23'N	87°40'W	14	200±100
West Hawk Lake, Manitoba, Canada	49°46'N	95°11'W	3.15	100±50
Wolfe Creek, Western Austr., Australia	19°10'S	127°47'E	0.875	<0.3
Zapadnaya, Ukraine	49°44'N	29°00'E	4	115±10
Zeleny Gai, Ukraine	48°42'N	35°54'E	2.5	>140
Zhamanshin, Kazakhstan	48°24'N	60°48'E	13.5	0.87±0.1

Sources: Garvin, J. B., Grieve, R. A. F., & Schnetzler, C. C., 1995, *Meteoritics* 30, 509. Graham, A. L., Bevan, A. W. R., & Hutchison, R., 1985, *Catalogue of meteorites*, Univ. of Arizona Press, pp. 460. Grieve, R. A. F., 1987, *Annu. Rev. Earth Planet. Sci.*, 15, 245–270. Grieve, R. A. F., 1991, *Meteoritics* 26, 175–194. Gurov, E. P., & Gurova, E. P., 1995, *Meteoritics* 30, 515.

Earth's Atmosphere

The variation of temperature with altitude divides the terrestrial atmosphere into several different regions. The troposphere, which is closest to the surface, is convective with an average lapse rate of −6.5 K km^{-1}. The tropopause, which is the top of the troposphere, is at 12 km (216 K), but its altitude varies seasonally and with latitude from 10 to 15 km. For example, the tropical tropopause is at 15 km (195 K). The temperature inversion at the tropopause and the increase of temperature with altitude in the stratosphere are due to absorption of solar UV and IR radiation by O_3. The stratosphere contains ~5 ppmv H_2O and is extremely dry with respect to the troposphere (≤4% H_2O) because of the low temperature at the tropical tropopause. (Most of the air mass exchange between the troposphere and stratosphere occurs in the ascending branches of the tropical Hadley cells.) Above the stratosphere is the mesosphere, which lies between 50 km (the stratopause) and 85 km (the mesopause). The mesosphere is cooled by CO_2, O_3, O, and NO radiating energy to space, and temperature decreases with altitude in the mesosphere, reaching an average minimum temperature of ~190 K at the mesopause. The thermosphere, starting at ~85 km, is strongly heated by O_2 photolysis and ionization. Temperatures become isothermal at 1000 K by 250 km. The ionosphere, where the electron density is high enough to affect radio wave propagation, starts at ~60 km (in the mesosphere), and peak electron densities are in the 200–400 km region. The atmosphere can also be divided into well mixed (homosphere) and diffusively unmixed (heterosphere) regions. The homopause, where diffusive separation becomes important, is at 100–110 km.

Atmospheric circulation in the troposphere is driven by absorption of solar energy. The atmosphere absorbs about one-third, and the surface about two-thirds of all solar energy absorbed. The equatorial regions absorb about four times more energy than the polar regions; heat is transported from the equator to the poles by direct (Hadley and polar) and indirect (Ferrel) cells in each hemisphere. A region of ascending air at the equator, the interhemispheric tropical convergence zone (ITCZ), separates the two tropical Hadley cells. The alternating seasonal dominance of one tropical Hadley cell over the other causes the ITCZ to move north or south about its average position, and this oscillation is primarily responsible for air exchange between the northern and southern hemispheres. Zonal circulation in the northern hemisphere (mirrored in the southern hemisphere) is dominantly from west to east at midlatitudes and from east to west in the tropics. The subtropical jet stream, at an altitude close to the tropopause and at latitude ~30°, flows west to east at 25–50 m s^{-1}.

Characteristic magnitudes of atmospheric transport times are one hour for vertical mixing through the planetary boundary layer (the lowest 1–2 km tropospheric layer next to the surface), one month for vertical mixing up to the tropopause, three months for meridional transport from equator to pole (troposphere or stratosphere), one year for interhemispheric exchange in the troposphere and three years for interhemispheric exchange in the stratosphere, one to two years for stratospheric air exchange with the troposphere, and 50 years for tropospheric air exchange with the stratosphere. These transport times have been determined by measuring the distribution of radioactive fallout from atmospheric nuclear testing, as well as from the distribution of pollutants and inert tracers.

The dominant feature of tropospheric chemistry is the role of OH, the major oxidizer in the troposphere, in the destruction of different natural and anthropogenic trace gases such as CH_4, volatile organic compounds (VOCs), CO, SO_2, H_2S, organic sulfides, NO_x, NH_3, and O_3. Hydroxyl radicals are present at a global average concentration of ~10^6 OH per cm^3 and are mainly produced by reaction of electronically excited oxygen atoms with water vapor:

$$O(^1D) + H_2O \rightarrow OH + OH \qquad (1)$$

Tropospheric O_3, mixed downward from the stratosphere or produced via

$$NO_2 + h\nu \rightarrow NO + O(^3P) \quad (\lambda < 420 \text{ nm}) \qquad (2)$$
$$O(^3P) + O_2 + M \rightarrow O_3 + M \qquad (3)$$

is the source of the $O(^1D)$ atoms

$$O_3 + h\nu \rightarrow O_2 + O(^1D) \qquad (\lambda < 320 \text{ nm}) \qquad (4)$$

In reaction (3) and subsequent reactions, M is any third molecule or atom, such as N_2, O_2, or Ar, that is a collision partner to remove excess energy.

Other important facets of tropospheric chemistry include the sources, sinks, and radiative forcing of natural (H_2O, CO_2, CH_4, N_2O) and anthropogenic (chlorofluorocarbons) greenhouse gases; the sources, sinks, and fates of aerosols (e.g., sulfates, organic condensates, industrial particles); heterogeneous chemical reactions in and on cloud particles; the role of clouds in the atmospheric radiation balance via absorption, emission, and scattering of solar energy; O_3 and other photochemically produced air pollutants in urban areas; the production of even more toxic compounds from anthropogenic pollutants, such as organophosphorus insecticides; halogen atom reactions with VOCs; and the biogeochemical cycles of water, carbon, sulfur, nitrogen, and phosphorus. For example, it is now possible to measure the seasonal and annual trends in the abundance of O_2. These trends are due to fossil fuel combustion, CO_2 production by photosynthesis, and dissolution of CO_2 in the oceans.

Terrestrial stratospheric chemistry is closely linked to the ozone (O_3) layer at 15–35 km, which shields the earth's surface from harmful UV sunlight ($\lambda < 300$ nm) and dissipates the absorbed solar energy as heat. The abundance of O_3 in the stratosphere is regulated by a balance between O_3 production and destruction and O_3 transport between regions of net production and net destruction. Production and destruction of O_3 in the absence of other perturbing influences is described by the Chapman cycle:

$$O_2 + h\nu \rightarrow O + O \qquad (\lambda = 180\text{--}240 \text{ nm}) \qquad (5)$$
$$O + O_2 + M \rightarrow O_3 + M \qquad\qquad\qquad\qquad (6)$$
$$O_3 + h\nu \rightarrow O(^1D) + O_2 \qquad (\lambda = 200\text{--}300 \text{ nm}) \qquad (7)$$
$$O + O_3 \rightarrow O_2 + O_2 \qquad\qquad\qquad\qquad (8)$$

However, natural and anthropogenic trace gases in the earth's atmosphere catalyze ozone destruction more rapidly than reaction (8) in the Chapman cycle. Three important examples are the HO_x, NO_x, and halogen (Cl, Br, I) oxide cycles:

$$OH + O_3 \rightarrow HO_2 + O_2 \qquad\qquad\qquad (9)$$
$$HO_2 + O_3 \rightarrow OH + O_2 + O_2 \qquad\qquad (10)$$
$$O_3 + O_3 \rightarrow 3O_2 \qquad\qquad \text{Net reaction} \quad (11)$$

$$NO_2 + O \rightarrow NO + O_2 \qquad\qquad\qquad (12)$$
$$NO + O_3 \rightarrow NO_2 + O_2 \qquad\qquad\qquad (13)$$
$$O + O_3 \rightarrow 2O_2 \qquad\qquad \text{Net reaction} \quad (14)$$

$$Cl + O_3 \rightarrow ClO + O_2 \qquad\qquad\qquad (15)$$
$$ClO + O \rightarrow Cl + O_2 \qquad\qquad\qquad (16)$$
$$O + O_3 \rightarrow 2O_2 \qquad\qquad \text{Net reaction} \quad (17)$$

The OH radicals in reaction (9) are mainly produced by reaction (1) and the stratospheric H_2O is due either to tropospheric transport or to oxidation of CH_4 transported upward from the troposphere. The NO_x gases are produced from nitrous oxide (N_2O) transported upward from the troposphere:

$$O(^1D) + N_2O \rightarrow NO + NO \qquad\qquad (18)$$

whereas halogens and halogen oxide radicals are produced by solar UV photolysis of halocarbon (also known as chlorofluorocarbon or CFC) gases:

$$CFCl_3 + h\nu \rightarrow Cl + CFCl_2 \qquad (\lambda < 265 \text{ nm}) \qquad (19)$$

$$CFCl_2 + O_2 \rightarrow ClO + COFCl \qquad (20)$$

The halogen oxide cycles can be coupled to the NO_x cycle and each other. Modeling of O_3 chemistry shows that the NO_x, HO_x, ClO_x, and Chapman cycles account for 31–34%, 16–29%, 19–20%, and 20–25%, respectively, of O_3 destruction in the stratosphere. The relative importance of the Chapman reactions and the various catalytic cycles also varies with altitude and with the concentrations of NO_x, HO_x, and halogen oxide gases.

The halogen oxide-catalyzed O_3 depletion models are supported by observations of O_3 depletions over the Antarctic in an area known as the Antarctic ozone hole. Long-term observations (starting in 1956) show a continued decrease in springtime O_3 levels. During the dark polar winter, the polar vortex traps air over the Antarctic; stratospheric temperatures drop to ~190 K and polar stratospheric clouds composed of solid $HNO_3 \cdot 3H_2O$ form. The surfaces of the cloud particles are sites for heterogeneous reactions such as

$$HCl + ClONO_2 \rightarrow HNO_3 + Cl_2 \qquad (21)$$

that convert relatively inert Cl reservoir species into Cl-bearing species that dissociate into active Cl atoms and Cl-bearing radicals, e.g., via

$$Cl_2 + h\nu \rightarrow Cl + Cl \qquad (22)$$

once polar winter is over and the sun rises again. The active Cl atoms and radicals then catalytically destroy O_3, leading to the observed O_3 depletions in Antarctic spring. Mixing with the rest of the atmosphere is unable to restore O_3 to previous levels, and the depletions grow with time.

Sources: Special issue on tropospheric chemistry, *Science* 276, (No. 5315, 16 May 1997). Butcher, S. S., Charlson, R. J., Orians, G. H., & Wolfe, G. V. (eds.), 1992, *Global biogeochemical cycles*, Academic Press, New York, pp. 379. Chamberlain, J. W., & Hunten, D. M., 1987, *Theory of planetary atmospheres*, Academic Press, New York, pp. 481. Houghton, H. G., 1985, *Physical meteorology*, MIT Press, Cambridge, pp. 442. Houghton, J. T., 1986, *The physics of atmospheres*, Cambridge Univ. Press, Cambridge, pp. 203. Keeling, R. F., & Shertz, S. R., 1992, *Nature* 358, 723–727. Levine, J. S. (ed.), 1985, *The photochemistry of atmospheres*, Academic Press, New York, pp. 518. Warneck, P., 1988, *Chemistry of the natural atmosphere*, International Geophysics Series, Vol. 41, Academic Press, New York, pp. 757.

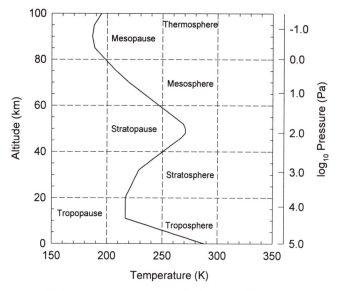

Figure 6.2 Temperature and pressure in the terrestrial atmosphere

Table 6.15 Temperature, Pressure, and Density in the Earth's Atmosphere

Altitude (km)	Temperature (K)	Pressure (Pa)	Density (kg m^{-3})
0	288.15	101,300	1.23
10	223.25	26,500	0.414
20	216.65	5,529	0.089
30	226.51	1,197	0.018
40	250.35	287.1	4.00(−3)
50	270.65	79.78	1.03(−3)
60	247.02	21.96	3.10(−4)
70	219.58	5.221	8.28(−5)
80	198.64	1.052	1.85(−5)
90	186.87	0.184	3.42(−6)
100	195.08	0.032	5.60(−7)

Exponents in parentheses.

Source: U.S. Standard Atmosphere, 1976, NOAA-S/T76-1562, NOAA, NASA, and U.S. Airforce, Washington, D.C., pp. 227.

Table 6.16 Chemical Composition of the Terrestrial Troposphere

Gas	Abundance[a]	Gas Source(s)	Gas Sink(s)	Sources
N_2	78.084%	denitrifying bacteria	nitrogen fixing bacteria	War88
O_2	20.946%	photosynthesis	respiration & decay	War88
H_2O	<4%, varies	evaporation	condensation	War88
Ar	9340 ppm	outgassing (^{40}K)	no known sinks	OP83
CO_2	350 ppm	combustion, biology	biology	KC84
Ne	18.18 ppm	outgassing	no known sinks	OP83
^4He	5.24 ppm	outgassing (U, Th)	escape	OP83
CH_4	1.7 ppm	biology & agriculture ($+1\%$ yr^{-1})	reaction with OH	Ehh88
Kr	1.14 ppm	outgassing	no known sinks	OP83
H_2	0.55 ppm	H_2O photolysis	H atom escape	War88
N_2O	~320 ppb	biology	photolysis (stratosphere)	War88
CO	125 ppb	photochemistry	photochemistry	War88
Xe	87 ppb	outgassing	no known sinks	OP83
O_3	~10–100 ppb	photochemistry	photochemistry	War88
HCl	~1 ppb	derived from sea salt	rainout	War88
Isoprene, etc.	~1–3 ppb	foliar emissions	photooxidation	War88
C_2H_6, etc.	~3–80 ppb	combustion, biomass burning, grasslands	photooxidation	War88
H_2O_2	~0.3–3 ppb	photochemistry	photochemistry	SSN92
C_2H_2, etc.	~0.2–3 ppb	combustion, biomass burning, oceans	photooxidation	War88
C_2H_4, etc.	~0.1–6 ppb	combustion, biomass burning, oceans	photooxidation	War88
C_6H_6 etc.	~0.1–1 ppb	anthropogenic	photooxidation	War88
NH_3	0.1–3 ppb	biology	wet & dry deposition	War88
HNO_3	~0.04–4 ppb	photochemistry (NO_x)	rainout	War88
CH_3Cl	612 ppt	ocean, biomass burning	reaction with OH	Pri88
OCS	500 ppt	biology	photodissociation	War88
NO_x	~30–300 ppt	combustion, biology	photooxidation	War88
CF_2Cl_2 (F12)	300 ppt	anthropogenic ($+5.1\%$ yr^{-1})	photolysis (stratosphere)	Pri88
$CFCl_3$ (F11)	178 ppt	anthropogenic ($+5.1\%$ yr^{-1})	photolysis (stratosphere)	Pri88
CH_3CCl_3	157 ppt	anthropogenic ($+4.4\%$ yr^{-1})	reaction with OH	PCS92
CCl_4	121 ppt	anthropogenic ($+1.3\%$ yr^{-1})	photolysis (stratosphere)	Pri88

continued

Table 6.16 *(continued)*

Gas	Abundance[a]	Gas Source(s)	Gas Sink(s)	Sources
CF_4 (F14)	69 ppt	anthropogenic (+2.0% yr^{-1})	photolysis (upper atm.)	Pri88
$CHClF_2$ (F22)	59 ppt	anthropogenic (+10.9% yr^{-1})	reaction with OH	Pri88
H_2S	30–100 ppt	biology	photooxidation	War88
$C_2Cl_3F_3$ (F113)	30–40 ppt	anthropogenic (+11.5% yr^{-1})	photolysis (stratosphere)	Pri88
CH_2Cl_2	30 ppt	anthropogenic	reaction with OH	War88
CH_2ClCH_2Cl	26 ppt	anthropogenic	reaction with OH	War88
CH_3Br	22 ppt	ocean, marine biota	reaction with OH	War88
SO_2	20–90 ppt	combustion	photooxidation	War88
$CHCl_3$	16 ppt	anthropogenic	reaction with OH	War88
CS_2	~15 ppt	anthropogenic	photooxidation	War88
$C_2Cl_2F_4$ (F114)	14 ppt	anthropogenic	photolysis (stratosphere)	War88
C_2H_5Cl	12 ppt	anthropogenic	reaction with OH	War88
$CHClCCl_2$	7.5 ppt	anthropogenic	reaction with OH	War88
$(CH_3)_2S$	5–60 ppt	biology	photooxidation	War88
C_2ClF_5 (F115)	4 ppt	anthropogenic	photolysis (stratosphere)	War88
C_2F_6 (F116)	4 ppt	anthropogenic	photolysis (upper atm)	War88
$CClF_3$ (F13)	3.3 ppt	anthropogenic	photolysis (stratosphere)	War88
CH_3I	~2 ppt	ocean, marine biota	photolysis (troposphere)	War88
$CHCl_2F$ (F21)	1.6 ppt	anthropogenic	reaction with OH	War88
$CClF_2Br$	1.2 ppt	anthropogenic (+20% yr^{-1})	photolysis (stratosphere)	Pri88

*Abundances by volume in dry air (non-urban troposphere).

Sources: **[Ehh88]** Ehhalt, D. H., 1988, in *The changing atmosphere* (Rowland F. S., & Isaksen, I. S. A., eds.), J. Wiley & Sons, pp. 25–32. **[KC84]** Keeling, C. D., & Carter, A. F., 1984, *J. Geophys. Res.* 89, 4615–4620. **[OP83]** Ozima, M., & Podosek, F. A., 1983, *Noble gas chemistry*, Cambridge Univ. Press, pp. 367. **[Pri88]** Prinn, R. G., 1988, in *The changing atmosphere* (Rowland, F. S., & Isaksen, I. S. A., eds.), J. Wiley & Sons, pp. 33–48. **[PCS92]** Prinn, R. G., Cunnold, D., Simmonds, P., Alyea, F., Boldi, R., Crawford, A., Fraser, P., Gutzler, D., Hartley, D., Rosen, R., & Rasmussen, R., 1992, *J. Geophys. Res.* 97, 2445–2461. **[SSN92]** Sigg, A., Staffelbach, T., & Neftel, A., 1992, *J. Atmos. Chem.* 14, 233–232. **[War88]** Warneck, P., 1988, *Chemistry of the natural atmosphere*, International Geophysics Series, Vol. 41, Academic Press, New York, pp. 757.

Table 6.17 Isotopic Composition of Noble Gases in the Terrestrial Atmosphere

Isotopic ratio	Observed Value	Sources
$^3\mathrm{He}/^4\mathrm{He}$	$(1.399\pm0.013)\times10^{-6}$	MAK70
$^{20}\mathrm{Ne}/^{22}\mathrm{Ne}$	9.800 ± 0.080	EEM65
$^{21}\mathrm{Ne}/^{22}\mathrm{Ne}$	$(2.899\pm0.025)\times10^{-2}$	EEM65
$^{36}\mathrm{Ar}/^{38}\mathrm{Ar}$	5.320 ± 0.002	Nie50
$^{40}\mathrm{Ar}/^{36}\mathrm{Ar}$	296.0 ± 0.5	Nie50
$^{78}\mathrm{Kr}/^{84}\mathrm{Kr}$	$(6.087\pm0.002)\times10^{-3}$	BDP73
$^{80}\mathrm{Kr}/^{84}\mathrm{Kr}$	$(3.960\pm0.002)\times10^{-2}$	BDP73
$^{82}\mathrm{Kr}/^{84}\mathrm{Kr}$	$(20.217\pm0.021)\times10^{-2}$	BDP73
$^{83}\mathrm{Kr}/^{84}\mathrm{Kr}$	$(20.136\pm0.021)\times10^{-2}$	BDP73
$^{86}\mathrm{Kr}/^{84}\mathrm{Kr}$	$(30.524\pm0.025)\times10^{-2}$	BDP73
$^{124}\mathrm{Xe}/^{132}\mathrm{Xe}$	$(3.537\pm0.0011)\times10^{-3}$	OP83
$^{126}\mathrm{Xe}/^{132}\mathrm{Xe}$	$(3.300\pm0.017)\times10^{-3}$	OP83
$^{128}\mathrm{Xe}/^{132}\mathrm{Xe}$	$(7.136\pm0.009)\times10^{-2}$	OP83
$^{129}\mathrm{Xe}/^{132}\mathrm{Xe}$	$(98.32\pm0.12)\times10^{-2}$	OP83
$^{130}\mathrm{Xe}/^{132}\mathrm{Xe}$	$(15.136\pm0.012)\times10^{-2}$	OP83
$^{131}\mathrm{Xe}/^{132}\mathrm{Xe}$	$(78.90\pm0.11)\times10^{-2}$	OP83
$^{134}\mathrm{Xe}/^{132}\mathrm{Xe}$	$(38.79\pm0.06)\times10^{-2}$	OP83
$^{136}\mathrm{Xe}/^{132}\mathrm{Xe}$	$(32.94\pm0.04)\times10^{-2}$	OP83

Sources: **[BDP73]** Basford, J. R., Dragon, J. C., Pepin, M. R., Coscio, M. R., & Murthy, V. R., 1973, *Proc. 4th. Lunar Sci. Conf.* 2, 1915–1955. **[EEM65]** Eberhardt, P., Eugster, O., & Marti, K., 1965, *Z. Naturforsch.* 20a, 623–624. **[MAK70]** Mamyrin, B. A., Anufriyev, G. S., Kamenskii, I. L, & Tolstikhin, I. A., 1970, *Geochem. Intl.* 7, 498–505. **[Nie50]** Nier, A. O., 1950, *Phys. Rev.* 77, 789–793. **[OP83]** Ozima, M., & Podosek, F. A., 1983, *Noble gas chemistry*, Cambridge Univ. Press, Cambridge, pp. 367.

Earth's Hydrosphere and Oceans

The total amount of water in Earth's hydrosphere, including the pore water in sediments and water in rocks, is about 1664×10^6 km^3 (1 km^3 H$_2$O = 10^{12} liters = 10^{12} kg = 1000 Tg = 1 Pg). Excluding pore waters and bound water in rocks, the exchangeable water reservoir is about 1410×10^6 km^3, of which the oceans comprise about 97%. The remaining 3% of the water is in glaciers and the polar ice caps (2.15%), soil moisture and ground water (0.62%), freshwater lakes, rivers, and streams (0.009%), saline lakes and inland seas (0.008%), and atmospheric water vapor (0.001%). Evaporation from the oceans is 425,000 km^3 yr^{-1} whereas precipitation onto the oceans is 385,000 km^3 yr^{-1}; the difference of 40,000 km^3 yr^{-1} is made up by runoff from the continents to the oceans. Precipitation on the continents is 111,000 km^3 yr^{-1}, whereas evaporation from the continents and transpiration from plants is 71,000 km^3 yr^{-1}; the difference is the runoff into the oceans. These global average fluxes are temporally and spatially variable. For example, precipitation is highest at the equator and drops off toward the poles, with a minimum at about 25° north and south of the equator where the oceanic central gyres are located. As a result, the residence time of atmospheric water vapor varies from 8 days near the equator to 12–15 days near the poles with a global average of 9 days.

The three major ocean basins (and their volumes in 10^6 km^3) are the Pacific (707.6), Atlantic (323.6), and Indian (291.0). The average depth of the global oceans is 3794 meters whereas the average elevation of the continents is 840 meters. The transition region between the continents and oceans is the continental margin. The landward part of the margin is the continental shelf, which is the submerged continuation of the land, and is variable in extent with an average width of 70 km. The continental slope begins at the continental shelf break, which is generally at a depth of 130 meters. The continental slope is the relatively narrow region where the topographic gradient changes from 1:1000 to 1:40 and the continent drops off into the ocean. The slope may form one side of a trench or may grade into the continental rise, which is a depositional feature formed by currents carrying sediments down the slope. The continental margin is the minimum (at ~1.8 km below sea level) in the earth's bimodal hypsometric curve. The greatest ocean depths (up to 11,035 meters in the Challenger Deep in the Marianas Trench) occur in ocean trenches.

The uppermost 75–200 meters of the ocean comprise the surface mixed layer, which has an average temperature of about 18°C. The thermocline with a characteristic depth of 1 km is the transition region between the surface mixed layer and the deep (or abyssal) ocean with an average temperature of 3.5°C and an average depth of 3794 meters (P ~386 bars). The abyssal regions contain about 95% of the ocean water. Because the oceans are heated from above, mixing between the warmer surface mixed layer and the deep ocean is fairly slow with a typical mixing time of 1–2 years. Vertical mixing across the 1 km deep thermocline takes ~2.5 years. Replacement of all water in the surface mixed layer by downwelling in the polar regions takes ~30 years. In contrast, vertical mixing across the surface mixed layer takes about 10–20 hours and air-sea exchange of a soluble gas, such as CO_2, proceeds at a typical rate of 5 m of seawater per day (the piston velocity).

The surface circulation of the oceans is driven by the winds and is principally in the upper few hundred meters. Trade winds drive currents from east to west in the equatorial regions, and when the currents encounter land they divide north and south flowing along the eastern edges of continents. The poleward currents are deflected by the Coriolis force, e.g., the northward-flowing Gulf Stream crosses the North Atlantic to northern Europe. Cold surface currents flow along the western edges of the continents and return to equatorial latitudes. The cyclic patterns of surface currents at about 25° north and south latitude are the central gyres. About 50% of the heat transport from equator to poles is by oceanic surface circulation. Thermocline, or thermohaline, circulation in the oceans occurs via downwelling at high latitudes, and transport along constant density (isopycnal) contours to equatorial regions where upwelling occurs. The characteristic transport times are ~130 years and have been measured by tracing the penetration of $^{14}CO_2$ and HTO (from atmospheric nuclear testing) in the oceans. Thermocline circulation is important for removing CO_2 from the atmosphere. Abyssal circulation below the thermocline is very slow and is constrained by the topography of the ocean floor. The two main sources of abyssal ocean water are the Weddell Sea (Antarctic) and the Norwegian-Greenland Sea in the North Atlantic.

The salinity (S) of the oceans is defined as grams NaCl per kilogram seawater (in parts per thousand, ‰) because mass is conserved as pressure and temperature change in the oceans. Salinity is related to chlorinity (Cl) by $S(‰) = 1.80655\ Cl(‰)$. The alkalinity of seawater is defined as

$$[Alk] = [HCO_3^-] + 2[CO_3^{2-}] + [B(OH)_4^-] + [OH^-] - [H^+] \tag{1}$$

where alkalinity is in milliequivalents (meq) per kg and the square brackets are concentrations in millimoles kg^{-1}. The average salinity, density, pH, and alkalinity of seawater are 34.7‰, 1.025 g cm^{-3}, 8.1, and 2.4 meq kg^{-1}, respectively. Salinity variations of 33–38‰ are observed in the oceans and have several causes. Higher salinity regions occur in the central gyres ~25° north and south of the equator, where the evaporation rate is larger than the precipitation rate. Lower salinity regions occur during summer in the polar oceans when some of the ice sheets melt.

The elements Cl, Na, S, Mg, Ca, and K dominate the composition of seawater and have constant ratios relative to each other, despite variations in salinity. These six elements and others with the same behavior (e.g., Br, B, F) are conservative elements. Variations in their concentrations can be explained solely by either subtraction or addition of pure water to seawater. Calcium is only conservative to first approximation because it is slightly depleted in surface water relative to the deep ocean because of carbonate shell formation by organisms.

Conservative elements are fairly nonreactive and have oceanic residence times (t_{res} ~ mass in oceans/river input rate) that are significantly longer than the residence time of ocean water itself, which is about 34,000 years. For example, the residence times of Cl, Br, Na, Ca, and F are 120 Ma, 100 Ma, 75 Ma, 1.1 Ma, and 0.5 Ma, respectively. The long residence times lead to uniform relative concentrations throughout the oceans.

Other elements are nonconservative and are classified by their depth-dependent concentration profiles. Nutrient elements (e.g., C, N, P) and elements such as Ca and Si that are used to build shells are depleted in surface waters and are enriched in the deep oceans where they are released by decomposition of sinking detritus. Conversely, O_2 and other biologically produced compounds are enriched in surface waters. Atmospheric inputs such as elements in dust (Fe), radionuclides from atmospheric nuclear explosions (T, ^{14}C, ^{90}Sr, ^{137}Cs, Pu), anthropogenic pollutants (Pb), and photochemical products (NO) are also enriched in surface waters. Some species have midwater maxima as a result of input from hydrothermal vents (^3He, ^{222}Rn, CH_4) or thermocline circulation (HTO). Fluxes out of sediments lead to enrichments in abyssal waters (^{228}Ra), whereas precipitation onto settling particles leads to depletions for other species (^{210}Pb, ^{230}Th) in the deep ocean. Measuring the concentrations, speciation, and vertical profiles for many elements in seawater is an active research area.

The composition of seawater is regulated by the sources and sinks for different elements; seawater is not simply the result of evaporation of river water. River input is the major source of most elements; ion exchange of river-borne clays is a major source for Ca and a sink for Na. Other sinks include seaspray, porewaters, and evaporites for Na and Cl; hydrothermal activity for Mg; pyrite burial for sulfate; and carbonate formation for Ca.

The oceans account for about 50% of biological productivity on Earth with a net primary productivity (NPP) of ~51 Pg C yr^{-1} versus 60 Pg C yr^{-1} on land. The highest productivity is in upwelling regions where nutrients are brought up from the deep oceans (420 g C m^{-2} yr^{-1}) and in coastal zones (250 g C m^{-2} yr^{-1}) where nutrients are washed off the continents. However, total oceanic productivity is dominated by the open ocean (130 g C m^{-2} yr^{-1}) because of its much larger area. Schematically, productivity in the oceans is represented by the Redfield-Ketchum-Richards equation:

$$106 \ CO_2 + 16 \ HNO_3 + H_3PO_4 + 122 \ H_2O + h\nu \rightarrow$$
$$(CH_2O)_{106}(NH_3)_{16}(H_3PO_4) + 138 \ O_2 \qquad (2)$$

About 0.5% of the organic matter produced in the oceans is incorporated into sediments; the rest is remineralized.

Finally, geochemical and isotopic analyses of oceanic sediments provide a variety of data on Earth's past climate and chemistry. For example, variations in the $^{16}O/^{18}O$ ratio of sedimentary carbonates record paleotemperatures; variations in the $^{34}S/^{32}S$ ratio of gypsum sediments record changes in the sulfate content of seawater and in biological reduction of sulfate to sulfide; changes in the $^{87}Sr/^{86}Sr$ ratio of carbonates track the relative weathering rate of continental rocks; and boron isotopic ratios of carbonates track oceanic pH over time.

Sources and further reading: Butcher, S. S., Charlson, R. J., Orians, G. H., & Wolfe, G. V. (eds.), 1992, *Global biogeochemical cycles*, Academic Press, New York, pp. 379. Holland, H. D., 1978, *The chemistry of the atmosphere and oceans*, Wiley-Interscience, New York, pp. 351. Holland, H. D., 1984, *The chemical evolution of the atmosphere and oceans*, Princeton Univ. Press, Princeton, pp. 582. Li, Y.-H., 1991, *Geochim. Cosmochim. Acta*, 55, 3223–3240. Quinby-Hunt, M. S., & Turekian, K. K., 1983, *EOS Trans. AGU* 64, 130–132. Riley, J. P., Skirrow, G., & Chester, R. (eds.), 1975–1995, *Chemical oceanography*, Academic Press, London. Schlesinger, W. H., 1997, *Biogeochemistry*, 2nd ed., Academic Press, San Diego, pp. 588. Stumm, W., & Morgan, J. J., 1996, *Aquatic chemistry*, 3rd. ed., Wiley, New York, pp. 1022.

Table 6.18 Major Elements in River Water (mg/liter)

Species	Africa	N. America	S. America	Asia	Australia	Europe	Global
Na^+	11	9	4	9.3	2.9	5.4	6.3
Mg^{2+}	3.8	5	1.5	5.6	2.7	5.6	4.1
K^+	2	1.4	2	2	1.4	1.7	2.3
Ca^{2+}	12.5	21	7.2	18.4	3.9	31.1	15
HCO_3^-	43	68	31	79	31.6	95	58.4
NO_3^-	0.8	1	0.7	0.7	0.05	3.7	1
SO_4^{2-}	13.5	20	4.8	8.4	2.6	24	11.2
Cl^-	12.1	8	4.9	8.7	10	6.9	7.8
SiO_2	23.2	9	11.9	11.7	3.9	7.5	13.1
Fe	1.3	0.16	1.4	0.01	0.3	0.8	0.37

Values are approximate continental means. The composition of rivers on a given continent varies depending on geological setting of bedrock. In addition, compositions of individual rivers may show seasonal changes in composition.

Sources: Berner, E. K., & Berner, R. A., 1996, *Global environment. Water, air, and geochemical cycles,* Prentice Hall, Upper Saddle River, NJ, pp. 376. Schlesinger, W. H., 1997, *Biogeochemistry,* Academic Press, San Diego, pp. 588.

Table 6.19 Mean Chemical Composition of Terrestrial Oceans

Element	Abundance	Element	Abundance	Element	Abundance
He	6.8–7.6 ppt	Ni	480–530 ppt	Ba	11.7–15 ppb
Li	178 ppb	Cu	120–210 ppt	La	4.5–5.6 ppt
Be	0.21 ppt	Zn	320–390 ppt	Ce	1.7–3.5 ppt
B	4.4 ppm	Ga	1.7–20 ppt	Pr	0.6–0.87 ppt
C	96.8 ppm CO_2	Ge	4.3–5 ppt	Nd	4.2 ppt
N	16.5 ppm N_2	As	1.7 ppb	Sm	0.84 ppt
	1.86 ppm NO_3	Se	130–155 ppt	Eu	0.15–0.21 ppt
O	4.8 ppm O_2	Br	67 ppm	Gd	0.8–1.3 ppt
F	1.3 ppm	Kr	200–310 ppt	Tb	0.17–0.21 ppt
Ne	120–161 ppt	Rb	124 ppb	Dy	1.1–1.5 ppt
Na	1.0781 wt%	Sr	7.8 ppm	Ho	0.28–0.45 ppt
Mg	1280 ppm	Y	13 ppt	Er	0.92–1.3 ppt
Al	0.3–1 ppb	Zr	17–30 ppt	Tm	0.13–0.25 ppt
Si	2.8 ppm	Nb	1–10 ppt	Yb	0.9–1.5 ppt
P	65–71 ppb	Mo	11 ppb	Lu	0.14–0.32 ppt
S	898 ppm	Ru	~1 ppt	Hf	3.4 ppt
Cl	1.9353 wt%	Rh	≤0.005 ppt	Ta	≤2.5 ppt
Ar	4.3–623 ppb	Pd	≤0.074 ppt	W	0.1 ppb
K	399 ppm	Ag	2.7 ppt	Re	4–8 ppt
Ca	415 ppm	Cd	70–79 ppt	Au	0.03–4.9 ppt
Sc	0.67 ppt	In	0.1–0.2 ppt	Hg	0.42–6 ppt
Ti	0.01 ppb	Sn	0.5–0.6 ppt	Tl	12–14 ppt
V	1.2–2.15 ppb	Sb	0.15 ppb	Pb	2–2.7 ppt
Cr	253–330 ppt	Te	0.07 ppt	Bi	0.0042–10 ppt
Mn	10–72 ppt	I	59 ppb	Th	0.06 ppt
Fe	40–250 ppt	Xe	50–65.7 ppt	Pa	2×10^{-7} ppt
Co	1.2–2 ppt	Cs	0.3 ppb	U	3.2 ppb

1 wt% = 10 g/kg = 10,000 ppm
1 ppm = 1 mg/kg = 10^{-3} g/kg = 10^{-6} g/g = 10^{-4} wt%
1 ppb = 1 μg/kg = 10^{-6} g/kg = 10^{-9} g/g = 10^{-7} wt%
1 ppt = 1 ng/kg = 10^{-9} g/kg = 10^{-12} g/g = 10^{-10} wt%

Sources: Quinby-Hunt, M. S., & Turekian, K. K., 1983, *EOS Trans. AGU* 64, 130–132. Li, Y.-H., 1991, *Geochim. Cosmochim. Acta* 55, 3223–3240.

6.2 The Moon

The earth and the moon orbit about their barycenter, which is ~4671 km from the earth's center, in the middle of the lower mantle. At present, the mean Earth–Moon distance is ~384,400 km. Over time, conservation of angular momentum in the Earth–Moon system causes the moon to move further away from Earth and causes Earth's rotation to slow down. The secular increase in the length of day is ~0.002^s per century. Studies of diurnal and tidally related growth rings in fossil corals show that the length of day and month were significantly shorter ~365 Ma ago in the mid-Devonian when days were ~21.9^h long and there were ~400 days per year.

Over time, tidal friction slowed the moon's rotation rate which is now synchronous with its orbital period. However, ~59% of the lunar surface can be seen from Earth because of lunar librations. Librations in longitude arise because the moon rotates at a nearly constant rate but has a variable orbital velocity that is greatest at perigee and smallest at apogee. As a result, the center of the sub-earth tidal bulge on the moon is sometimes ahead and sometimes behind the earth's position.

The moon is 27% as big as Earth, but only 1.2% as massive because of its lower density (3.344 g cm^{-3}). The average crustal thickness is ~61 km, but the farside crust is ~67 km thick whereas the nearside crust is ~55 km thick. The moon's low density constrains its core (if any) to <4% of its total mass. At present the evidence for a lunar core is weak. The moon's average heat flow is 29 mW m^{-2}, less than half that of Earth. Seismic activity has been measured, but it is much less than on Earth. There is no evidence for volcanism on the moon at present.

The two major types of terrain on the lunar surface are the maria and the highlands. The maria are younger dark plains formed by basalt flows that flooded impact basins. Because they are younger than the highlands, the maria are less heavily cratered. The maria are concentrated on the near side of the moon and cover ~16% of its surface. The highlands are the brighter areas of the moon and cover ~84% of its surface. The *Apollo* samples from maria and highlands regions show that, to a first approximation, there are two major types of lunar rocks: mare basalts and highland rocks. The highland rocks can be subdivided into three categories: (1) ferroan anorthosites, represented by the *Apollo 16* samples and lunar meteorites, (2) Mg-rich rocks, represented by the norites, troctolites, and dunites, and (3) potassium, rare earth element, phosphorus-rich (KREEP) rocks.

Various types of lunar soils and glasses were also returned by the *Apollo* missions. Representative compositions of the major rock types are tabulated in this section.

The moon is covered by a regolith that is 2 to 8 meters thick in the maria and >15 meters thick in the highlands. The top layer of the regolith is a good insulator because of its high porosity, so diurnal temperature changes only penetrate to a depth of <1 meter. The mean daytime and nighttime temperatures on the moon are 107°C and −153°C, respectively. Day and night both last for two weeks. Because of the moon's small axial tilt, it has no seasons. However, sunlight is nearly always horizontal to the lunar poles, so permanently cold, dark regions exist there. This situation has led to suggestions that large ice deposits (e.g., from impacts of water-bearing bodies) may stably exist at the lunar poles. Some radar observations by the *Clementine* spacecraft and some neutron spectrometer observations by the *Lunar Prospector* spacecraft have been interpreted in terms of ice deposits, but radar observations from the *Aricebo* radio telescope find no evidence of polar ice deposits.

Scientific models for the origin of the moon fall into four major categories: (1) fission from the earth, proposed by George Darwin in 1879, (2) gravitational capture (Gerstenkorn, 1955), (3) coaccretion of the earth and the moon (Schmidt, 1959), and (4) the giant impact of a Mars-sized planetesimal on the earth, proposed by Hartmann and Davis in 1975 and Cameron and Ward in 1976, and studied in detail by Cameron, Benz, and colleagues. The giant impact hypothesis is attractive because it accounts for the angular momentum of the Earth–Moon system and it is generally consistent with models of planetary accretion from smaller planetesimals.

Sources and further reading: Many lunar science results are reported in *Proceedings of the Lunar and Planetary Science Conference,* Lunar & Planetary Institute, Houston (1969–1992). Special issue on *Clementine* mission, *Science* 266 (16 Dec. 1994). Cameron, A. G. W., & Ward, W. R., 1976, Lunar Sci. Conf. VII, 120–122. Hartmann, W. K., & Davis, D. R., 1975, *Icarus* 24, 504–515. Darwin, G. H., 1879, *Phil. Trans. Roy. Soc.* 170, part 2, 447–538. Gerstenkorn, H., 1955, *Zeitschr. Astrophys.* 36, 245–274. Hartmann, W. K., Phillips, R. J., & Taylor, G. J. (eds.), 1986, *Origin of the moon*, Lunar and Planetary Institute, Houston, pp. 781. Heiken, G. H., Vaniman, D. T., & French, B. M. (eds.), 1991, *Lunar sourcebook*, Cambridge University Press, Cambridge, pp. 736. Neumann, G. A., Zuber, M. T., Smith, D. E., & Lemoine, F. G., 1996, *J. Geophys. Res.* 101, 16841–16843. Schmidt, O. Yu., 1959, *A theory of the origin of the earth*, Lawrence and Wishart, London, pp. 139. Smith, D. E., Zuber, M. T., Neumann, G. A., & Lemoine, F. G., 1997, *J. Geophys. Res.* 102, 1591–1611. Stacey, N. J. S., Campbell, D. B., & Ford, P. G., 1997, *Science* 276, 1527–1530. Taylor, S. R., 1982, *Planetary science: A lunar perspective*, Lunar and Planetary Institute, Houston, pp. 481.

Table 6.20 Spacecraft Missions to the Moon

Mission	Launch Date	Remarks
Pioneer 0 (USA)	17 Aug. 1958	attempted flyby; launch failure
Luna 1958A (USSR)	23 Sept. 1958	attempted impact; launch failure
Pioneer 1 (USA)	11 Oct. 1958	attempted lunar orbit; missed Moon
Luna 1958B (USSR)	12 Oct. 1958	attempted lunar impact; launch failure
Pioneer 2 (USA)	8 Nov. 1958	attempted lunar orbit; launch failure
Luna 1958C (USSR)	4 Dec. 1958	attempted lunar impact; launch failure
Pioneer 3 (USA)	6 Dec. 1958	attempted flyby; launch failure
Luna 1 (USSR)	2 Jan. 1959	lunar flyby; confirmed solar wind
Pioneer 4 (USA)	3 Mar. 1959	lunar flyby
Luna 1959A (USSR)	16 June 1959	attempted lunar impact; launch failure
Luna 2 (USSR)	12 Sept. 1959	first spacecraft to impact Moon
Luna 3 (USSR)	4 Oct. 1959	lunar flyby, first image of Moon's far side
Pioneer P-3 (USA)	26 Nov. 1959	attempted lunar orbit; launch failure
Luna 1960A (USSR)	15 Apr. 1960	attempted flyby; launch failure
Luna 1960B (USSR)	18 Apr. 1960	attempted flyby; launch failure
Pioneer P-30 (USA)	25 Sept. 1960	attempted orbit; launch failure
Pioneer P-31 (USA)	15 Dec. 1960	attempted orbit; launch failure
Ranger 1 (USA)	23 Aug. 1961	test flight
Ranger 2 (USA)	18 Nov. 1961	test flight
Ranger 3 (USA)	26 Jan. 1962	attempted impact, missed moon
Ranger 4 (USA)	23 Apr. 1962	first US spacecraft impacting the moon
Ranger 5 (USA)	18 Oct. 1962	flyby; attempted lander
Sputnik 33 (USSR)	4 Jan. 1963	attempted lander
Lunar 1963B (USSR)	2 Feb. 1963	attempted lander; launch failure
Luna 4 (USSR)	2 Apr. 1963	attempted lander; missed Moon
Ranger 6 (USA)	30 Jan. 1964	impact; camera failure
Luna 1964A (USSR)	21 Mar. 1964	attempted lander; launch failure
Luna 1964B (USSR)	20 Apr. 1964	attempted lander; launch failure
Zond 1964A (USSR)	4 June 1964	attempted lander; launch failure
Ranger 7 (USA)	28 July 1964	close range photographs, impact

continued

Table 6.20 *(continued)*

Mission	Launch Date	Remarks
Ranger 8 (USA)	17 Feb. 1965	high resolution photographs; impact
Cosmos 60 (USSR)	12 Mar. 1965	attempted lander
Ranger 9 (USA)	21 Mar. 1965	photographs; impact
Luna 1965A (USSR)	10 Apr. 1965	attempted lander; launch failure
Luna 5 (USSR)	9 May 1965	impact; attempted soft landing
Luna 6 (USSR)	8 June 1965	attempted lander; missed Moon
Zond 3 (USSR)	18 July 1965	flyby; pictures from far side
Luna 7 (USSR)	4 Oct. 1965	impact; attempted soft landing
Luna 8 (USSR)	3 Dec. 1965	impact; attempted soft landing?
Luna 9 (USSR)	31 Jan. 1966	first unmanned lander; photographs from surface
Cosmos 111 (USSR)	1 Mar. 1966	attempted orbiter
Luna 10 (USSR)	31 Mar. 1966	orbiter; γ-ray data, still in orbit
Luna 1966A (USSR)	30 Apr. 1966	attempted orbiter; launch failure
Surveyor 1 (USA)	30 May 1966	first soft landing of U.S. spacecraft on Moon
Explorer 33 (USA)	1 Jul. 1966	attempted orbiter
Lunar Orbiter 1 (USA)	10 Aug. 1966	orbiter; pictures of far side; impact on command
Luna 11 (USSR)	24 Aug. 1966	orbiter; photography; still in orbit
Surveyor 2 (USA)	20 Sept. 1966	attempted lander
Luna 12 (USSR)	22 Oct. 1966	orbiter; still in orbit
Lunar Orbiter 2 (USA)	6 Nov. 1966	orbiter; pictures of far side; impact on command
Luna 13 (USSR)	21 Dec. 1966	lander
Lunar Orbiter 3 (USA)	4 Feb. 1967	orbiter; pictures of far side; impact on command
Surveyor 3 (USA)	17 Apr. 1967	lander
Lunar Orbiter 4 (USA)	8 May 1967	orbited at polar inclination; impact on command
Surveyor 4 (USA)	14 July 1967	attempted lander
Explorer 35 (USA)	19 July 1967	orbiter; measured magnetic field and particle data
Lunar Orbiter 5 (USA)	1 Aug. 1967	orbiter; photography
Surveyor 5 (USA)	8 Sept. 1967	lander
Zond 1967A (USSR)	28 Sept. 1967	attempted test flight; launch failure
Surveyor 6 (USA)	7 Nov. 1967	lander and take-off

continued

Table 6.20 *(continued)*

Mission	Launch Date	Remarks
Zond 1967B (USSR)	22 Nov. 1967	attempted test flight; launch failure
Surveyor 7 (USA)	7 Jan. 1968	lander
Luna 1968A (USSR)	7 Feb. 1968	attempted orbiter; launch failure
Zond 4 (USSR)	2 Mar. 1968	attempted flyby
Luna 14 (USSR)	7 Apr. 1968	orbiter
Zond 1968A (USSR)	23 Apr. 1968	attempted test flight?; launch failure
Zond 5 (USSR)	15 Sept. 1968	lunar flyby and return to Earth
Zond 6 (USSR)	10 Nov. 1968	lunar flyby and return to Earth, first returned film
Apollo 8 (USA)	21 Dec. 1968	manned orbital flight (Frank Borman, James A. Lovell, William Anders)
Zond 1969A (USSR)	20 Jan. 1969	attempted flyby and return; launch failure
Luna 1969A (USSR)	19 Feb. 1969	attempted rover; launch failure
Zond L1S-1 (USSR)	21 Feb. 1969	attempted orbiter; launch failure
Luna 1969B (USSR)	15 Apr. 1969	attempted sample return; launch failure
Apollo 10 (USA)	18 May 1969	manned orbital flight (Thomas Stafford, Eugene A. Cernan, John W. Young)
Luna 1969C (USSR)	14 June 1969	attempted sample return?; launch failure
Zond L1S-2 (USSR)	3 July 1969	attempted orbiter; launch failure
Luna 15 (USSR)	13 July 1969	attempted sample return; impact
Apollo 11 (USA)	16 July 1969	first manned lunar landing on 20 July 1969 at Mare Tranquillitatis (Edwin E. Aldrin, Neil A. Armstrong, Michael Collins)
Zond 7 (USSR)	7 Aug. 1969	flyby and return to Earth; returned film
Cosmos 300 (USSR)	23 Sept. 1969	attempted sample return?
Cosmos 305 (USSR)	22 Oct. 1969	attempted sample return?
Apollo 12 (USA)	14 Nov. 1969	manned landing at Oceanus Procellarum (Alan L. Bean, Charles Conrad, Richard F. Gordon)
Luna 1970A (USSR)	6 Feb. 1970	attempted sample return; launch failure
Luna 1970B (USSR)	19 Feb. 1970	attempted orbiter; launch failure
Apollo 13 (USA)	11 Apr. 1970	manned flyby; aborted landing (Fred W. Haise, James A. Lowell, John L. Swigert)
Luna 16 (USSR)	12 Sept. 1970	first unmanned sample return, Mare Fecunditatis

continued

Table 6.20 *(continued)*

Mission	Launch Date	Remarks
Zond 8 (USSR)	20 Oct. 1970	flyby and return, film
Luna 17 (USSR)	10 Nov. 1970	lunar rover Lunokhod 1
Apollo 14 (USA)	31 Jan. 1971	manned landing at Fra Mauro highlands (Edgar D. Mitchell, Stuart A. Roosa, Alan B. Shepard)
Apollo 15 (USA)	26 July 1971	manned landing at Palus Putredinis, Hadley-Apenninus region (James B. Irwin, David R. Scott, Alfred M. Worden)
Luna 18 (USSR)	2 Sept. 1971	attempted sample return
Luna 19 (USSR)	28 Sept. 1971	orbiter; still in orbit
Luna 20 (USSR)	14 Feb. 1972	sample return from Apollonius highlands, Crisium basin rim
Apollo 16 (USA)	16 Apr. 1972	manned landing at Descartes crater (Charles M. Duke, Thomas K. Mattingly, John W. Young)
Soyuz L3 (USSR)	23 Nov. 1972	attempted orbiter; launch failure
Apollo 17 (USA)	7 Dec. 1972	manned landing at Taurus-Littrow valley (Eugene A. Cernan, Ronald B. Evans, Harrison H. Schmitt)
Luna 21 (USSR)	8 Jan 1973	lunar rover Lunokhod 2
Explorer 49 (RAE-B)	10 June 1973	orbiter; radio astronomy
Luna 22 (USSR)	29 May 1974	orbiter
Luna 23 (USSR)	28 Oct. 1974	attempted sample return
Luna 1975A (USSR)	16 Oct. 1975	attempted sample return?
Luna 24 (USSR)	9 Aug. 1976	sample return from Mare Crisium
Galileo (USA-EU)	18 Oct. 1989	primary mission to Jupiter, approaches Earth & Moon on 8 Dec. 1990 and 8 Dec. 1992; imaging
Muses-A (Japan)	24 Jan. 1990	two attempted orbiters; lost contact
Clementine (USA)	25 Jan. 1994	orbiter; topographic surface mapping
Lunar Prospector (USA)	24 Sept. 1997	in orbit

Sources: National Space Science Data Center, Greenbelt, MD. Heiken, G. H., Vaniman, D. T., & French, B. M. (eds.), 1991, *Lunar sourcebook,* Cambridge Univ. Press, Cambridge, pp. 736. Lewis, J. S., 1995, *Physics and chemistry of the solar system,* Academic Press, San Diego, pp. 556. Wilhelms, D. E., 1984, in *The geology of the terrestrial planets,* NASA-SP 469, pp. 107–205.

Table 6.21 Some Physical Parameters of the Moon

Property	Value	Property	Value
Mean radius (km)	1737.103 ±0.015	Sidereal rotation period, sidereal month ($\oplus$ days)	27.321661
Mean equatorial radius (km)	1738.139 ±0.065	Synodic month, new moon to new moon ($\oplus$ days)	29.530589
Mean polar radius (km)	1735.972 ±0.200	Eccentricity of orbit	0.05490
Surface area (km^2)	3.792×10^8	Inclination of orbit to ecliptic	5.145°
Volume (km^3)	2.196×10^{10}	Mean orbital velocity (km s^{-1})	1.023
Mass kg	7.349×10^{22}	Inclination of equator to orbit	6.68°
Bulk density (g cm^{-3})	3.344	Mean distance from Earth (km)	3.84401×10^5
GM (m^{3}s^{-2})	4.9028×10^{12}		$= 60.27\ R_{\oplus}$
Gravity GM/R^2 (m s^{-2})	1.624	$T_{surface}$ (K)	120–390
Escape velocity (km s^{-1})	2.376	$T_{blackbody}$ (K)	277
$J_2 \times 10^6$	202.7	Mean crustal thickness (km)	61
Moment of inertia, C/(MR^2)	0.394	on farside (km)	67
Moment of inertia difference		on nearside (km)	55
$\beta = (C-A)/B$	631.7×10^{-6}	Crustal density (g cm^{-3})	2.97±0.07
$\gamma = (B-A)/C$	227.9×10^{-6}	Heat flow (mW m^{-2})	22–31

Sources: Heiken, G. H., Vaniman, D. V., & French, B. M. (eds.), 1991, *Lunar sourcebook*, Cambridge Univ. Press, Cambridge, pp. 736. Smith, D. E., Zuber, M. T., Neumann, G. A., & Lemoine, F. G., 1997, *J. Geophys. Res.* 102, 1591–1611.

Table 6.22 Composition of the Lunar Atmosphere

Species	Number density (cm^{-3})	Species	Number density (cm^{-3})
H	<17	Ar	4×10^4
He	$(2-40) \times 10^3$	K	16
Li	<0.01	Ca	<6
O	<500	Ti	<2
Na	70		

Source: Sprague, A. L., & Hunten, D. M., 1995, in *Volatiles in the earth and the solar system*, (Farley, K. A., ed.), AIP Conf. Proc. 341, pp. 200–208.

Table 6.23 Composition of the Moon, Lunar Bulk Silicates, and the Lunar Highland Crust

	Unit	Bulk Moon		Bulk Silicates				Highland Crust
		[Tay82]	[An77]	[ON91]	[JD89]	[RSW86]	[WPB77]	[Tay82]
H	ppm	...	2.34	...	...	...	...	...
Li	ppm	0.83	9.27	1.9	...	...	...	...
Be	ppm	0.18	0.198	...	...	...	...	...
B	ppm	0.54	0.0139	...	...	...	...	...
C	ppm	...	10.5	...	...	...	...	...
N	ppm	...	0.277	...	...	...	...	...
O	wt%	...	44.11	...	...	...	42.60	...
F	ppm	...	32.0	1.3	...	...	...	...
Na	ppm	600	960	260	...	450	1520	3300
Mg	wt%	19.3	18.50	20.8	22.4	22.23	12.80	4.10
Al	wt%	3.17	6.21	2.04	1.96	1.97	8.63	13.0
Si	wt%	20.3	19.83	20.5	19.9	20.21	18.70	21.00
P	ppm	...	573	43	...	...	...	...
S	ppm	...	4150	800	...	...	1930	...
Cl	ppm	...	0.746	...	...	...	...	...
K	ppm	83	102	31	...	...	178	600
Ca	wt%	3.22	6.78	2.31	2.1	2.17	9.14	11.3
Sc	ppm	19	42.6	15.4	...	14	60.9	...
Ti	ppm	1800	3600	1220	1100	1800	4670	3360
V	ppm	150	362	81	...	79	315	21
Cr	ppm	4200	1280	3140	...	2200	2030	680
Mn	ppm	1200	352	1310	1200	1200	914	...
Fe	wt%	10.6	3.09	9.9	10.6	9.51	7.0	5.10
Co	ppm	...	256	220	...	95	...	...
Ni	ppm	...	5430	4720	...	2487	914	...
Cu	ppm	...	7.35	3.3	...	...	...	...
Zn	ppm	...	21.2	1.9	...	...	...	...
Ga	ppm	...	0.703	0.24	...	...	...	...
Ge	ppm	...	1.77	0.52	...	...	...	...

continued

Table 6.23 *(continued)*

	Unit	Bulk Moon [Tay82]	[An77]	Bulk Silicates [ON91]	[JD89]	[RSW86]	[WPB77]	Highland Crust [Tay82]
As	ppm	...	0.959	0.082	...	...	...	...
Se	ppm	...	1.39	...	...	...	...	...
Br	ppb	...	4.05	...	...	...	...	...
Rb	ppm	0.28	0.352	0.12	...	...	0.406	1.7
Sr	ppm	30	63.9	...	...	...	66	120
Y	ppm	5.1	11.6	...	...	...	17.3	13.4
Zr	ppm	14	69.2	...	...	...	47.7	63
Nb	ppm	1.1	3.52	...	...	...	3.35	4.5
Mo	ppm	...	1.04	0.068	...	...	...	...
Ru	ppm	...	5.22	...	...	...	...	...
Rh	ppm	...	1.12	...	...	...	...	...
Pd	ppb	...	266	...	...	...	...	...
Ag	ppb	...	10.2	...	...	...	...	...
Cd	ppb	...	0.618	...	...	...	...	...
In	ppb	...	0.08	0.4	...	...	...	...
Sn	ppb	...	90.5	34	...	...	...	...
Sb	ppb	...	8.09	2.8	...	...	...	...
Te	ppb	...	0.213	...	...	...	...	...
I	ppb	...	0.511	...	...	...	...	...
Cs	ppb	12	35.2	4.8	...	...	20.3	70
Ba	ppm	8.8	17.9	...	...	...	24.4	66
La	ppm	0.90	16.7	...	...	...	2.54	5.3
Ce	ppm	2.34	4.47	...	...	...	...	13
Pr	ppm	0.34	0.565	...	...	...	...	1.8
Nd	ppm	1.74	3.09	...	...	...	...	7.4
Sm	ppm	0.57	0.916	...	...	...	...	2.0
Eu	ppm	0.21	0.352	...	...	...	0.579	1.0
Gd	ppm	0.75	1.26	...	...	...	...	2.3
Tb	ppm	0.14	0.234	...	...	...	...	0.35

continued

Table 6.23 *(continued)*

	Unit	Bulk Moon [Tay82]	Bulk Moon [An77]	Bulk Silicates [ON91]	Bulk Silicates [JD89]	Bulk Silicates [RSW86]	Bulk Silicates [WPB77]	Highland Crust [Tay82]
Dy	ppm	0.93	1.59	...	...	...	...	2.3
Ho	ppm	0.21	0.352	...	...	...	...	0.53
Er	ppm	0.61	1.02	...	...	...	...	1.51
Tm	ppm	0.088	0.154	...	...	...	...	0.22
Yb	ppm	0.61	1.01	...	...	...	...	1.4
Lu	ppm	0.093	0.17	...	...	...	...	0.21
Hf	ppm	0.42	1.01	...	...	...	...	1.4
Ta	ppm	...	0.102	...	...	...	0.132	...
W	ppb	740	799	41	...	...	...	...
Re	ppb	...	266	16	...	...	...	...
Os	ppb	...	3.83	...	...	...	...	...
Ir	ppb	...	3.73	210	...	...	...	...
Pt	ppb	...	7.35	...	...	...	...	...
Au	ppb	...	76.7	...	...	...	...	...
Hg	ppb	...	0.298	...	...	...	...	...
Tl	ppb	...	0.128	...	...	...	...	...
Bi	ppb	...	0.111	...	...	...	...	...
Th	ppb	125	224	...	...	...	223	900
U	ppb	33	62.8	19	...	...	60.9	240

Sources: **[An77]** Anders, E., 1977, *Phil. Trans. Roy. Soc. Lond.* A285, 23–40. **[JD89]** Jones, J. H., & Drake, M. J., 1989, *Geochim. Cosmochim. Acta* 53, 513–528. **[ON91]** O'Neill, H. S. C., 1991, *Geochim. Cosmochim. Acta* 55, 1135–1157. **[RSW86]** Ringwood, A. E., Seifert, S., & Wänke, H., 1986, *Earth Planet. Sci. Lett.* 81, 105–117. **[Tay82]** Taylor, S. R., 1982, *Phys. Earth Planet. Interiors* 29, 233–241. **[WPB77]** Wänke, H., Palme, H, Baddenhausen, H., Kruse, H., & Spettel, B., 1977, *Phil. Trans. Roy. Soc. Lond.* A285, 41–48.

Table 6.24 Compositions of Some Typical Lunar Rocks

Sample ID:	Highland Rocks					Mare Basalts		
	KREEP basalt	ferroan-anorthosite	norite	gabbro-norite	trocto-lite	high Ti	low Ti	high Al
	15382	15415	77215	61224	76535	70215	12064	14053
Na$_2$O mass%	0.87	0.364	0.40	0.91	0.23	0.36	0.28	...
MgO mass%	7.83	0.26	12.5	12.8	20.0	8.4	6.5	8.5
Al$_2$O$_3$ mass%	16.9	35.6	15.0	13.2	19.9	8.8	10.7	13.6
SiO$_2$ mass%	52.5	44.5	51.1	50.7	43.0	37.8	46.3	46.4
K$_2$O mass%	0.53	0.0147	0.173	0.017	0.028	0.05	0.07	0.10
CaO mass%	9.43	20.4	9.1	11.6	10.8	10.7	11.8	11.2
TiO$_2$ mass%	1.90	0.02	0.33	0.40	0.05	13.0	4.0	2.6
FeO mass%	9.02	0.21	9.9	9.91	5.0	19.7	19.9	16.8
Li ppm	...	...	12.3	...	3.0	...	...	...
C ppm	...	9	...	...	...	...	...	...
F ppm	...	...	...	...	9	...	...	...
P ppm	2400	40	480	...	100	...	...	...
S ppm	...	<100	...	...	<100	...	...	...
Cl ppm	...	150	...	...	<0.8	...	...	...
Sc ppm	19	0.42	15.6	20.8	1.94	85.9	63.1	55
V ppm	60	...	...	...	...	50	119	...
Cr ppm	1780	20	2470	1990	730	2800	2500	...
Mn ppm	940	45	1270	1230	500	2100	2100	2000
Co ppm	17	0.21	28.0	23.6	27.6	...	...	...
Ni ppm	23	9	2	8.3	44	3	6.9	14
Cu ppm	...	58	...	...	...	...	...	...
Zn ppm	2.6	0.26	3.1	4.0	1.2	...	...	...
Ga ppm	...	3.1	4.4	3.0	...	...	...	...
Ge ppm	0.047	1.2	0.0152	4.3	1.7	...	...	...
As ppb	...	4.1	...	...	...	...	...	...
Se ppb	72	0.23	77	...	4.1	...	...	...
Br ppb	142	2.3	42	...	3.2	...	...	...
Rb ppm	15.4	0.17	4.38	...	0.22	0.36	1.05	2.19
Sr ppm	189	188	104	160	114	121	135	98
Zr ppm	1.068	...	175	...	17	192	114	215
Pd ppb	<0.6	...	...	...	...	...	...	...
Ag ppb	0.44	1.73	0.62	...	0.12	...	...	...
Cd ppb	87	0.57	4.9	4.1	0.60	...	...	...

continued

Table 6.24 Compositions of Some Typical Lunar Rocks

	Highland Rocks					Mare Basalts		
	KREEP basalt	**ferroan-anorthosite**	**norite**	**gabbro-norite**	**trocto-lite**	**high Ti**	**low Ti**	**high Al**
Sample ID:	15382	15415	77215	61224	76535	70215	12064	14053
In ppb	2.66	0.18	0.30	<0.6	...	...	...	...
Sn ppb	0.17	...	...	...	...	...	...	...
Sb ppb	...	0.067	0.121	...	0.014	...	...	...
Te ppb	1.0	2.1	1.0	...	0.28	...	...	...
I ppb	...	...	...	...	1.1	...	...	...
Cs ppb	725	26	180	...	14	...	...	...
Ba ppm	702	6.2	176	32	33	56.9	70	146
La ppm	73.8	0.15	8.5	1.47	1.51	5.22	6.76	13.0
Ce ppm	215	0.33	24.6	4.3	3.8	16.5	17.5	34.5
Nd ppm	115	0.19	14.2	<9	2.3	16.7	16	21.9
Sm ppm	31.4	0.053	4.0	0.87	0.61	6.69	5.51	6.56
Eu ppm	2.75	0.81	1.05	1.43	0.70	1.37	1.16	1.21
Gd ppm	42.9	0.056	6.6	...	0.73	10.4	7.2	8.59
Tb ppm	6.2	0.0085	0.89	0.22	...	...	1.27	...
Dy ppm	...	0.054	7.1	...	0.80	12.2	9.03	10.5
Ho ppm	...	...	...	...	...	...	1.72	...
Er ppm	...	...	...	...	...	7.4	6	6.51
Yb ppm	21.6	0.034	4.5	1.06	0.56	7.04	4.59	6
Lu ppm	3.07	0.0043	0.68	0.16	0.079	1.03	0.67	...
Hf ppm	29.9	0.014	3.4	0.55	0.41	6.33	3.9	9.8
Ta ppm	3.1	...	0.37	0.16	...	...	...	...
W ppb	...	26	...	...	...	...	...	...
Re ppb	0.0089	0.00084	0.008	0.013	0.0012	...	...	...
Os ppb	0.018	...	...	...	...	...	...	...
Ir ppb	0.0132	<0.01	0.071	0.148	0.0054	...	...	...
Au ppb	0.0033	0.117	0.045	0.079	0.0025	...	...	...
Tl ppb	3.2	0.09	0.61	...	0.012	...	...	...
Pb ppm	5.94	0.23	1.08	...	...	...	...	...
Bi ppb	290	0.097	0.13	...	0.037	...	...	...
Th ppm	10.3	0.004	1.84	0.19	0.19	0.34	0.84	2.1
U ppm	3.37	0.002	0.60	<0.6	0.054	0.13	0.22	0.59

Sources: Basaltic Volcanism Study Project: *Basaltic volcanism on the terrestrial planets*, 1981, Pergamon Press, pp. 1286. Heiken, G. H., Vaniman, D. T., & French, B. M. (eds.), 1991, *Lunar sourcebook*, Cambridge Univ. Press, Cambridge, pp. 736.

Table 6.25 Geologic Time Scale for the Moon

Time (Gyrs ago)	Period	Marking Event
4.54		accretion of the Moon
4.6–4.4		accretion & melting, magma ocean
4.4–4.2	Pre-Nectarian	cooling and primary differentiation, crust formation
4.2–3.9	Nectarian	intense bombardment, impact basins, highlands
3.9–3.2	Imbrian	volcanism, mare basalt formation
3.2–0.9	Eratosthenian	continuing bombardment,
0.9–present	Copernian	crater & regolith formation

Table 6.26 Locations, Sizes, and Ages of Lunar Maria

Mare	Sea	Latitude	Longitude	Basin Diameter (km)	Period of Volcanism (Gyrs ago)
Mare Crisium	Sea of crises	18°N	58°E	1060	3.2–3.3
Mare Fecunditatis	Sea of fertility	4°S	53°E	690	3.3–3.4
Mare Humorum	Sea of moisture	23°S	38°W	820	...
Mare Imbrium	Sea of rain	36°N	16°W	1500	3.1–3.4
Mare Nectaris	Sea of nectar	14°S	34°E	860	...
Mare Nubium	Sea of clouds	19°S	14°W	690	...
Mare Orientale	Eastern sea	19°S	95°W	930	...
Mare Serenitatis	Sea of serenity	30°N	17°E	880	3.7–3.9
Mare Smythii	Smyth's sea	3°S	80°E	840	...
Mare Tranquillitatis	Sea of tranquillity	7°N	30°E	775	3.6–3.9
Oceanus Procellarum	Ocean of storms	10°N	47°W	3200	3.0–3.3

Table 6.27 Diameters and Ages of Lunar Impact Basins

Basin	Basin Diameter (km)	Impact Date (Gyrs ago)
Mare Crisium/Mare Humorum Basin	1060/820	4.2/4.05
Mare Imbrium Basin	1500	3.95–3.84
Mare Nectaris Basin	860	4.25
Mare Orientale Basin	930	3.8–3.9
Mare Nectaris Basin	860	4.25
Mare Serenitatis Basin	880	3.98–4.28

Sources: Basaltic Volcanism Study Project: *Basaltic volcanism on the terrestrial planets*, 1981, Pergamon Press, pp. 1286. Chao. E. C. T., 1977, *Phil. Trans. Roy. Soc. Lond.* A285, 115–126. Ryder, G., & Spudis, P. D., 1979, in *Proc. conf. lunar highlands crust* (Papike, J. J., & Merrill, R. B., eds.), Pergamon Press, NY, pp. 353–375. Schaeffer, O. A., 1977, *Phil. Trans. Roy. Soc. Lond.* A285, 137–143. Tera, F., Papanastassiou, D. A., & Wasserburg, G. J., 1974, *Earth Planet. Sci. Lett.* 22, 1–21. Turner, G., 1977, *Phil. Trans. Roy. Soc. Lond.* A285, 97–103. Wilhelms, D. E., 1984, in *The geology of the terrestrial planets*, NASA-SP 469, pp. 107–205.

Table 6.28 Locations and Sizes of Lunar Impact Craters (diameter 150 km and larger)

Name	Latitude	Longitude	Diameter (km)
Hertzsprung	2.6°N	129.2°W	591
Apollo	36.1°S	151.8°W	537
Korolev	4.0°S	157.4°W	437
Birkhoff	58.7°N	146.1°W	345
Poincaré	56.7°S	163.6°E	319
Planck	57.9°S	136.8°E	314
Mendeleev	5.7°N	140.9°E	313
Schrödinger	75.0°S	132.4°E	312
Lorentz	32.6°N	95.3°W	312
Bailly	66.5°S	69.1°W	287
Milne	31.4°S	112.2°E	272
Gagarin	20.2°S	149.2°E	265
Deslandres	33.1°S	4.8°W	256
D'Alembert	50.8°N	163.9°E	248
Clavius	58.8°S	14.1°W	245
Leibnitz	38.3°S	179.2°E	245
Harkhebi	39.6°N	98.3°E	237
Van de Graaff	27.4°S	172.2°E	233
Lagrange	32.3°S	72.8°W	225
Pasteur	11.9°S	104.6°E	224
Galois	14.2°S	151.9°W	222
Campbell	45.3°N	151.4°E	219
Bel'kovich	61.1°N	90.2°E	214
Landau	41.6°N	118.1°W	214
Schwarzschild	70.1°N	121.2°E	212
Oppenheimer	35.2°S	166.3°W	208
Schickard	44.3°S	55.3°W	206
Janssen	45.4°S	40.3°E	199
Einstein	16.3°N	88.7°W	198
Mandel'shtam	5.4°N	162.4°E	197
Poczobutt	57.1°N	98.8°W	195
Maginus	50.5°S	6.3°W	194
Zeeman	75.2°S	133.6°W	190
Humboldt	27.0°S	80.9°E	189
Petavius	25.1°S	60.4°E	188

continued

Table 6.28 *(continued)*

Name	Latitude	Longitude	Diameter (km)
Vertregt	19.8°S	171.1°E	187
Tsiolkovskiy	21.2°S	128.9°E	185
Fabry	42.9°N	100.7°E	184
Fermi	19.3°S	122.6°E	183
Compton	55.3°N	103.8°E	182
Tsander	6.2°N	149.3°W	181
Schiller	51.9°S	39.0°W	180
Von Kármán	44.8°S	175.9°E	180
Mach	18.5°N	149.3°W	180
Chebyshev	33.7°S	133.1°W	178
Rozhdestvenskiy	85.2°N	155.4°W	177
Gauss	35.7°N	79.0°E	177
Grimaldi	5.5°S	68.3°W	172
Rowland	57.4°N	162.5°W	171
Sommerfeld	65.2°N	162.4°W	169
Hecataeus	21.8°S	79.4°E	167
Hausen	65.0°S	88.1°W	167
Heaviside	10.4°S	167.1°E	165
J. Herschel	62.0°N	42.0°W	165
Ptolemaeus	9.3°S	1.9°W	164
Struve	22.4°N	77.1°W	164
Joliot	25.8°N	93.1°E	164
Riemann	38.9°N	86.8°E	163
Roche	42.3°S	136.5°E	160
Keeler	10.2°S	161.9°E	160
Lippmann	56.0°S	114.9°W	160
Brouwer	36.2°S	126.0°W	158
Longomontanus	49.6°S	21.8°W	157
W. Bond	65.4°N	4.5°W	156
Ashbrook	81.4°S	112.5°W	156
Hilbert	17.9°S	108.2°E	151
Curie	22.9°S	91.0°E	151
Fersman	18.7°N	126.0°W	151
Hedin	2.0°N	76.5°W	150
Zwicky	15.4°S	168.1°E	150

Sources: IAU recommendations published by the U.S. Geological Survey, Branch of Astrogeology, Flagstaff, AZ.

MARS AND SATELLITES

7.1 Mars

Mars has an elliptical orbit (e~0.093), which leads to ~45% more insolation at perihelion than aphelion. As a consequence, the seasons are more accentuated than on Earth. Seasonal temperatures range from 148 K in polar winter to 290 K in southern summer. The seasonal temperature changes cause CO_2 condensation into and sublimation from the polar caps, producing an annual pressure change of 37% (2.4 mbar) relative to the global mean pressure of 6.36 mbar. Annual pressure variations due to CO_2 condensation and sublimation were monitored by the *Viking* landers. Another seasonal effect is the global dust storms that generally occur in late southern spring near perihelion. At present, Mars has an obliquity of 25.1°, but calculations predict changes with a period of 120,000 years. In the past, long term climate changes resulted as the obliquity varied from 15° to 30° and the insolation in the polar regions varied from 35% greater to 40% less than the present value.

Mars is about 50% farther away from the sun than Earth is, so the average solar flux at Mars is ~43% of that received by Earth. The equilibrium temperatures for Mars and Earth are 216 K and 253 K, respectively, whereas the global average surface temperatures are 220 K and 288 K, respectively, because greenhouse warming is smaller on Mars than on Earth.

Mars is 53% as big as the Earth, but is only 11% as massive because of its low density (3.934 g cm^{-3}). Bulk compositional models based on either the chemical or the oxygen isotopic composition of the SNC (shergottite, nakhlite, chassignite) meteorites, which are believed to come from Mars, predict that Mars has an Fe-FeS core ~21% of its total mass and ~49% of its total radius. The other 78% of the planet are the silicate mantle and crust, which is 20–150 km thick in the Tharsis bulge. The upper limit on the dipole magnetic field, if any, is <10^{-4} that of Earth. The lack of a magnetic field may indicate either complete solidification of the Martian core or very slow motions in any liquid inner core. As a result of the very weak (or nonexistent) magnetic field, the solar wind may be eroding the Martian atmosphere at a substantial rate. The high $^{15}N/^{14}N$ ratio in Martian N_2 also shows that significant amounts of nitrogen have been lost from Mars over geologic time. (The unusual, but not unique, ^{15}N

Mars

enrichment in some SNC meteorites, also supports the idea that Mars is the parent body for the SNC meteorites.)

The Martian atmosphere is dominantly CO_2 and, as on Venus, is continually converted to O_2 and CO by solar UV light. However, the observed abundances of CO_2 (95.3%), O_2 (0.13%), and CO (0.08%) cannot be explained simply by the direct recombination of CO and O atoms to CO_2, because this reaction is too slow to maintain the high CO_2 and low CO and O_2 abundances. Instead, OH radicals produced from atmospheric water vapor by UV photolysis or by reaction with electronically excited O atoms

$$H_2O + h\nu \rightarrow H + OH \tag{1}$$
$$O(^1D) + H_2O \rightarrow OH + OH \tag{2}$$

enter into catalytic cycles such as

$$OH + CO \rightarrow CO_2 + H \tag{3}$$
$$H + O_2 + M \rightarrow HO_2 + M \tag{4}$$
$$HO_2 + O \rightarrow OH + O_2 \tag{5}$$
$$CO + O \rightarrow CO_2 \qquad \text{Net reaction} \tag{6}$$

that recombine CO and O atoms to CO_2. Another catalytic cycle involves photolysis of hydrogen peroxide:

$$H_2O_2 + h\nu \rightarrow OH + OH \tag{7}$$

Reactions (3) and (4) repeated twice give the net reaction

$$2\,CO + O_2 \rightarrow 2\,CO_2 \tag{8}$$

which also regenerates CO_2. The HO_x radicals also regulate the O_3 level in the Martian atmosphere via catalytic cycles such as

$$H + O_3 \rightarrow OH + O_2 \tag{9}$$
$$O + OH \rightarrow O_2 + H \tag{10}$$
$$O + O_3 \rightarrow O_2 + O_2 \qquad \text{Net reaction} \tag{11}$$

analogous to HO_x catalytic O_3 destruction in the terrestrial stratosphere.

An interesting feature of the Martian atmosphere is that solar UV photolysis of atmospheric gases occurs all the way down to the surface, so H_2O_2, as well as other reactive species, are able to react with surface rocks and may be present in the soil. Peroxide chemistry is probably responsible for the results of the life detection experiments on the two *Viking* landers.

The Martian surface is divided into heavily cratered highlands in the southern hemisphere and lightly cratered plains in the northern hemisphere. The dividing line is roughly a great circle at 30° to the equator, with the highlands covering 60% of the surface. The *Viking 1* landing site in Chryse Planitia, the *Viking 2* landing site in Utopia Planitia, and the *Mars Pathfinder* landing site in Ares Valles are in the northern hemisphere plains regions. *Viking 1* and *2* are on opposite sides of Mars and *Mars Pathfinder* is ~850 km southeast of *Viking 1*.

The Tharsis bulge, or rise, dominates the western hemisphere and is the largest volcanic province on Mars. The Olympus Mons shield volcano in Tharsis is ~600 km wide at the base (which has 6 km high cliffs), reaches 27 km in altitude, and has a summit caldera 80 km across. It is the largest volcano in the solar system. Several other large shield volcanoes (Arsia Mons, Pavonis Mons, Ascraeus Mons) are also found in Tharsis; the Alba Patera volcano is north of Tharsis. Volcanic shields and domes are also prominent in the Elysium region in the eastern hemisphere and along the eastern side of the Hellas Basin in the southern hemisphere.

The Valles Marineris canyon system, discovered by and named for the *Mariner 9* orbiter, stretches eastward from Tharsis between 30°W and 110°W longitude. Valles Marineris is about 4000 km long, up to 600 km wide, and reaches maximum depths of 7–9 km. The formation of Valles Marineris may be connected to the formation of the Tharsis bulge.

Many channels and valleys on Mars look like they were formed by flowing water. At present, liquid water is thermodynamically unstable on the surface because of the low pressure. The channels are evidence that water existed on Mars in the past. The amount of water needed to form the observed features has also been estimated. Other geologic features may be evidence of present–day ground ice and ancient glaciation. Geologists estimate the amount of water on Mars to be equivalent to a global layer 0.5–2 km deep, almost all of it buried beneath the surface as ice, absorbed water, and chemically bound water in the Martian regolith. Water ice is also the dominant ice in the permanent north polar cap, but CO_2 ice dominates the permanent south polar cap.

The geologic evidence for flowing water on Mars in the past has led to theoretical models of how to warm Mars. Greenhouse warming of CO_2 is insufficient by itself, because CO_2 clouds condense in the middle atmosphere at the large CO_2 pressures needed. Instead, other greenhouse gases, such as SO_2, which is emitted from volcanoes on Earth and Io, are needed.

The question of a warmer, wetter early Mars is closely connected to the question of life on Mars. The popular perception of Mars as a haven for life dates back to the late 19th century. In 1878, Schiaparelli, the director of the observatory in Milan, Italy, reported seeing canali (channels) on

Mars during the favorable opposition in 1877. Schiaparelli's work stimulated Percival Lowell to found the Lowell Observatory in Flagstaff, Arizona, so he could observe Mars during the next favorable opposition in 1894. Lowell claimed that he observed extensive systems of canals, presumably constructed by an ancient civilization to bring water from the polar caps to their cities at lower latitudes. At about the same time, H. G. Wells wrote his novel *War of the Worlds,* in which the Martians invaded Earth. During the early 20th century, Edgar Rice Burroughs, the creator of Tarzan, also wrote a series of novels about ancient Martian civilizations. The ideas of life on Mars thus became firmly rooted in the public imagination.

However, spacecraft and Earth-based observations of Mars show that Lowell's canals were illusions that do not exist. Furthermore, the life detection experiments on *Viking 1* and *2* showed no evidence for organic carbon or for life in their Martian soil samples. In addition, the UV light with wavelengths down to 195 nm that reaches the Martian surface can effectively sterilize the upper layers of Martian soil. The short–wavelength UV light also forms peroxides, such as H_2O_2, which are strong oxidizing agents that destroy organic material.

Interest in life on Mars was recently revived when scientists reported evidence for "microfossils" in one of the SNC meteorites (denoted ALH84001) believed to come from Mars. However, the geochemical and isotopic measurements that suggest microbial activity can be interpreted in alternative ways, and the putative "microfossils" may be inorganic artifacts. More detailed studies of the ALH84001 meteorite are currently under way, and may resolve this issue.

Finally, Mars has two satellites, Phobos and Deimos. They are small and irregular in shape with low albedos, and they may be composed of carbonaceous material. See section 7.2 for background information and physical properties of Phobos and Deimos.

Sources and further reading: Scientific Results of the *Viking* Project, *J. Geophys. Res.* 82, 3959–4681 (30 Sept. 1977 issue). Reports of the *Mars Pathfinder* mission, *Science* 278, 1734–1776 (5 Dec. 1997 issue). Reports of the *Mars Global Surveyor* mission, *Science* 279, 1671–1698 (13 Mar. 1998 issue). Barth, C. A., 1985, in *The photochemistry of atmospheres* (Levine, J. S., ed.), Academic Press, Orlando, pp. 337–392. Kieffer, H. H., Jakosky, B. M., Snyder, C. W., & Matthews, M. S. (eds.) 1992, *Mars*, University of Arizona Press, Tucson, pp. 1536. Kliore, A., (ed.) 1982, *The Mars reference atmosphere*, Pergamon Press, Oxford, pp. 107. Lodders, K., & Fegley, B., Jr., 1997, *Icarus* 126, 373–394. Lowell, P., 1906, *Mars and its canals*, Macmillan, New York, pp. 393. Schiaparelli , G. V., 1877, *Memoria della el. di scienze fisiche, Atti della R. Academia dei Lincei*, Mem. 1, Ser. 3, Vol. 2, pp. 308–439. Sullivan, W., 1966, *We are not alone*, Signet, New York, pp. 319.

Table 7.1 Spacecraft Missions to Mars

Mission	Launch Date	Remarks
Mars 1960A (USSR)	10 Oct. 1960	attempted flyby; launch failure
Mars 1960B (USSR)	14 Oct. 1960	attempted flyby; launch failure
Mars 1962A (USSR)	24 Oct. 1962	attempted flyby; failed to leave Earth's orbit
Mars 1 (USSR)	1 Nov. 1962	flyby; lost contact in transit
Mars 1962B (USSR)	4 Nov. 1962	attempted lander; failed to leave Earth's orbit
Mariner 3 (USA)	5 Nov. 1964	attempted flyby
Mariner 4 (USA)	28 Nov. 1964	flyby, imaging
Zond 2 (USSR)	30 Nov. 1964	flyby, lost contact in transit
Mariner 6 (USA)	25 Feb. 1969	flyby, imaging, atmospheric measurements
Mariner 7 (USA)	27 Mar. 1969	flyby, imaging, atmospheric measurements
Mars 1969A (USSR)	27 Mar. 1969	attempted lander; launch failure
Mars 1969B (USSR)	2 Apr. 1969	attempted lander; launch failure
Mariner 8 (USA)	8 May 1971	attempted flyby; launch failure
Cosmos 419 (USSR)	10 May 1971	attempted orbiter/lander
Mars 2 (USSR)	19 May 1971	orbiter; lander crashed on surface
Mars 3 (USSR)	28 May 1971	orbiter; lander lost contact
Mariner 9 (USA)	30 May 1971	orbiter; imaging of Mars, Phobos, & Deimos
Mars 4 (USSR)	21 July 1973	flyby imaging; attempted orbiter
Mars 5 (USSR)	25 July 1973	orbiter; imaging
Mars 6 (USSR)	5 Aug. 1973	orbiter; lander lost contact, some data
Mars 7 (USSR)	9 Aug. 1973	orbiter; attempted lander
Viking 1 (USA)	20 Aug. 1975	orbiter and lander in Chryse Planitia
Viking 2 (USA)	9 Sept. 1975	orbiter and lander in Utopia Planitia
Phobos 1 (USSR)	7 July 1988	attempted Mars orbiter & Phobos landers
Phobos 2 (USSR)	12 July 1988	Mars orbiter, some imaging before failure; Phobos lander failed
Mars Observer (USA)	25 Sept. 1992	orbiter; contact lost during Mars' orbit entry
Mars Global Surveyor (USA)	7 Nov. 1996	orbiter; operating
Mars 96 (Russia)	16 Nov. 1996	attempted orbiter and landers; launch failure
Mars Pathfinder (USA)	4 Dec. 1996	lander and rover, Ares Vallis
Planet-B, Nozomi (Japan)	4 July 1998	orbiter, atmospheric probe; arrives in 1999

Sources: National Space Science Data Center, Greenbelt, MD. Kieffer, H. H., Jakosky, B. M., Snyder, C. W., & Matthews, M. S. (eds.), *Mars*, Univ. of Arizona Press, Tucson, pp. 1498. Lewis, J. S., *Physics and chemistry of the solar system*, Academic Press, San Diego, pp. 556.

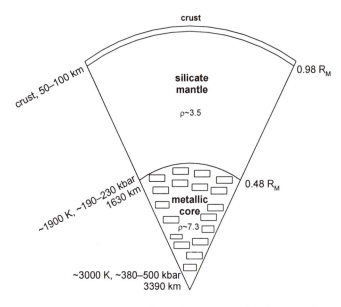

Figure 7.1 Interior of Mars based on geochemical models

Table 7.2 Some Physical Properties of Mars

Property	Value	Property	Value
Mean radius (km) *	3389.92±0.04	Sidereal revolution period ($\oplus$ days)	686.980
Equatorial radius (km) *	3396.9±0.4	Mean synodic period ($\oplus$ days)	779.94
Polar radius (km)	3374.9	Eccentricity of orbit	0.0933
Surface area (10^6 km^2)	144	Inclination of orbit to ecliptic	1.85°
S. polar cap (10^6 km^2) †	0.088	Mean orbital velocity (km s^{-1})	24.13
N. polar cap (10^6 km^2) †	0.837	Inclination of equator to orbit	25.189°
Volume (km^3)	1.6318×10^{11}	Sidereal rotation period ($\oplus$ hours)	24.6229
Oblateness ($R_{eq.} - R_{pol.})/R_{eq.}$	6.476×10^{-3}	Solar constant 1.52 AU (W m^{-2})	588.98
Mass (kg)	6.4185×10^{23}	Mean surface temperature (K)	214
Mean density (g cm^{-3})	3.9335±0.0004	Surface temperature range (K)	~140–300
GM (m^{3}s^{-2})	4.2828×10^{13}	Mean surface pressure (mbar)	6.36
Equatorial gravity (m s^{-2})	3.711	Annual pressure range (mbar)	2.4

continued

Table 7.2 *(continued)*

Property	Value	Property	Value
Polar gravity (m s^{-2})	3.758	Mean atmospheric mass (kg)	2.46×10^{16}
Eq. escape velocity (km s^{-1})	5.027	Mean scale height at surface (km)	11.0
C/MR^2	0.3662 ± 0.0017	Mean atm. lapse rate (K km^{-1})	~2.5
J$_2 \times 10^6$	1960.454	N. seasonal polar cap (kg) ‡	3.5×10^{15}
Mag. dipole moment (Tm^{-3})	$\leq 8 \times 10^{11}$	S. seasonal polar cap (kg) ‡	8.1×10^{15}

*The mean radius is for a sphere of equal volume. The equatorial radius is (1+b)/2 where the three ellipsoidal radii are a=3394.5±0.3 km, b=3399.2±0.3 km, and c=3376.1±0.4 km. The polar radius is calculated from the oblateness.

† Areas of perennial polar caps

‡ Based on seasonal pressure variations

Sources: Barth, C. A., 1985, in *The photochemistry of atmospheres* (Levine, J. S., ed.), Academic Press, New York, pp. 337–392. Christensen, E. J., & Balmino, G., 1979, *J. Geophys. Res.* 84, 7943–7953. Folkner, W. M., Yoder, C. F., Yuan, D. N., Standish, E. M., & Oreston, R. A., 1997, *Science* 278, 1749–1752. Kieffer, H. H., Jakosky, B. M., Snyder, C. W., Matthews, M. S. (eds.), 1992, *Mars*, Univ. of Arizona Press, Tucson, pp. 1498. Jordan, J. F., & Lorell, J., 1975, *Icarus* 25, 146–165.

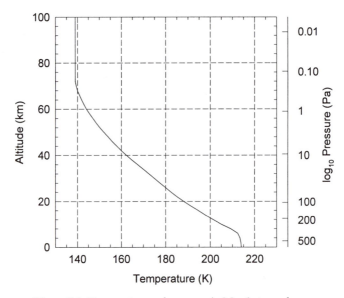

Figure 7.2 Temperature and pressure in Mars' atmosphere

Table 7.3 Temperature, Pressure, and Density in Mars' Atmosphere

Altitude (km)	Temperature (K)	Pressure (Pa)	Density (kg m^{-3})
0	214.0	636	1.55(–2)
10	205.0	254	6.47(–3)
20	188.2	0.947	2.63(–5)
30	175.0	0.328	9.78(–6)
40	162.4	0.106	3.41(–6)
50	152.2	3.15(–2)	1.08(–6)
60	144.2	8.79(–3)	3.18(–7)
70	139.5	2.33(–3)	8.72(–8)
80	139.0	6.09(–4)	2.29(–8)
90	139.0	1.60(–4)	6.01(–9)
100	139.0	4.24(–5)	1.59(–9)

Exponents in parentheses.

Source: Barth, C. A., 1985, in *The photochemistry of atmospheres* (Levine, J. S., ed.), Academic Press, New York, pp. 337–392.

Table 7.4 Chemical Composition of the Atmosphere of Mars

Gas	Abundance[a]	Gas Source(s)	Gas Sink(s)	Sources
CO_2	95.32%	outgassing & evaporation	condensation	Kui52, OBR77
N_2	2.7%	outgassing	escape as N	OBR77
Ar	1.6%	outgassing (^{40}K), primordial	—	OBR77
O_2	0.13%	CO_2 photolysis	photoreduction	Bar72, CT72, TL83
CO	0.08%	CO_2 photolysis	photooxidation	CMB90, KWW77, KCC69
H_2O*	0.03%	evaporation & desorption	condensation & adsorption	JH92, KMS64
NO	~100 ppm (at 120 km)	photochemistry (N_2, CO_2)	photochemistry	NM77
Ne	2.5 ppm	outgassing, primordial	—	OBR77
HDO	0.85±0.02 ppm	evaporation & desorption	condensation & adsorption	BML89, OMD88
Kr	0.3 ppm	outgassing, primordial	—	OBR77
Xe	0.08 ppm	outgassing, primordial	—	OBR77
O_3*	~(0.04–0.2) ppm	photochemistry (CO_2)	photochemistry	Bar85, BH71

[a] The mixing ratios, but not the column densities, of noncondensible gases are seasonally variable as a result of the annual condensation and sublimation of CO_2.

* Spatially and temporally variable

Sources: [Bar72] Barker, E. S., 1972, *Nature* 238, 447–448. [Bar85] Barth, C. A., 1985, in *The photochemistry of atmospheres,* (Levine, J. S., ed.), Academic Press, New York, pp. 337–392. [BH71] Barth, C. A., & Hord, C. W., 1971, *Science* 173, 197–201. [BML89] Bjoraker, G. L., Mumma, M. J., & Larson, H. P., 1989, in *Abstracts of the fourth international conference on Mars*, pp. 69–70. [BOR76] Biemann, K., Owen, T., Rushneck, D. R., Lafleur, A. L., & Howarth, D. W., 1976, *Science* 194, 76–78. [CMB90] Clancy, R. T., Muhleman, D. O., & Berge, G. L., 1990, *J. Geophys. Res.* 95, 14543–14554. [CT72] Carleton, N. P., & Traub, W. A., 1972, *Science* 177, 988–992. [JH92] Jakosky, B. M., & Haberle, R. M., 1992, in *Mars,* (Kieffer, H., Jakosky, B., Snyder, C., & Matthews, M. S., eds.) Univ. of Arizona Press, Tucson, pp. 969–1016. [KCC69] Kaplan, L. D., Connes, J., & Connes, P., 1969, *ApJ.* 157, L187–L192. [KMS64] Kaplan, L. D., Münch, G., & Spinrad, H., 1964, *ApJ.* 139, 1–15. [Kui52] Kuiper, G. P., 1952, in *Atmospheres of the earth and planets,* (Kuiper, G. P., ed.), Univ. of Chicago Press, Chicago, pp. 306–405. [KWW77] Kakar, R. K., Walters, J. W., & Wilson, W. J., 1977, *Science* 196, 1090–1091. [NM77] Nier, A. O., & McElroy, M. B., 1977, *J. Geophys. Res.* 82, 4341–4349. [OBR77] Owen, T., Biemann, K.,

Rushneck, D. R., Biller, J. E., Howarth, D. W., & Lafleur, A. L., 1977, *J. Geophys. Res.* 82, 4635–4639. **[OMD88]** Owen, T., Maillard, J. P., DeBergh, C., & Lutz, B. L., 1988, *Science* 240, 1767–1770. **[TL83]** Trauger, J. T., & Lunine, J. I., 1983, *Icarus* 55, 272–281.

Table 7.5 Isotopic Composition of the Atmosphere of Mars

Isotopic Ratio	Observed Value [a]	Notes	Sources
D/H	$(9\pm4)\times10^{-4}$	IR spectroscopy	OMD88
	$(7.8\pm0.3)\times10^{-4}$	IR spectroscopy	BML89
$^{12}C/^{13}C$	90 ± 5	*Viking* MS	NM77
$^{14}N/^{15}N$	170 ± 15	*Viking* MS	NM77
$^{16}O/^{17}O$	2655 ± 25	IR spectroscopy	BML89
$^{16}O/^{18}O$	490 ± 25	*Viking* MS	NM77
	545 ± 20	IR spectroscopy	BML89
$^{36}Ar/^{38}Ar$	5.5 ± 1.5	*Viking* MS	BOR76
$^{40}Ar/^{36}Ar$	3000 ± 500	*Viking* MS	OBR77
$^{129}Xe/^{132}Xe$	2.5^{+2}_{-1}	*Viking* MS	OBR77

[a] Isotopic compositions inferred from SNC meteorites are not given here; only direct observations of isotopic compositions of the Martian atmosphere are listed.
MS: mass spectrometer

For sources, see Table 7.4.

Table 7.6 Elemental Analyses of the Martian Surface

mass%	Mars 5 (1973)	Viking 1 (1975) Chryse	Viking 2 (1975) Utopia	Phobos 2 (1988)	Pathfinder (1997) Rocks	Soil
Na_2O	...	...	...	...	2.5±0.7	2.3±1.5
MgO	...	6	(6)*	10±5	4.3±1.3	7.7±1.2
Al_2O_3	9.5±4	7.3	(7)*	9.5±4	10.1±1.1	8.5±0.9
SiO_2	30±6	44	43	41±9	56.6±3.4	49.5±2.6
SO_3	...	6.7	7.9	...	2.4±1.2	5.5±1.3
Cl	...	0.8	0.4	...	0.5±0.2	0.6±0.2
K_2O	0.4±0.1	<0.5	<0.5	0.4±0.1	0.6±0.1	0.3±0.2
CaO	...	5.7	5.7	8.4±4	6.6±1.2	6.5±1.1
TiO_2	...	0.62	0.54	1.7±0.8	0.9±0.2	1.2±0.2
Fe_2O_3	20±6	17.5	17.3	12.9±4	14.9±1.5	17.8±1.9
Th ppm	2.1±0.5	...	...	1.9±0.6	...	...
U ppm	0.6±0.1	...	...	0.5±0.1	...	...
Br ppm	...	~80	present	...	...	...
Rb ppm	...	≤30	≤30	...	...	...
Sr ppm	...	60±30	100±40	...	...	...
other	...	2	2	...	...	...
total	60	91	90	84	99.4	99.9

* Mg and Al for Utopia are taken as equal to Chryse data.

Sources: Clark, B. C., Baird, A. K., Weldon, R. J., Tsusaki, D. M., Schnabel, L., & Candelaria, M. P., 1982, J. Geophys. Res. 87, 10059–10067. Clark, B. C., & van Hart, D. C., 1981, Icarus 45, 370–378. Rieder, R, Economou, T., Wänke, H., Turkevich, A., Crisp, J., Brückner, J., Dreibus, G., & McSween, H. Y., 1997, Science 278, 1771–1776. Surkov, Yu. A., Moskaleva, L. P., Zolotov, M. Yu., Kharnykova, V. P. Manvelyan, O. S., Smirnov, G. G., & Manvelyan, O. S., 1989, Nature 241, 595–598. Surkov, Yu. A., Moskaleva, L. P., Zolotov, M. Yu., Kharynkova, V. P., Manvelyan, O. S., Smirnov, G. G., & Golovin, A. V., 1994, Geochem. Intern. 31, 50–58. Toulmin, P., Baird, A. K., Clark, B. C., Keil., K., Rose, H. J., Christian, R. P., Evans, P. H., & Kelliher, W. C., 1977, J. Geophys. Res. 82, 4625–4634.

Table 7.7 Model Mantle and Core Compositions of Mars

Compound	[LF97]	[WD94]	[Goe83]	[Wei81]	[MA79]	[MG78]	[An72]
Silicate portion (mass%)							
Na_2O	0.98	0.5	1.4	1.4	0.10	0.50	0.8
MgO	29.71	30.1	32.08	31.2	29.78	32.68	27.2
Al_2O_3	2.89	2.9	3.26	3.2	6.39	3.09	3.1
SiO_2	45.39	44.4	45.07	43.9	41.6	39.41	39.8
P_2O_5	0.17	0.17	...	...	...	...	0.3
K_2O	0.11	0.04	0.12	0.14	0.01	0.11	0.07
CaO	2.35	2.4	3.03	3.0	5.16	2.69	2.5
TiO_2	0.14	0.13	...	0.16	0.33	0.62	0.1
Cr_2O_3	0.68	0.8	...	...	0.65	...	0.6
MnO	0.37	0.5	...	...	0.15	...	0.2
FeO	17.22	17.9	15.07	16.7	15.85	20.11	24.2
Fe_2O_3	0	0	0	0	0	0.79	0
H_2O	...	0.004	...	0.44	0.001	...	0.9
High-pressure norm (%)							
pyroxene	42.63	37.75	34.28	30.5	21.55	12.06	18.8
olivine	50.91	51.87	62.32	65.8	48.98	73.21	71
garnet	4.82	8.61	3.43	3.1	25.86	11.41	7.2
oxides	0	0	0	0	2.05	2.14	0
other *	1.64	1.37	...	0.3	1.58	1.2	1.8
ρ (STP) (g cm^{-3}) †	3.50	3.52	3.46	3.48	3.58	3.57	3.61
Core (mass%)							
Fe	61.48	53.1	45.60	1.5	82.02	74.77	39.0
Co	0.38	...	...	...	0.37	...	...
Ni	7.67	8.0	6.20	5.8	7.99	...	9.3
Fe_3P	1.55	0	0	0	0	0	0
FeS	28.97	38.9	48.20	92.7	9.62	25.23	51.7
ρ (STP) (g cm^{-3}) †	7.27	7.04	6.81	5.92	7.85	7.34	6.74
Mantle & core							
silicates mass%	79.37	78.3	...	74.3	81.0	85.0	87.8
core mass%	20.63	21.7	...	25.7	19.0	15.0	11.2
bulk ρ(STP) (g cm^{-3})†	3.92	3.95	...	3.89	3.99	3.87	3.85
r_{core}/R_{total}	0.481	0.496	...	0.552	0.458	0.430	0.425
$C/(MR^2)$	0.367	0.367	...	0.371	0.368	0.375	0.380

* other normative minerals include ilmenite, chromite, whitlockite.

† densities recalculated by [LF97]

Sources: [An72] Anderson, D. L., 1972, *J. Geophys. Res.* 77, 789–795. [Goe83] Goettel, K., 1983, *Annu. Rep. Dir., Geophys. Lab. Washington 1982–1983*, pp. 363–366. [LF97] Lodders, K., & Fegley, B., 1997, *Icarus* 126, 373–394. [MA79] Morgan, J. W., & Anders, E., 1979, *Geochim. Cosmochim. Acta* 43, 1601–1610. [MG78] McGetchin, T. R., & Smyth, J. R., 1978, *Icarus* 34, 512–536. [WD94] Wänke, H., & Dreibus, G., 1994, *Phil. Trans. R. Soc. Lond.* A349, 285–293 and references therein. [Wei81] model after Weidenschilling, S. J., in Basaltic Volcanism Study Project: *Basaltic volcanism on the terrestrial planets*, 1981, Pergamon Press, pp. 1286.

Table 7.8 Model Elemental Abundances in Mars (mantle, crust, and core)

Element	Unit	[LF97]	[MA79]	Element	Unit	[LF97]	[MA79]
Li	ppm	1.69	1.94	Pd	ppm	0.86	0.74
B	ppm	0.5	0.00226	Ag	ppb	60	37.9
C	ppm	<2960	16.3	Cd	ppb	80	0.62
N	ppm	<180	0.148	In	ppb	16	0.077
O	mass%	33.75	34.11	Sn	ppb	915	258
F	ppm	33	19.1	Sb	ppb	76	23.4
Na	ppm	5770	574	Te	ppb	430	13.8
Mg	mass%	14.16	14.55	I	ppb	100	0.48
Al	mass%	1.21	2.73	Cs	ppb	120	21
Si	mass%	16.83	15.74	Ba	ppm	4.3	8.0
P	ppm	1100	1608	La	ppm	0.320	0.75
S	ppm	22000	6660	Ce	ppm	0.890	1.99
Cl	ppm	120	0.71	Pr	ppm	0.130	0.25
K	ppm	730	62	Nd	ppm	0.670	1.38
Ca	mass%	1.33	2.98	Sm	ppm	0.200	0.41
Sc	ppm	8.4	19	Eu	ppm	0.078	0.157
Ti	ppm	650	1580	Gd	ppm	0.310	0.56
V	ppm	77	162	Tb	ppm	0.055	0.105
Cr	ppm	3680	3620	Dy	ppm	0.360	0.71
Mn	ppm	2250	940	Ho	ppm	0.078	0.156
Fe	mass%	27.24	26.72	Er	ppm	0.240	0.455
Co	ppm	795	702	Tm	ppm	0.040	0.068
Ni	mass%	1.58	1.52	Yb	ppm	0.220	0.451
Cu	ppm	87	210	Lu	ppm	0.035	0.076
Zn	ppm	66	33.9	Hf	ppm	0.180	0.451
Ga	ppm	6.3	1.97	Ta	ppm	0.023	0.0455
Ge	ppm	15	5	W	ppm	0.16	0.356
As	ppm	2.0	2.68	Re	ppm	0.07	0.118
Se	ppm	8.5	2.22	Os	ppm	0.82	1.71
Br	ppb	740	3.83	Ir	ppm	0.76	1.66
Rb	ppm	2.7	0.209	Pt	ppm	1.40	3.28
Sr	ppm	10.7	28.4	Au	ppm	0.21	0.214
Y	ppm	2.2	5.19	Tl	ppb	14	0.14
Zr	ppm	6.5	30.8	Pb	ppb	460	0.056
Mo	ppm	1.7	1.57	Bi	ppb	25	0.106
Ru	ppm	1.1	4.65	Th	ppb	44	101
Rh	ppm	0.22	0.50	U	ppb	12.6	28

Sources: **[LF97]** Lodders, K., & Fegley, B., 1997, *Icarus* 126, 373–394. **[MA79]** Morgan, J. W., & Anders, E., 1979, *Geochim. Cosmochim. Acta* 43, 1601–1610.

7.2 Phobos and Deimos

The Martian moons Phobos and Deimos were discovered by Asaph Hall at the U.S. Naval Observatory in 1877. Phobos, the larger moon, has a short orbital period of 0.32 days. Phobos may impact Mars in less than 100 Myrs because its orbital semimajor axis is decreasing over time as a result of tidal torque. In contrast, the semimajor axis of Deimos, which is farther away from Mars than Phobos, is increasing, so Deimos is moving away from Mars over time.

Both moons are irregularly shaped and have smooth surfaces with filled-in craters. The surface of Phobos is heavily cratered and shows linear grooves. The largest crater, Stickney, on Phobos is about 10 km in diameter, which is comparable to Phobos' mean radius of 11.1 km.

Phobos has a visual albedo of 0.06. Deimos shows higher reflectivity (~30% brighter) in some areas, but the overall albedo is only 0.07.

Both moons have low densities, suggesting similarities in composition. They may consist of porous carbonaceous chondrite or interplanetary dust material. Hydrous material is apparently absent on their surfaces, but hydrated interiors cannot be ruled out.

Phobos and Mars (but not Deimos) were subjects of the *Phobos I & II* missions in 1989; unfortunately, only 37 images of Phobos and limited data from other instruments were obtained before the missions failed.

Table 7.9 Some Physical Properties of Mars' Moons Phobos and Deimos

Property	Phobos	Deimos
Semimajor axis (km)	9378	23460
Eccentricity of orbit	0.015	0.0005
Orbital inclination	1.02°	1.82°
Size (km)	13.5×10.8×9.4	7.5×6.1×5.5
Mass (kg)	$9.6×10^{15}$	$1.9×10^{15}$
Observed density (g cm^{-3})	1.90±0.1	1.76
Geometric albedo	0.06	0.07

Sources and further reading: Reports from Phobos missions in *Nature* 341 (no. 6243, Oct. 1989 issue) and *Planet. Space Sci.* 39 (no. 1/2, Jan./Feb. 1991 issue). Burns, J. A., & Matthews, M. S. (eds.), 1986, *Satellites*, Univ. Arizona Press, Tucson, pp. 1021.

Jupiter (vertical tab)

8

JUPITER, RINGS, AND SATELLITES

Jupiter has a mass of ~$10^{-3}M_\odot$ and is the largest planet in the solar system. It has been explored by five spacecraft: *Pioneer 10* (4 Dec. 1973 flyby), *Pioneer 11* (3 Dec. 1974 flyby), *Voyager 1* (5 Mar. 1979 flyby), *Voyager 2* (9 July 1979 flyby), and *Galileo* (7 Dec. 1995 entry probe; and an ongoing multiyear orbiter mission).

Jupiter's low density and large mass, as well as its atmospheric composition, indicate that it is dominantly H_2 and He, with composition close to the sun's; interior structure models reinforce this conclusion. The observed D/H ratio, which would have been increased by accretion of large amounts of ices (e.g., H_2O, CH_4, NH_3) during Jupiter's formation, also indicates that Jupiter's composition is close to that of the sun and is only slightly enriched in elements heavier than He.

Jupiter emits ~1.67 times as much energy as it absorbs from the sun. Theoretical models suggest that this large heat flux must be transported by convective mixing. Remote sensing by the *Voyager* spacecraft and in situ measurements by the *Galileo* entry probe show that the troposphere is convective down to at least the 20 bar level. Observations of the abundances of CO, GeH_4 (germane), PH_3 (phosphine), and AsH_3 (arsine), at orders of magnitude greater than expected at the level of observation, indicate that convective mixing extends downward to at least kilobar levels where larger abundances of these gases are produced.

At great depth, the molecular H_2-He is transformed to metallic H-He, which is electrically conductive, and supports the dynamo currents that generate Jupiter's intense magnetic field. The rotation rate of the magnetic field is $9^h55^m29^s$. The magnetic dipole is tilted 9.6° from the spin axis and is offset ~0.1 R_J from Jupiter's center. The Jovian magnetosphere extends out to ~10 R_J between the orbits of Europa (~9.4 R_J) and Ganymede (~15.0 R_J). The radio emission from charged particles in the Jovian magnetosphere was the first evidence for the Jovian magnetic field. Io's interactions with the magnetosphere modulate some of the radio emissions and also lead to the formation of Io's plasma torus.

Jupiter's atmosphere is dominated by alternating dark belts, bright zones, the Great Red Spot (GRS), and vigorous storm-like features. The GRS has apparently existed on Jupiter since at least 1664 when it (or a similar feature) was observed by Robert Hooke. Detailed records of the Great Red Spot's appearance and activity extend over 150 years. The GRS is believed to be a long-lived, intense storm system. The red color is possibly due to formation of red P by photolysis of PH_3 gas.

The brilliant colors in Jupiter's atmosphere are probably due to inorganic sulfur and/or phosphorus compounds instead of organic matter. Theoretical models predict a trimodal system of condensation clouds: the visible clouds of solid NH_3, a lower layer of ammonium hydrosulfide (NH_4HS) clouds, and an even lower layer of aqueous water clouds (~6 bar level). This trimodal structure was not seen by the *Galileo* entry probe, which apparently entered a relatively clear, low-humidity region.

During the period 16 July through 24 July 1994, Jupiter was hit by over 20 fragments of comet Shoemaker-Levy 9 (SL9), which had been broken up during a previous close approach to Jupiter in July 1992. The impacting fragments ranged in size from <1 km to a few km and impacted Jupiter at ~60 km s^{-1}; the larger fragments produced large, hot impact plumes and had energies equivalent to millions of megatons of TNT. The impacts led to the formation of S_2, OCS, CS_2, CO, HCN, and other gases, lifted tropospheric gas into the upper atmosphere, and left brown-colored scars in Jupiter's clouds. Only small amounts of water vapor were detected in the impacts, suggesting that SL9 was water-poor and perhaps not a typical comet.

Jupiter's ring system was discovered by *Voyager 1* in March 1979. The Jovian ring system consists of the halo, the main ring, and the gossamer ring. The halo extends from ~1.3 to 1.7 R_J, at the inner edge of the main ring. The halo reaches a thickness of ~20,000 km, comparable to Saturn's E ring. The main ring is ~6400 km wide and extends out to ~1.8 R_J. Although it is the most prominent part of the Jovian ring system, the optical depth of the main ring is ~10^{-6} to ~10^{-5}, which is much fainter than Saturn's main rings. The gossamer ring is ~30 times fainter than the main ring and extends from ~1.8 to 2.9 R_J. Metis and Adrastea are in the main ring and Amalthea is in the gossamer ring. The ring particles are submicron- to meter-size and are made of silicates and carbonaceous material.

Jupiter's 16 satellites fall into four different groups: (1) four small, inner, irregularly shaped satellites (Metis, Adrastea, Amalthea, and Thebe) at 1.8–3.1 R_J, (2) the four Galilean satellites (Io, Europa, Ganymede, and Callisto) at 5.9–26.4 R_J, (3) four prograde outer satellites (Leda, Himalia, Lysithea, and Elara) at 155–164 R_J, and (4) four retrograde outer satellites

(Ananke, Carme, Pasiphae, and Sinope) at 297–332 R_J. The Galilean satellites, Amalthea, and possibly the other three inner satellites rotate synchronously. The Galilean satellites and the inner satellites have nearly circular (e <0.01) orbits in Jupiter's equatorial plane (i <1°). The prograde outer satellites have high eccentricity (e ~ 0.11–0.21), high inclination (i ~ 28°) orbits; and the retrograde outer satellites have high eccentricity (e ~ 0.2–0.4), very high inclination (i ~147–163°) orbits. The orbital periods of Io, Europa, and Ganymede are 1:2:4 because of the Laplace resonance between their orbits. The orbits of the four prograde outer satellites cross and are unstable on timescales <1 Ga. All the outer satellites are probably captured bodies. In contrast, the eight inner satellites probably formed in a subnebula around Jupiter; the radial density decrease of the Galilean satellites (Io ~ 3.5 g cm^{-3}, Europa ~ 3.0 g cm^{-3}, Ganymede ~ 1.9 g cm^{-3}, and Callisto ~ 1.8 g cm^{-3}) supports this concept.

Water ice is present on the surfaces of Europa, Ganymede, and Callisto. Recently, oxygen and ozone have been discovered on Europa and Ganymede. The O_2 and O_3 are trapped in water ice and probably formed by photochemical and charged particle reactions. Ozone has also been detected on Saturn's satellites Rhea and Dione, and is presumably produced by similar processes.

Sources and further reading: Papers on the *Galileo* spacecraft instruments in *Space Sci. Rev.* 60 (no. 1–4, 1992). Reports of the *Galileo* probe results in *Science* 272, 837–860 (no. 5263, 10 May 1996 issue). Belton, M. J. S., West, R. A., Rahe, J., & Pereyda, M. (eds.), 1989, *Time variable phenomena in the Jovian system*, NASA, Washington, D.C., pp. 409. Burns, J. A., Showalter, M. R., & Morfill, G. E., 1984, in *Planetary rings* (Greenberg, R., & Brahic, A., eds.), University of Arizona Press, Tucson, pp. 200–272. Fegley, B., Jr., & Lodders, K., 1994, *Icarus* 110, 117–154. Gehrels, T. (ed.), 1976, *Jupiter*, Univ. of Arizona Press, Tucson, pp. 760. Hall, D. T., Strobel, D. F., Feldman, P. D., McGrath, M. A., & Weaver, H. A., 1995, *Nature* 373, 677–679. Noll, K. S., Feldman, P., & Weaver, H. A. (eds.), 1996, *The collision of comet Shoemaker-Levy 9 and Jupiter*, Cambridge University Press, Cambridge, pp. 650. Noll, K. S., Johnson, R. E., Lane, A. L., Domingue, D. L., & Weaver, H. A., 1996, *Science* 273, 341–343. Peek, B. M., 1981, *The planet Jupiter*, 2nd. ed., Faber & Faber, London, pp. 450. Showalter, M. R., Burns, J. A., Cuzzi, J. N., & Pollack, J. B., 1987, *Icarus* 69, 458–498. Showalter, M. R., Burns, J. A., Cuzzi, J. N., & Pollack, J. B., 1985, *Nature* 316, 526–528. Spencer, J. R., Calvin, W. M., & Person, M. J., 1995, *J. Geophys. Res.* 100, 19049–19056.

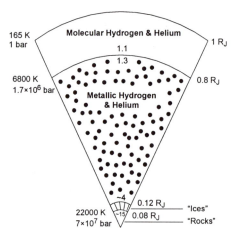

Figure 8.1 Internal structure of Jupiter from adiabatic models.
The numbers at the phase boundaries are densities in g cm^{-3}.

Table 8.1 Some Physical Properties of Jupiter

Property	Value	Property	Value
Equatorial radius at 1 bar (km)	71492±4	Rotational period (hours)	9.925
Polar radius at 1 bar (km)	66854±10	GM (10^{15} m^{3}s^{-2})	126.6865
Oblateness (R_{eq}–$R_{pol.}$)/$R_{eq.}$	0.06487±0.0002	Equatorial gravity (m s^{-2})	23.12±0.01
Mass (10^{24} kg)	1898.6	Polar gravity (m s^{-2})	27.01±0.01
Mass (Earth masses)	317.83	J_2 10^5	1469.7±1
Rocky core mass (% of total)	2.61	J_4 10^6	–584±5
Mean Density (g cm^{-3})	1.326	J_6 10^4	0.31±0.20
Temperature at 1 bar (K)	165±5	C/MR^2	0.254
Effective temperature (K)	124.4±0.3	Escape velocity (km s^{-1})	60.236
Internal energy flux (W m^{-2})	5.44±0.43	Mean molec. wt. (g mol^{-1})	2.306
Intern. power/mass (10^{-11}J s^{-1}kg^{-1})	17.6±1.4	Scale height (km)	24.35
Therm. emis./absorbed solar energy	1.67±0.09	Magnetic axis offset	9.6°
Magnetic dipole moment (Tesla R$_{Jup}$^{3})	4.3×10^{-4}	Solar wind–magnetopause boundary (R$_{Jup}$)	70

Sources: Guillot, T., Chabrier, G., Morel, P. & Gautier, D., 1994, *Icarus* 112, 354–367. Hanel, R. A., Conrath, B. J., Hearth, L. W., Kunde, V. G., & Pirragalia, J. A., 1981, *J. Geophys. Res.* 86, 8705–8712. Hubbard, W. B., Podolak, J. B., & Stevenson, D. J., 1996, in *Neptune and Triton* (Cruikshank, D. P., ed.), Univ. of Arizona Press, Tucson, pp. 109–138. Lindal, G. F., 1992, *Astron. J.* 103, 967–982. Stevenson, D. J., 1982, *Annu. Rev. Earth Planet. Sci.* 10, 257–295.

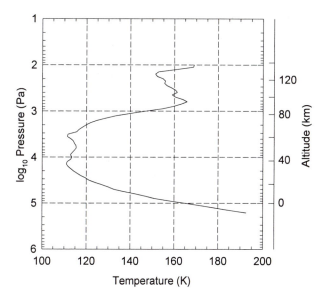

Figure 8.2 Temperature and pressure in Jupiter's atmosphere

Table 8.2 Temperature, Pressure, and Density in Jupiter's Atmosphere

Altitude (km)	Temperature (K)	Pressure (Pa)	Density (kg m⁻³)
0	165.0	100,000	0.173
20	123.5	38,522	8.88(−2)
40	112.1	11,972	3.04(−2)
60	111.7	3,630	9.25(−3)
80	137.4	118	2.45(−4)
100	160.4	49.6	8.80(−5)
120	155.4	21.0	3.85(−5)
140	169.4	8.94	1.50(−5)

Exponents in parentheses.

Source: Lindal, G. F., 1992, *Astron. J.* 103, 967–982.

Table 8.3 Chemical Composition of the Atmosphere of Jupiter

Gas	Abundance	Comments	Sources
H_2	86.2±2.6%	*Galileo* probe mass spectrometer (GPMS) & He detector (HAD)	VZH96, NAC98
^4He	13.6±2.6%	GPMS and HAD	VZH96, NAC98
CH_4	$(1.81\pm0.34)\times10^{-3}$	$CH_4/H_2 = (2.1\pm0.4)\times10^{-3}$, ~3 × solar C/H_2 ratio, GPMS value	NAC98
NH_3	$(2.6\pm0.4)\times10^{-4}$	NH_3/H_2 ~1.5 × solar N/H_2 ratio, no value yet from GPMS	BLK86
H_2O	≤(570±260) ppm	$H_2O/H_2 \leq(6\pm3)\times10^{-4}$, ≤0.35 × solar O/H_2 ratio, GPMS, 19 bar level	NAC98
HD	45±12 ppm	$D/H = (2.6\pm0.7)\times10^{-5}$, ~0.2 × terrestrial SMOW, GPMS value	NAC98
H_2S	67±4 ppm	$H_2S/H_2 = 77\pm5$ ppm, GPMS, > 16 bar	NAC98
^{20}Ne	≤26 ppm	$^{20}Ne/H_2 = \leq30$ ppm, GPMS	NAC98
$^{13}CH_4$	19±1 ppm	$^{12}C/^{13}C = 92.6\pm4.3$, terrestrial within error, GPMS value	NAC98
^3He	22.6±0.7ppm	$^3He/^4He = (1.66\pm0.05)\times10^{-4}$	NAC98
^{36}Ar	≤9.06 ppm	$^{36}Ar/H_2 = \leq10.5$ ppm, GPMS	NAC98
C_2H_6	5.8±1.5 ppm	from CH_4 photolysis in stratosphere, varies with altitude and latitude	NKT86
$^{15}NH_3$	~2 ppm	$^{14}N/^{15}N$ ~125 from [DEC85]	DEC85, TKR80
PH_3	0.7±0.1 ppm	mixed upward from deep atmosphere, photolyzed in stratosphere	BLK86
C_2H_2	0.11±0.03 ppm	from CH_4 photolysis in stratosphere, varies with altitude & latitude	NKT86
CH_3D	0.20±0.04 ppm	formed by D/H exchange of CH_4 + HD in deep atmosphere	BT73, BT78, BLK86
$^{13}CCH_6$	~58 ppb	for $^{12}C/^{13}C$ ~ 94	WBJ91
$^{13}CCH_2$	~10 ppb	for $C_2H_2/^{13}CCH_2$ ~ 10	DLS85
C_2H_4	7±3 ppb	in N. polar auroral zone, 0.4 ppb in equatorial region	KCR85, KEM89
^{84}Kr	≤3.2 ppb	$^{84}Kr/H_2$ ≤3.7 ppb, GPMS	NAC98
^{132}Xe	≤3.8 ppb	$^{132}Xe/H_2 \leq 4.5$ ppb, GPMS	NAC98

continued

Table 8.3 *(continued)*

Gas	Abundance	Comments	Sources
CH_3C_2H	2.5^{+2}_{-1} ppb	in N. polar auroral zone	KCR85
HCN	<12 ppb	deep atmospheric source?	BGL95
C_6H_6	2^{+2}_{-1} ppb	in N. polar auroral zone	KCR85
CO	1.6±0.3 ppb	mixed upward from deep atmosphere	BLK86, NKG88
GeH_4	$0.7^{+0.4}_{-0.2}$ ppb	mixed upward from deep atmosphere, destroyed in stratosphere	BLK86
C_4H_2	0.3±0.2 ppb	midlatitude region	GY83
AsH_3	0.22±0.11 ppb	mixed upward from deep atmosphere, destroyed in stratosphere	NGK89, NLG90
H_3^+	...	in auroral regions	BDM92, DMC89
C_3H_8	<0.6 ppm	in N. polar auroral zone	KCR85

Note: GMPS is *Galileo* Probe Mass Spectrometer; HAD is Helium Abundance Detector

Sources: **[BDM92]** Billebaud, F., Drossart, P., Maillard, J. P., Caldwell, J., & Kim, S., 1992, *Icarus* 96, 281–283. **[BGL95]** Bézard, B., Griffith, C., Lacy, J., & Oates, T., 1995, *Icarus* 118, 384–391. **[BLK86]** Bjoraker, G. L., Larson, H. P., & Kunde, V. G., 1986, *Icarus* 66, 579–609. **[BT73]** Beer, R., & Taylor, F. W., 1973, *ApJ.* 179, 309–327. **[BT78]** Beer, R., & Taylor, F. W., 1978, *ApJ.* 219, 763–767. **[DEC85]** Drossart, P., Encrenaz, T., & Combes, M., 1985, *Astron. & Astrophys.* 146, 181–184. **[DLS85]** Drossart, P., Lacy, J., Serabyn, E., Tikunaga, A., Bézard, B., & Encrenaz, T., 1985, *Astron. & Astrophys.* 149, L10–L12. **[DMC89]** Drossart, P., Maillard, J. P., Caldwell, J., Kim, S. J., Watson, J. K. G., Majewski, W. A., Tennyson, J., Miller, S., Atreya, S. K., Clarke, J. T., Waite, J. H., & Wagener, R., 1989, *Nature,* 340, 539–541. **[GY83]** Gladstone, G. R., & Yung, Y. I., 1983, *ApJ.* 266, 415–424. **[KCR85]** Kim, S. J., Caldwell, J., Rivolo, A. R., Wagener, R., & Orton, G. S., 1985, *Icarus* 64, 233–248. **[KEM89]** Kostiuk, T., Espenak, F., Mumma, M. J., & Romani, P., 1989, *Infrared Physics* 29, 199–204. **[NAC98]** Niemann, H. B., Atreya, S. K., Carignan, G. R., Donahue, T. M., Haberman, J. A., Harpold, D. N., Hartle, R. E., Hunten, D. M., Kasprzak, W. T., Mahaffy, P. R., Owen, T. C., & Way, S. H., 1998, *J. Geophys. Res.* 103, *in press.* **[NGK89]** Noll, K. S., Geballe, T. R., & Knacke, R. F., 1989, *ApJ.* 338, L71–L74. **[NKG88]** Noll, K. S., Knacke, R. F., Geballe, T. R., & Tokunaga, A. T., 1988, *ApJ.* 324, 1210–1218. **[NKT86]** Noll, K. S., Knacke, R. F., Tokunaga, A. T., Lacy, J. H., Beck, S., & Serabyn, E., 1986, *Icarus* 65, 257–263. **[NLG90]** Noll, K. S., Larson, H. P., & Geballe, T. R., 1990, *Icarus* 83, 494–499. **[TKR80]** Tokunaga, A. T., Knacke, R. F., & Ridgway, S. T., 1980, *Icarus* 44, 93–101. **[VZH96]** von Zahn, U., & Hunten, D. M., 1996, *Science* 272, 849–851. **[WBJ91]** Wiedemann, G., Bjoraker, G. L., & Jennings, D. E., 1991, *ApJ.* 383, L29–L32.

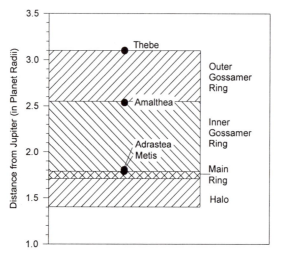

Figure 8.2 Jupiter's satellites and rings. Not shown are the Galilean satellites and moons Leda, Himalia, Lysisthea, Elara, Anake, Carme, Parsiphae, and Sinope, which are more than five Jupiter radii away from Jupiter.

Table 8.4 Jupiter's Rings

Ring Name	a (km)	$a/R_{Jup.}$	Width (km)	Thickness (km)
Halo inner edge	100,000	1.41	7,000	~20,000
Halo outer edge	122,000	1.71		
Main ring inner edge	123,000	1.72	7,000	≤100
Main ring outer edge	129,130	1.81		
Inner gossamer ring inner edge	127,800	1.79	54,000	2,000
Gossamer ring transition	182,000	2.55		
Outer gossamer ring outer edge	222,000	3.1	39,000	4,000

a: semimajor axis

Sources: Burns, J. A., Showalter, M. R., & Morfill, G. E., 1984, in *Planetary Rings* (Greenberg, R., & Brahic, A., eds.), Univ. of Arizona Press, Tucson, pp. 200–272. Showalter, M. R., Burns, J. A., Cuzzi, J. N., & Pollack, J. B., 1985, *Nature* 316, 526–528. Showalter, M. R., Burns, J. A., Cuzzi, J. N., & Pollack, J. B., 1987, *Icarus* 69, 458–498, and additional data kindly provided by J. A. Burns.

Table 8.5 Some Physical Properties of the Galilean Satellites

Property	Io	Europa	Ganymede	Callisto
Semimajor axis (R_{Jup})	5.897	8.969	14.98	26.34
Mean radius (km)	1821.3±5	1560±10	2634±10	2400±10
Volume (km^3)	2.531×10^{10}	1.59×10^{10}	7.655×10^{10}	5.812×10^{10}
Mass (10^{23} kg)	0.89319 ±0.00012	0.47910 ±0.00012	1.48167 ±0.0002	1.0766 ±0.001
Mean density (g cm^{-3})	3.5294±0.0013	3.018±0.035	1.936±0.022	1.851
GM ($km^3 s^{-2}$)	5959.91±0.28	3202.86±0.072	9886.6±0.5	7180
$g = GM/R^2$ (m s^{-2})	1.797	1.31	1.425	1.24
C/MR^2	0.378	0.26–0.33	0.3105±0.0028	0.406±0.039
$J_2 \times 10^6$	...	389±39	126.9±6.7	47.7±11.5
$J_3 \times 10^6$	...	0.1±7.1	...	...
Albedo, p_{vis}	0.60	0.64	0.43	0.19
$T_{subsolar}$ (K)	~135	140	156	168
Surf. heat flux (W m^{-2})	~2.5	...	...	...
Magnetic dipole (Tm^3)	2.1×10^{13}	$\leq 3.8 \times 10^{12}$	1.4×10^{13}	...
Magnetic dipole offset from spin axis	...	...	~10°	...

Sources: Anderson, J. D., Lau, E. L., Sjorgren, W. L., Schubert, G., & Moore, W. B., 1996, *Nature* 384, 541–544. Anderson, J. D., Lau, E. L., Sjorgren, W. L., Schubert, G., & Moore, W. B., 1997, *Science* 276, 1236–1239. Anderson, J. D., Lau, E. L., Sjorgren, W. L., Schubert, G., & Moore, W. B., *Nature* 387, 264–266. Anderson, J. D., Sjorgren, W. L., & Schubert, G., 1996, *Science* 272, 709–712. Burns, J. A., & Matthews, M. S. (eds.), 1986, *Satellites*, Univ. Arizona Press, Tucson, pp. 1021. Kivelson, M. G., Khurana, K. K., Russell, C. T., Walker, R. J., Warnecke, J., Coroniti, F. V., Polanskey, C., Southwood, D. J., & Schubert, G., 1996, *Nature* 384, 537–541. Kivelson, M. G., Khurana, K. K., Walker, R. J., Russell, C. T., Linker, J. A., Southwood, D. J., & Poanskey, C., 1996, *Science* 273, 337–340. Kivelson, M. G., Khurana, K. K., Joy, S., Russell, C. T., Southwood, D. J., Walker, R. J., & Polanskey, C., 1997, *Science* 276, 1239–1241. McKinnon, W. B., 1997, *Icarus* 130, 540–543. Schubert, G., Zhang, K., Kivelson, M. G., & Anderson, J. D., 1996, *Nature* 384, 544–545.

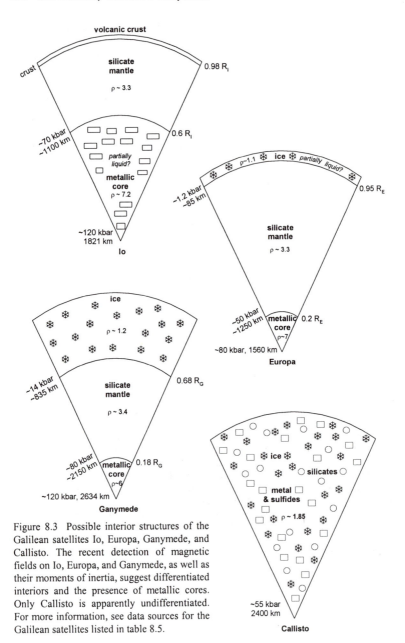

Figure 8.3 Possible interior structures of the Galilean satellites Io, Europa, Ganymede, and Callisto. The recent detection of magnetic fields on Io, Europa, and Ganymede, as well as their moments of inertia, suggest differentiated interiors and the presence of metallic cores. Only Callisto is apparently undifferentiated. For more information, see data sources for the Galilean satellites listed in table 8.5.

9

SATURN, RINGS, AND SATELLITES

9.1 Saturn

Saturn is the most distant planet visible with the naked eye. It has the most spectacular ring system and the most satellites of any of the gas giant planets. Spacecraft flybys of the Saturnian system were done by *Pioneer 11* (1 Sept. 1979), *Voyager 1* (12 Nov. 1980), and *Voyager 2* (25 Aug. 1981). The *Cassini* spacecraft, launched on 15 Oct. 1997, will provide more information on Saturn, its satellites, and rings; in particular, the *Huygens* entry probe will explore the atmosphere and surface of Titan.

Saturn's bulk density of 0.687 g cm^{-3} is less than that of Jupiter, because Saturn's smaller size leads to less self-compression. However, interior structure models and spectroscopy of Saturn's atmosphere suggest that Saturn is more enriched than Jupiter in elements heavier than He. For example, the CH_4/H_2, PH_3/H_2, and AsH_3/H_2 molar ratios in Saturn's atmosphere are ~6, ~2, and ~6.5 times the solar C/H_2, P/H_2, and As/H_2 ratios, respectively. The NH_3/H_2 molar ratio in Saturn's atmosphere is approximately 0.2–1 times the solar N/H_2 ratio, but refers to the NH_3 abundance above the cloud tops, which has been depleted by condensation to form the clouds. Likewise, H_2O is depleted by condensation in deeper atmospheric regions, as on Jupiter. The H_2O observed in Saturn's upper atmosphere plausibly comes from the rings and/or icy satellites.

However, Saturn's atmospheric He/H_2 molar ratio of 0.034 is about 4.6 times smaller than the Jovian atmospheric He/H_2 molar ratio of 0.158. The difference is believed to result from separation and gravitational settling of He from H_2 in Saturn's interior, rather than from a bulk He depletion on Saturn. In fact, it is difficult to explain Saturn's internal heat flux, which is 1.78 times larger than the absorbed solar flux, without the additional energy from He phase separation and sedimentation.

Saturn's magnetic rotation period is 10.65 hours, and the rapid rotation causes an oblateness of ~10%. The dipole magnetic field is tilted <1° from Saturn's spin axis, offset by ~0.04 R_S from Saturn's center, and probably generated by dynamo action in metallic hydrogen in the interior.

Saturn has 18 satellites: Titan, the largest satellite; six intermediate-size satellites (Mimas, Enceladus, Tethys, Dione, Rhea, and Iapetus); and

11 minor, irregularly shaped satellites, with mean radii ranging from 141 km (Hyperion) to 10 km (Pan). The six intermediate-size satellites, Janus, and Epimethus rotate synchronously; Hyperion rotates chaotically; Phoebe rotates in ~0.4 days; and the rotation rates of Titan, and the other satellites are unknown. Titan's pole position and obliquity are also unknown. Except for Phoebe, the most distant satellite, the satellites have prograde orbits around Saturn. Two of the minor satellites, Janus and Epimetheus, are co-orbital. They periodically pass one other, exchange their slightly different orbital periods (16.66 and 16.67 hours), and exchange their leading and trailing positions. Three other minor satellites are Lagrangian satellites that orbit ~60° ahead or behind larger satellites. Helene is at a Lagrangian point in Dione's orbit, and Telesto and Calypso are at the two Lagrangian points in Tethys' orbit. The densities of the satellites range from ~0.64 g cm^{-3} for Janus and Epimetheus, to ~1–1.1 g cm^{-3} for Mimas, Enceladus, and Tethys, from 1.2 to 1.4 g cm^{-3} for the three other intermediate-size satellites, and up to 1.88 g cm^{-3} for Titan. These densities indicate compositions ranging from porous water ice with little rock, to ice-rock mixtures with ~50% rock. Reflection spectroscopy shows water ice on the surfaces of several satellites. Photometry shows that Iapetus has an albedo of ~0.5 on its trailing hemisphere and an albedo of ~0.05 over most of its leading hemisphere. Trapped O_3 was recently discovered on the surfaces of Rhea and Dione. Water vapor was recently observed in Titan's atmosphere but water ice has not yet been observed on its surface.

Saturn's rings were observed by Galileo in 1610 (who called them "ears"). In 1659, Huygens correctly identified the "ears" as rings. In 1675, Cassini observed the gap in the rings that bears his name. The ring system orbits Saturn from 1.1 R_S to 7.96 R_S and is made up of seven rings of varying width, thickness, and optical depth. The ring particles are made of ice and plausibly also contain some rocky and carbonaceous material. The ring particles range in size from sub-micron to several meters with different sizes and size distributions found in different rings. Most of the mass in the main rings (C, B, A) is in 1–10 meter-size particles, whereas the E, F, and G rings are mainly composed of microscopic particles.

Sources and further reading: Descriptions of the instruments on *Voyager 1 and 2* are in *Space Sci. Rev.* 21, 103–376 (1977). Alexander, A. F. O'D., 1962, *The planet Saturn*, Faber & Faber, London, pp. 474. Gehrels, T., & Matthews, M. S. (eds.), 1985, *Saturn*, Univ. o Arizona Press, Tucson, pp. 968. Greenberg, R., & Brahic, A. (eds.), 1984, *Planetary rings* Univ. of Arizona Press, Tucson, pp. 784. Noll, K. S., Roush, T. L., Cruikshank, D. P., John son. R. E., & Pendleton, Y. J., 1997, *Nature* 388, 45–47.

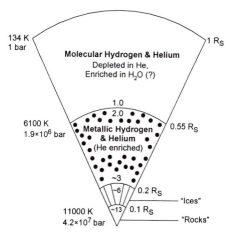

Figure 9.1 Internal structure of Saturn. The numbers at the phase boundaries are densities in g cm^{-3}. The interior structure is similar to that of Jupiter.

Table 9.1 Some Physical Properties of Saturn

Property	Value	Property	Value
Equatorial radius at 1 bar (km)	60268±4	Rotational period (hours)	10.65
Polar radius at 1 bar (km)	54364±10	GM (10^{15} m^{3}s^{-2})	37.9313
Oblateness ($R_{eq.}-R_{pol.})/R_{eq.}$	0.09796±2E-4	Equatorial gravity (m s^{-2})	8.96±0.01
Mass (10^{24} kg)	568.46	Polar gravity (m s^{-2})	12.14±0.01
Mass (Earth masses)	95.16	$J_2 \, 10^5$	1633.2±10
Rocky core mass (% of total)	10.3	$J_4 \, 10^6$	−919±40
Mean density (g cm^{-3})	0.6873	$J_6 \, 10^4$	1.04±0.50
Temperature at 1 bar (K)	134±4	C/MR^2	0.210
Effective temperature (K)	95.0±0.4	Escape velocity (km s^{-1})	35.478
Internal energy flux (W m^{-2})	2.01±0.14	Mean molec. wt. (g mol^{-1})	2.07
Intern. power/mass (10^{-11}J s^{-1}kg^{-1})	15.2±1.1	Scale height (km)	51.54
Therm. emis./absorbed solar energy	1.78±0.09	Magnetic axis offset	0.8°
Magnetic dipole moment ($\text{Tesla } R_{Sat}^3$)	0.21×10^{-4}	Solar wind–magnetopause boundary (R_{Sat})	22

Sources: Guillot, T., Chabrier, G., Morel, P., & Gautier, D., 1994, *Icarus* 112, 354–367. Hanel, R. A., Conrath, B. J., Kunde, V. G., Pearl, J. C., & Pirragalia, J. A., 1983, *Icarus* 53, 262–285. Hubbard, W. B., Podolak, J. B., & Stevenson, D. J., 1996, in *Neptune and Triton* (Cruikshank, D. P., ed.), Univ. of Arizona Press, Tucson, pp. 109–138. Lindal, G. F., 1992, *Astron. J.* 103, 967–982. Lindal, G. F., Lyons, J. R., Sweetnam, D. N., Eshleman, V. R., Hinson, D. P., & Tyler, G. L., 1987, *J. Geophys. Res.* 92, 14987–15001. Stevenson, D. J., 1982, *Annu. Rev. Earth Planet. Sci.* 10, 257–295.

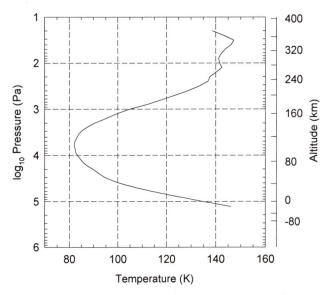

Figure 9.2 Temperature and pressure in Saturn's atmosphere

Table 9.2 Temperature, Pressure, and Density in Saturn's Atmosphere

Altitude (km)	Temperature (K)	Pressure (Pa)	Density (kg m⁻³)
0	134.8	100,000	0.196
40	101.8	41,472	0.107
80	85.4	13,565	4.19(–2)
120	83.4	3,947	1.25(–2)
160	99.6	1,253	3.32(–3)
200	123.6	500	1.07(–3)
240	137.2	232	4.46(–4)
280	142.0	112	2.08(–4)
320	143.3	54.3	9.99(–5)
360	144.4	27.1	4.95(–5)
400	130.5	10.9	2.20(–5)

Exponents in parentheses.

Sources: Lindal, G. F., 1992, *Astron. J.* 103, 967–982. Lindal, G. F., Sweetnam, D. N., & Eshleman, V. R., 1985, *Astron. J.* 90, 1136–1146.

Table 9.3 Chemical Composition of the Atmosphere of Saturn

Gas	Abundance	Comments	Sources
H_2	$96.3 \pm 2.4\%$	abundance from *Voyager* IRIS & radio occultation results, many spectroscopic studies of the pressure-induced dipole and quadrupole lines of H_2, and the ortho-para ratio	CGH84
He	$3.25 \pm 2.4\%$	*Voyager* IRIS & radio occultation	CGH84
CH_4	$4.5^{+2.4}_{-1.9} \times 10^{-3}$	abundance from *Voyager* IRIS is ~6 × solar C/H_2 ratio [GLS91], CH_4 is photolyzed to hydrocarbons in stratosphere	CGM84, PLC84
NH_3	$(0.5–2.0) \times 10^{-4}$	abundance from *Voyager* IRIS is ~(0.2–1.0) × solar N/H_2 ratio [GLS90], NH_3 undergoes condensation & photolysis in upper troposphere & stratosphere	CGM84, DM85, PLC84
HD	110 ± 58 ppm	discovered by [TRM77], R(1) rotational line observed by ISO [GND96], $D/H = (15–35) \times 10^{-6}$	MS78
$^{13}CH_4$	~51 ppm	$^{12}C/^{13}C \sim 89$	CMD77
C_2H_6	7.0 ± 1.5 ppm	abundance from [NKT86]. C_2H_6 due to CH_4 photolysis in stratosphere, abundance varies with altitude and latitude	NKT86
PH_3	1.4 ± 0.8 ppm	due to vertical mixing from deep atmosphere, photolyzed in stratosphere	CGM84, LFS80, NL90, PLC84
CH_3D	0.39 ± 0.25 ppm	formed by D/H exchange of CH_4 + HD in the deep atmosphere	CGM84, FL78, NL90
C_2H_2	0.30 ± 0.10 ppm	due to CH_4 photolysis in stratosphere	CGM84, NKT86
AsH_3	3 ± 1 ppb	due to vertical mixing from deep atmosphere, destroyed in stratosphere	BDL89, NL90, NLG90
H_2O	2–20 ppb	ISO observations of upper atmosphere	FLD97
CO	1.0 ± 0.3 ppb	due to vertical mixing from deep atmosphere, destroyed in upper atmosphere	NL90, NKG86
GeH_4	0.4 ± 0.4 ppb	due to vertical mixing from deep atmosphere, destroyed in stratosphere	BDL89, NL90, NKG88
CO_2	0.3 ppb	ISO observations of upper atmosphere	FLD97
C_3H_4	...	tentative detection, no abundance given	HCF81
C_3H_8	...	tentative detection, no abundance given	HCF81
H_2S *	<0.2 ppm	upper limit of 1 cm amagat	OME77
HCN *	<4 ppb	upper limit of 0.025 cm amagat	TBG81
SiH_4 *	<4 ppb	upper limit of 0.025 cm amagat	LFS80

* Converted to a mixing ratio using a H_2 column abundance of 70 km amagat [Tra77].

1 ppt μm = 10^{-4} g cm^{-2} = 0.124 cm amagat. 1 amagat = 2.69×10^{19} molecules cm^{-3}

continued

Table 9.3 *(continued)*

Sources: **[BDL89]** Bézard, B., DeBergh, C., Crisp, D., & Maillard, J. P., 1990, *ApJ.* 346, 509–513. **[CGH84]** Conrath, B. J., Gautier, D., Hanel, R. A., & Hoernstein, J. S., 1984, *ApJ.* 282, 807–815. **[CGM84]** Courtin, R., Gautier, D., Marten, Bézard, B., & Hanel, R., 1984, *ApJ.* 287, 899–916. **[CMD77]** Combes, M., Maillard, J. P., & DeBergh, C., 1977, *Astron. & Astrophys.* 61, 531–537. **[DM85]** DePater, I., & Massie, S. T., 1985, *Icarus*, 62, 143–171. **[FL78]** Fink, U., & Larson, H. P., 1978, *Science* 201, 343–345. **[FLD97]** Feuchtgruber, H., Lellouch, E., de Graauw, T., Bézard, B., Encrenaz, T., & Griffin, M., 1997, *Nature* 389, 159–162. **[GLS90]** Grevesse, N., Lambert, D. L., Sauval, A. J., van Dishoeck, E. F., Farmer, C. B., & Norton, R. H., 1990, *Astron. & Astrophys.* 232, 225–230. **[GLS91]** Grevesse, N., Lambert, D. L., Sauval, A. J., van Dishoeck, E. F., Farmer, C. B., & Norton, R. H., 1991, *Astron. & Astrophys.* 242, 488–495. **[GND96]** Griffin, M. J., Naylor, D. A., Davis, G. R., Ade, P. A. R., Oldham, P. G., Swinyard, B. M., Gautier, D., Lellouch, E., Orton, G. S., Encrenaz, T., de Graauw, T., Furniss, I., Smith, H., Armand, C., Burgdorf, M., Di Giorgio, A., Ewart, D., Gry, C., King, K. J., Lim, T., Molinari, S., Price, M., Sidher, S., Smith, A., Texier, D., Trams, N., Unger, S. J., & Salama, A., 1996, *Astron. & Astrophys.* 315, L389–L392. **[HCF81]** Hanel, R. A., Conrath, B. J., Flasar, F. M., Kunde, V. G., Maguire, W., Pearl, J., Pirragalia, J., Samuelson, R., Hearth, L., Allison, M., Cruikshank, D., Gautier, D., Horn, L., Koppany, R., & Ponnamperuma, C., 1981, *Science* 212, 192–200. **[LFS80]** Larson, H. P., Fink, U., Smith, H. A., & Davis, D. S., 1980, *ApJ.* 240, 327–337. **[MS78]** Macy, W., & Smith, W. H., 1978, *ApJ.* 222, L73–L75. **[NKG86]** Noll, K. S., Knacke, R. F., Geballe, T. R., & Tokunaga, A. T.,1986, *ApJ.* 309, L91–L94. **[NKG88]** Noll, K. S., Knacke, R. F., Geballe, T. R., & Tokunaga, A. T., 1988, *Icarus* 75, 409–422. **[NKT86]** Noll, K. S., Knacke, R. F., Tokunaga, A. T., Lacy, J. H., Beck, S., & Serabyn, E.,1986, *Icarus* 65, 257–263. **[NL90]** Noll, K. S., & Larson, H. P., 1990, *Icarus* 89, 168–189. **[NLG90]** Noll, K. S., Larson, H. P., & Geballe, T. R., 1990, *Icarus* 83, 494–499. **[OME77]** Owen, T., McKellar, A. R. W., Encrenaz, T., Lecaheux. J., DeBergh, C., & Maillard, J. P., 1977, *Astron. & Astrophys.* 54, 291–295. **[PLC84]** Prinn, R. G., Larson, H. P., Caldwell, J. J., & Gautier, D., 1984, in *Saturn* (Gehrels, T., & Matthews, M. S., eds.) Univ. of Arizona Press, Tucson, pp. 88–149. **[TBG81]** Tokunaga, A. T., Beck, S. C., Geballe, T. R., Lacy, J. H., & Serabyn, E., 1981, *Icarus* 48, 283–289. **[Tra77]** Trafton, L. M., 1977, *Icarus* 31, 369–384. **[TRM77]** Trauger, J. T., Roesler, F. L., & Mickelson, M. E., 1977, *BAAS* 9, 516.

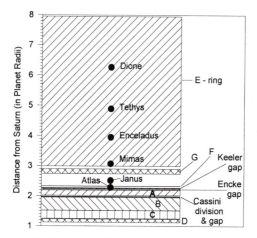

Figure 9.3 Saturn's satellites and rings. Not shown are the satellites Titan, Rhea, and Hyperion, which are farther from Saturn.

Table 9.4 Saturn's Rings

Ring Name	a (km)	a/R_Sat.	Width (km)	Mass (kg)
D inner edge	69970	1.110	4540	
D outer edge	74510	1.235		
C inner edge	74510	1.235	17500	$\sim1.1\times10^{21}$
C outer edge	92000	1.525		
B inner edge	92000	1.525	25500	$\sim2.8\times10^{22}$
B outer edge	117580	1.949		
Cassini Division center	119000	1.972	4500	$\sim5.7\times10^{20}$
Cassini Gap center	119900	1.988	246	
A inner edge	122170	2.025	14700	$\sim6.3\times10^{21}$
A outer edge	136780	2.267		
Encke Gap center	135706	2.214	325	
Keeler Gap center	136526	2.263	~35	
F center	140300	2.324	30–500	
G center	170000	2.818	1000	
E inner edge	180000	2.984	300000	
E outer edge	480000	7.956		

a: semimajor axis

Sources: Burns, J. A., Showalter, M. R., & Morfill, G. E., 1984, in *Planetary rings* (Greenberg, R., & Brahic, A., eds.), Univ. of Arizona Press, Tucson, pp. 200–272. Cuzzi, J. N., Lissauer, J. J., Esposito, L. W., Holberg, J. B., Marouf, E. A., Tyler, G. L., & Boischot, A., 1984, in *Planetary rings* (Greenberg, R., & Brahic, A., eds.), Univ. of Arizona Press, Tucson, pp. 73–199. Tyler, G. L., Eshleman, V. R., Anderson, J. D., Levy, G. S., Lindal, G. F., Wood, G. E., & Croft, T. A., 1982, *Science* 215, 553–558. Zebker, R. A., & Tyler, G. L., 1984, *Science* 223, 396–398.

9.2 Titan

Titan, which is about 6% larger than Mercury, is the only planet-size moon among the 18 Saturnian satellites. Titan's density is comparable to that of Jupiter's large moons Ganymede and Callisto indicating compositions of ~50 mass% rock and ~50 mass% ices (e.g., H_2O, CH_4, and NH_3 ice on Titan).

Titan is exceptional because it has an extensive atmosphere, which was discovered in 1944 when Kuiper observed CH_4 absorption in Titan's spectrum. From *Voyager* data, we now know that most of Titan's atmosphere is N_2 and Ar, with a CH_4/N_2 molar ratio of 2–10%. Titan's atmosphere will be more closely investigated by the *Huygens* probe released from the *Cassini* spacecraft, which was launched on 15 October 1997 and is to arrive at the Saturnian system in June of 2004. Scientific experiments on the *Huygens* probe include a gas chromatograph and mass spectrometer for chemical analyses of gases and aerosol particles in Titan's atmosphere, an atmospheric structure instrument, a Doppler wind experiment, an imager and spectral radiometer, and an aerosol particle collector and pyrolyser. Additional equipment on the *Cassini* spacecraft, such as ultraviolet, visible, and infrared spectrometers, and radar will also be used to study Titan.

Table 9.5 Some Physical Parameters of Saturn's Moon Titan

Property	Value	Property	Value
Mass; M (kg)	1.3455×10^{23}	$T_{surface}$ (K)	94
Radius; R (km)	2575	$T_{blackbody}$ (K)	90
Observed density (g cm^{-3})	1.881	$P_{surface}$ (bar)	1.5
GM (m^{3}s^{-2})	8.978×10^{12}	Mean molecular weight (g mol^{-1})	~28.6
Gravity (m s^{-2})	1.354	Pressure scale height (km)	~20.2
Moment of inertia; C/(MR^2)	...	Solar constant (W m^{-2})	15.04
Escape velocity; $v_{esc.}$ (km s^{-1})	2.641		

Sources and further reading: Hunten, D. M., Tomasko, M. G., Flasar, F. M., Samuelson, R. E., Strobel, D. F., & Stevenson, D. J., 1984, in *Saturn* (Gehrels, T., & Matthews, M. S., eds.), Univ. of Arizona Press, Tucson, pp. 671–759. Kuiper, G. P., 1944, *ApJ.* 100, 378–383. Lunine, J. I., Atreya, S. K., & Pollack, J. B., 1989, in *Origin and evolution of planetary and satellite atmospheres* (Atreya, S. K., Pollack, J. B., & Matthews, M. S., eds.), Univ. of Arizona Press, Tucson, pp. 605–665. Muhleman, D. O., Grossman, A. W., Butler, B. J., & Slade, M. A., 1990, *Science* 248, 975–980. Yung, Y. L., Allen, M., & Pinto, J. P., 1984, *ApJ. Suppl. Ser.* 55, 465–506.

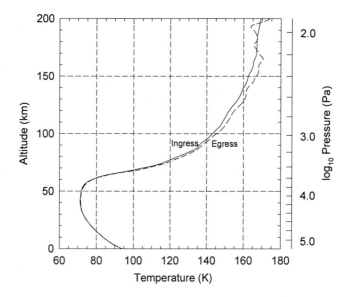

Figure 9.4 Temperature and pressure in Titan's atmosphere

Table 9.6 Temperature, Pressure, and Density in Titan's Atmosphere

Altitude (km)	Temperature (K)	Pressure (Pa)	Density (kg m⁻³)
0	94.0	149,526	5.55
20	77.0	49,404	2.20
40	71.2	14,426	0.687
60	77.5	4,278	0.186
80	124.4	1,806	4.89(–2)
100	143.1	962	2.26(–2)
120	152.1	546	1.21(–2)
140	160.1	322	6.79(–3)
160	165.3	195	3.95(–3)
180	167.2	120	2.42(–3)
200	169.4	75	1.49(–3)

Data taken during ingress. Exponents in parentheses.

Source: Lindal, G. F., Wood, G. E., Hotz, H. B., Sweetnam, D. N., Eshleman, V. R., & Tyler, G. L., 1983, *Icarus* 53, 348–363.

Table 9.7   Chemical Composition of the Atmosphere of Titan

Gas	Abundance	Comments	Sources
N_2	65–98 %	abundance indirectly inferred from *Voyager* IRIS & radio occultation data that constrain mean molec. weight of atmosphere, directly detected by *Voyager* UVS in upper atmosphere	ELT83, HTF84, LWH83
Ar	≤25 %	upper limit from deduced mean molecular weight of atmosphere, UVS data show Ar/N_2 < 6 % at 3900 km	BSS81, HTF84
CH_4	2–10 %	indirectly inferred from *Voyager* IRIS & radio occultation data, about 2% at tropopause	ELT83, HCF81, LWH83
H_2	0.2%	directly measured by *Voyager* IRIS	HTF84, Tra72
CO	60–150 ppm	Earth-based IR spectroscopy	HTF84, LDO83
	10^{+10}_{-5} ppm	Earth-based IR spectroscopy	NGK96
CH_3D	110^{+70}_{-60} ppm	*Voyager* IRIS gives $D/H = (1.5^{+1.4}_{-0.5}) \times 10^{-4}$	CBG89b, DLO88, KC82, OLD86
C_2H_6	13–20 ppm	*Voyager* IRIS, uniformly mixed over disk	CBG89a, CBG91, HCF81, KAH81
C_3H_8	0.5–4 ppm	*Voyager* IRIS, uniformly mixed over disk	CBG89a, CBG91, HTF84
C_2H_2	2–5 ppm	*Voyager* IRIS, uniformly mixed over disk	CBG89a, CBG91, HCF81, KAH81
C_2H_4	0.09–3 ppm	*Voyager* IRIS, polar/equatorial ratio ~30	CBG89a, CBG91, HCF81, KAH81
HCN	0.2–2 ppm	*Voyager* IRIS & Earth-based mm wavelength observations, polar/equatorial ratio ≤2 [CBG91]	CBG89a, CBG91, KAH81, TBM90
HC_3N	80–250 ppb	*Voyager* IRIS, abundances in N. polar region, no detection in equatorial region	CBG89a, CBG91
CH_3C_2H	4–60 ppb	*Voyager* IRIS, polar/equatorial ratio ~2–5	CBG89a, CBG91, KAH81, MHJ81
C_4H_2	1–40 ppb	*Voyager* IRIS, polar/equatorial ratio ~16	CBG89a, CBG91, KAH81
C_2N_2	5–16 ppb	*Voyager* IRIS, abundances in N. polar region, no detection in equatorial region	CBG89a, CBG91
CO_2	1.5–14 ppb	*Voyager* IRIS, polar/equatorial ratio ~0.5	CBG89a, CBG91, HTF84

continued

Table 9.7 *(continued)*

Sources: **[BSS81]** Broatfoot, A. L., Sandel, B. R., Shemansky, D. E., Holberg, J. B., Smith, G. R., Strobel, D. F., Kumar, S., McConnell, J. C., Hunten, D. M., Atreya, S. K., Donahue, T. M., Moos, H. W., Bertaux, J. L., Blamont, J. E., Pomphray, R. B., & Linick, S., 1981, *Science* 212, 206–211. **[CBG89a]** Coustenis, A, Bézard, B., & Gautier, D., 1989, *Icarus* 80, 54–76. **[CBG89b]** Coustenis, A, Bézard, B., & Gautier, D., 1989, *Icarus* 82, 67–80. **[CBG91]** Coustenis, A., Bézard, B., Gautier, D., Marten, A, & Samuelson, R., 1991, *Icarus* 89, 152–167. **[DLO88]** DeBergh, C., Lutz, B. L., Owen, T., & Chauville, J., 1988, *ApJ.* 329, 951–955. **[ELT83]** Eshleman, V. R., Lindal, G. F., & Tyler, G. L., 1983, *Science* 221, 53–55. **[HCF81]** Hanel, R. A., Conrath, B. J., Flasar, F. M., Kunde, V. G., Maguire, W., Pearl, J., Pirragalia, J., Samuelson, R., Hearth, L., Allison, M., Cruikshank, D., Gautier, D., Horn, L., Koppany, R., & Ponnamperuma, C., 1981, *Science* 212, 192–200. **[HTF84]** Hunten, D. M., Tomasko, M. G., Flasar, F. M., Samuelson, R. E., Strobel, D. F., Stevenson, D. J., 1984, in *Saturn* (Gehrels, T., & Matthews, M. S., eds.), Univ. of Arizona Press, Tucson, pp. 671–759. **[KAH81]** Kunde, V. G., Aikin, A. C., Hanel, R. A., Jennings, D. E., Maguire, W. C., & Samuelson, R. E., 1981, *Nature* 292, 686–688. **[KC82]** Kim, S. J., & Caldwell, J., 1982, *Icarus* 52, 473–482. **[LDO83]** Lutz, B. L., DeBergh, C., & Owen, T., 1983, *Science* 220, 1374–1375. **[LWH83]** Lindal, G. F., Wood, G. E., Hotz, H. B., Sweetnam, D. N., Eshleman, V. R., & Tyler, G. L., 1983, *Icarus* 53, 348–363. **[MHJ81]** Maguire, W. C., Hanel, R. A., Jennings, D. E., Kunde, V. G., & Samuelson, R. E., 1981, *Nature* 292, 683–686. **[NGK96]** Noll, K. S., Geballe, T. R., Knacke, R. F., & Pendleton, Y. J., 1996, *Icarus* 124, 625–631. **[OLD86]** Owen, T., Lutz, B. L., & DeBergh, C., 1986, *Nature* 320, 244–246. **[TBM90]** Tanguy, L., Bézard, B., Marten, A., Gautier, D., Gérard, E., Paubert, G., & Lecacheux, A., 1990, *Icarus* 85, 43–57. **[Tra72]** Trafton, L. M., 1972, *ApJ.* 175, 285–293.

10

URANUS, RINGS, AND SATELLITES

Uranus was discovered by Sir William Herschel on 13 March 1781. Observations of Uranus and its satellite and ring systems prior to the *Voyager 2* encounter on 26 January 1986 are summarized by [Alx65, Hu82, Ber84, BM86, GB84]. A post-*Voyager* summary of the atmosphere and interior of Uranus, its rings, satellites, and magnetosphere is in [BMM91].

The rotational and orbital axes of Uranus are nearly aligned because it has an obliquity of 97.9°. This was originally inferred from the orbital plane of its satellite system. As a result, we see Uranus pole-on every 42 years. Uranus' mean orbital distance is ~19.2 AU, and its orbital period is ~84 years. Uranus rotates in a retrograde, that is, east to west, direction. The magnetic rotation period determined by *Voyager 2* is 17.24 hours. However, at midlatitudes, atmospheric rotation is nearly 200 m s^{-1} faster than that of the planetary magnetic field.

Unlike the other gas giant planets, Uranus apparently has a weak, or nonexistent, internal heat source. The upper limit on the internal heat flux is 14% of the absorbed solar flux. This is consistent with heating from decay of radionuclides in the rocky material inside Uranus.

The bulk density of 1.318 g cm^{-3} shows that Uranus does not have solar composition. Instead, elements heavier than He comprise about 75–90% of its total mass. The molar CH_4/H_2 ratio of ~2%, about 24 times higher than the solar C/H_2 ratio, also suggests a large heavy element enrichment. In contrast, microwave observations indicate an apparent NH_3 depletion of 0.005–0.01 times the solar N/H_2 ratio in the 150–200 K region of the atmosphere. Temporal and latitudinal variations of the NH_3/H_2 ratio in this region of the atmosphere may be due to atmospheric circulation patterns. The NH_3 depletion may be due to the lack of nitrogen on Uranus, the loss of NH_3 in water- and NH_4SH-cloud layers deeper in Uranus' atmosphere, or other factors. Within error, the He/H_2 ratio of ~0.18 is the same as the solar value and suggests that the Uranian atmosphere is not depleted in He (as on Saturn). The D/H ratio derived from observations of CH_3D and the temperature dependent fractionation of D between HD and CH_4 is ~0.7×10^{-4}, intermediate between values of (0.2–0.5)×10^{-4} on Jupiter and Saturn and ~1.6×10^{-4} for Earth. The temperature gradient in the upper troposphere suggests local equilibrium of the ortho (parallel) and para (antiparallel) nuclear spin states of hydrogen. The ortho and para modifications of H_2 have different properties at the low temperatures in Uranus' upper troposphere and affect atmospheric structure and dynamics.

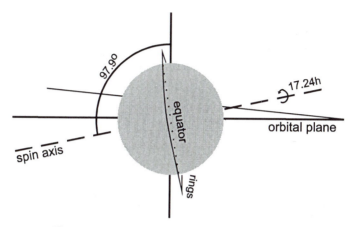

Figure 10.1 Uranus' inclination to its orbital plane

Uranus has a regular system of five major satellites (Ariel, Umbriel, Titania, Oberon, and Miranda), discovered prior to the *Voyager 2* encounter, and 10 minor satellites, all lying inside Miranda's orbit, discovered by *Voyager 2*. Two new distant moons were discovered more recently by B. J. Gladman and colleagues [GNB98] and other minor satellites may also exist. Except for Miranda ($\rho \sim 1.2$ g cm^{-3}), the major satellites have bulk densities of ~1.5–1.7 g cm^{-3}, implying >50% silicates by mass plus ice. Miranda's lower density implies a lower silicate mass fraction. Prior Earth-based spectroscopy showed water ice on the surfaces of all five major satellites. The topography and range of geologic features observed on the major satellites show that cryovolcanism and impacts have modified their surfaces to different extents.

The rings of Uranus were discovered during the stellar occultation of SAO 158687 on 10 March 1977. Earth-based occultation and *Voyager 2* results show 11 narrow rings that circle Uranus at roughly three distances: about 42,500 km, 48,000 km, and 51,000 km. The narrowest rings have widths of ~5 km; the widest is ~100 km. The narrowness of the rings is believed to arise from gravitational interactions with nearby shepherding satellites. The rings particles are ~1 m in size and are very dark with an albedo of only a few percent. Orbital periods are ~8 hours. A continuous distribution of dust fills the entire ring plane.

The *Voyager 2* flyby provided the conclusive evidence for Uranus' magnetic field, which can be modeled as a dipole offset $0.3\ R_U$ from the center of Uranus and tilted ~60° from the spin axis. The surface field strength varies from 0.1 to 1 gauss. The nature of the magnetic field suggests that it is generated by dynamo activity in a region of Uranus that extends out to ~70% of the planet's radius. In turn, this implies a convective, partially fluid interior.

Sources and further reading: **[Alx65]** Alexander, A. F. O'D, 1965, *The planet Uranus: A history of observation, theory and discovery*, Faber and Faber, London, pp. 316. **[Ber84]** Bergstralh, J. T. (ed.), 1984, *Uranus and Neptune*, NASA-CP 2330. **[BM86]** Burns, J. A., & Matthews, M. S. (eds.) 1986, *Satellites*, Univ. of Arizona Press, Tucson, pp. 598. **[BMM91]** Bergstralh, J. T, Miner, E. D., & Matthews, M. S. (eds.) 1991, *Uranus*, Univ. of Arizona Press, Tucson, pp. 1076. **[GB84]** Greenberg, R., & Brahic, A. (eds.) 1984, *Planetary Rings*, Univ. of Arizona Press, Tucson, pp. 784. **[GNB98]** Gladman, B. J., Nicholson, P. D., Burns, J. A., Kavelaars, J. J., Marsden, B. G., Williams, G. V., & Offutt, W. B., 1998, *Nature* 392, 897–899. **[Hu82]** Hunt, G. E. (ed.), 1982, *Uranus and the outer planets*, Cambridge Univ. Press, pp. 307.

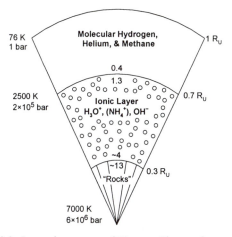

Figure 10.2   Internal structure of Uranus. The numbers at the phase boundaries are densities in g cm^{-3}. The interior structure of Uranus is a guide for constructing models of Neptune.

Table 10.1   Some Physical Properties of Uranus

Property	Value	Property	Value
Equatorial radius at 1 bar (km)	25559±4	Rotational period (hours)	17.24±0.01
Polar radius at 1 bar (km)	24973±20	GM (10^{15} m^{3}s^{-2})	5.79395
Oblateness ($(R_{eq.} - R_{pol.})/R_{eq.}$)	0.02293±0.0008	Equatorial gravity (m s^{-2})	8.69
Mass (10^{24} kg)	86.832	Polar gravity (m s^{-2})	9.19
Mass (Earth masses)	14.54	J_2 10^6	3516±3
Ice & rock mass (% of total)	76–92	J_4 10^6	−31.9±5
Mean density (g cm^{-3})	1.318	C/MR^2	0.225
Temperature at 1 bar (K)	76±2	Escape velocity (km s^{-1})	21.267
Effective temperature (K)	59.1±0.3	Solar constant (W m^{-2})	3.71
Internal energy flux (W m^{-2})	0.042±0.047	Mean molec. wt. (g mol^{-1})	2.64
Intern. power/mass (10^{-11}J s^{-1}kg^{-1})	0.392±0.441	Scale height (km)	27.05
Therm. emis./absorbed solar energy	1.06±0.08	Magnetic axis offset	58.6°
Magnetic dipole moment (Tesla R_{Ura}^3)	0.23×10^{-4}	Solar wind–magnetopause boundary (R_{Ura})	18

Sources: Bergstralh, J. T, Miner, E. D., & Matthews. M. S. (eds.), 1991, *Uranus*, Univ. of Arizona Press, Tucson, pp. 1076. Lindal, G. F., 1992, *Astron. J.* 103, 967–982. Lindal, G. F., Lyons, J. R., Sweetnam, D. N., Eshleman, V. R., Hinson, D. P., & Tyler, G. L., 1987, *J. Geophys. Res.* 92, 14987–15001. Stevenson, D. J., 1982, *Annu. Rev. Earth Planet. Sci.* 10, 257–295. Hubbard, W. B., Podolak, J. B., & Stevenson, D. J., 1996, in *Neptune and Triton* (Cruikshank, D. P., ed.), Univ. of Arizona Press, Tucson, pp. 109–138.

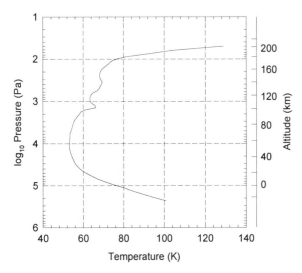

Figure 10.3 Temperature and pressure in Uranus' atmosphere

Table 10.2 Temperature, Pressure, and Density in Uranus' Atmosphere

Altitude (km)	Temperature (K)	Pressure (Pa)	Density (kg m⁻³)
0	76.4	100,000	0.365
20	60.9	48,867	0.223
40	54.5	20,893	0.107
60	53.4	8,525	4.44(−2)
80	55.1	3,504	1.77(−2)
100	62.3	1,518	6.82(−3)
120	63.4	721.9	3.17(−3)
140	68.8	353.8	1.43(−3)
160	68.0	174.3	7.14(−4)
180	76.1	88.5	3.24(−4)
200	104.7	51.7	1.39(−4)

Exponents in parenthesis.

Sources: Lindal, G. F., 1992, *Astron. J.* 103, 967–982. Lindal, G. F., Lyons, J. R., Sweetnam, D. N., Eshleman, V. R., Hinson, D. P., & Tyler, G. L., 1987, *J. Geophys. Res.* 92, 14987–15001.

Table 10.3 Chemical Composition of the Atmosphere of Uranus

Gas	Abundance	Comments	Sources
H_2	~82.5±3.3%	*Voyager* IRIS & radio occultation, by difference from sum of He + CH_4, many studies of the pressure-induced dipole & quadrupole lines and the ortho-para ratio	BB84, BB86, CS38, CGH87, FL79, Smi78, Tra87
He	15.2±3.3 %	*Voyager* IRIS & radio occultation	CGH87
CH_4	~2.3 %	CH_4 ~32 × solar C/H_2 ratio [GLS91], data from *Voyager* radio occultation data on lapse rate, abundance from vis/IR spectroscopy is 1–10%	Bai83, BB86, BF80, LLS87, LOC76, MGS78
HD	~148 ppm	based on D/H ~ 9 × 10^{-5}, no reliable observations of HD lines [SSS89]	CS83, MS78, SSS89, TR80
CH_3D	~8.3 ppm	based on $CH_3D/CH_4 = 3.6^{+3.6}_{-2.8} \times 10^{-4}$ [DLO86] and 2.3% CH_4	BMM91, DLO86
C_2H_6	~1–20 ppb	due to CH_4 photolysis, abundance varies with height & latitude, abundance from *Voyager* is a few times 10^{-8}	ASR91, HSB87, OAS87, OBC90
C_2H_2	~10 ppb	due to CH_4 photolysis, abundance varies with height & latitude	ASR91, CWF88, HSB87, OAS87, OBC90
H_2O	5–12 ppb	ISO observations	FLD97
H_2S *	<0.8 ppm	upper limit of 30 cm amagat	FL79
NH_3 *	<100 ppb	upper limit of 5 cm amagat from IR spectroscopy, abundance varies with height and latitude and is larger at lower levels, extensive microwave studies	FL79, GD84, GJO78, HM88
CO	<40 ppb	upper limit for stratosphere	MGO93, RLR92
CH_3CN	...	in stratosphere	RLR92
HCN	<15 ppb	in stratosphere	MGO93, RLR92
HC_3N	<0.8 ppb	in stratosphere	RLR92
CO_2	≤0.3 ppb	ISO observations	FLD97

* Converted to a mixing ratio using a H_2 column abundance of 400 km amagat.

Sources: **[ASR91]** Atreya, S. K., Sandel, B. R., & Romani, P. N., 1991, in *Uranus* (Bergstralh, J. T., Miner, E. D., & Matthews, M. S., eds.) Univ. of Arizona Press, Tucson, pp. 110–146. **[Bai83]** Baines, K. H., 1991, *Icarus* 56, 543–559. **[BB84]** Bergstralh, J. T., & Baines, K. H., 1984, in *Uranus and Neptune* (Bergstralh, J. T., ed.), NASA CP-2330, pp. 179–206. **[BB86]** Baines, K. H., & Bergstralh, J. T., 1986, *Icarus* 65, 406–441.

continued

Table 10.3 *(continued)*

[**BF80**] Benner, D. C., & Fink, U., 1980, *Icarus* 42, 343–353. [**BMM91**] Bergstralh, J. T., Miner, E. D., & Matthews, M. S. (eds.), 1991, *Uranus*, Univ. of Arizona Press, Tucson, pp. 1076. [**CGH87**] Conrath, B. J., Gautier, D., Hanel, R. A., Lindal, G., & Marten, A., 1987, *J. Geophys. Res.* 92, 15003–15010. [**CS83**] Cochran, W. D., & Smith, W. H., 1983, *ApJ.* 219, 756–762. [**CWF88**] Caldwell, J. T., Wagener, R., & Fricke, K. H., 1988, *Icarus* 74, 133–140. [**DLO86**] DeBergh, C., Lutz, B. L., Owen, T., Brault, J., & Chauville, J., 1986, *ApJ.* 311, 501–510. [**FL79**] Fink, U., & Larson, H. P., 1979, *ApJ.* 233, 1021–1040. [**FLD97**] Feuchtgruber, H., Lellouch, E., de Graauw, T., Bézard, B., Encrenaz, T., & Griffin, M., 1997, *Nature* 389, 159–162. [**GD84**] Gulkis, S., & DePater, I., 1984, in *Uranus and Neptune* (Bergstralh, J. T., ed.) NASA CP-2330, pp. 225–262. [**GJO78**] Gulkis, S., Janssen, M. A., & Olsen, E. T., 1978, *Icarus* 34, 10–19. [**GLS91**] Grevesse, N., Lambert, D. L., Sauval, A. J., van Dishoeck, E. F., Farmer, C. B., & Norton, R. H., 1991, *Astron. & Astrophys.* 242, 488–495. [**HSB87**] Herbert, F. L., Sandel, B. R., Broadfoot, A. L., Shemansky, D. E., Holberg, J. B., Yelle, R. V., Atreya, S. K., & Romani, P. N., 1987, *J. Geophys. Res.* 92, 15093–15109. [**HM88**] Hofstadter, M. D., & Muhleman, D. O., 1988, *Icarus* 81, 396–412. [**LLS87**] Lindal, G. F., Lyons, J. R., Sweetnam, D. N., Eshleman, V. R., Hinson, D. P., & Tyler, G. L., 1987, *J. Geophys. Res.* 92, 14987–15001. [**LOC76**] Lutz, B. L., Owen, T., & Cess, R. D., 1976, *ApJ.* 203, 541–551. [**MGO93**] Marten, A., Gautier, D., Owen, T., Sanders, D., Tilanus, R. T., Matthews, H., Atreya, S. K., Tilanus, R. P. J., & Deane, J. R., 1993, *ApJ.* 406, 285–297. [**MGS78**] Macy, W., Gelfand, J., & Smith, W. H., 1978, *Icarus* 26, 428–436. [**MS78**] Macy, W., & Smith, W. H., 1978, *ApJ.* 222, L73–L75. [**OAS87**] Ortin, G. S., Aitken, D. K., Smith, C., Roche, P. F., Caldwell, J., & Snyder, R., 1987, *Icarus* 70, 1–12. [**OBC90**] Orton, G. S. Baines, K. H., Caldwell, J., Romani, P., Tokunaga, A. T., & West, R. A., 1990, *Icarus* 85, 257–265. [**RLR92**] Rosenqvist, J., Lellouch, E. Romani, P. N., Paubert, G., & Encrenaz, T., 1992, *ApJ.* 392, L99–L102. [**Smi78**] Smith, W. H., 1978, *Icarus* 33, 210–216. [**SSS89**] Smith, W. H., Schempp, W. V., Simon, J., & Baines, K. H., 1989, *ApJ.* 336, 962–966. [**TR80**] Trafton, L. M., & Ramsay, D. A., 1980, *Icarus* 41, 423–429. [**Tra87**] Trafton, L. M., 1987, *Icarus* 70, 13–30.

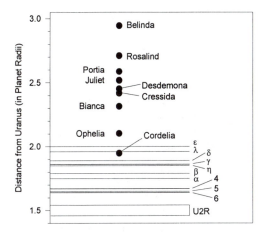

Figure 10.4 Uranus' satellites and rings. Not shown are the satellites Ariel, Miranda, Oberon, Puck, Titania, and Umbriel, which are farther from Uranus.

Table 10.4 Uranus' Rings

Ring Name	a (km)	$a/R_{\mathrm{Ura.}}$	e	i (deg.)
U2R inner edge	37000	1.45	...	...
U2R outer edge	39500	1.55	...	...
6	41837	1.64	1.013×10^{-3}	0.0616
5	42235	1.65	1.899×10^{-3}	0.0536
4	42571	1.67	1.059×10^{-3}	0.0323
α	44718	1.75	7.61×10^{-4}	0.0152
β	45661	1.79	4.42×10^{-4}	0.0051
η	47176	1.85	(4×10^{-6})	(0.0011)
γ	47627	1.86	1.09×10^{-4}	(0.0015)
δ	48300	1.89	4×10^{-6}	0.0011
λ	50024	1.96	(0.0)	(0.0)
ε	51149	2.00	7.936×10^{-3}	(0.0002)

a: semimajor axis; e: eccentricity; i: inclination
Two narrow and two broader rings of fine dust are located between the δ and ε rings.

Source: French, R. G., Nicholson, P. D., Porco, C. C., & Marouf, E. A., 1991, in *Uranus* (Bergstralh, J. T., Miner, E. D., Matthews, M. S., eds.), Univ. of Arizona Press, Tucson, pp. 327–409.

11

NEPTUNE, RINGS, AND SATELLITES

11.1 Neptune

During the six decades following the discovery of Uranus, it became clear that there were large discrepancies between the observed and calculated positions of the planet. Analyses of the discrepancies by Adams in England and LeVerrier in France led to predictions for the mass and orbit of a trans-Uranian planet. These calculations led to the discovery of Neptune by Galle and d'Arrest on 23 September 1846. Shortly thereafter, Lassell discovered Triton, Neptune's largest satellite on 10 October 1846. The account of Neptune's discovery is described by Grosser (1962).

Neptune's mean orbital distance is 30.07 AU and its orbit is nearly circular. In contrast, Pluto's orbit is highly eccentric and, as a consequence, Neptune is occasionally the outermost planet (e.g., from 21 Jan. 1979 to 14 Mar. 1999). However, a 3:2 orbital resonance between Pluto and Neptune prevents the close approach of the two planets.

Neptune is smaller than Uranus, but is also more massive and has a mean density of ~1.638 g cm^{-3}. Thus, Neptune, like Uranus, does not have solar composition and is enriched in elements heavier than He. A large enrichment of heavy elements is also suggested by the observed atmospheric composition. The atmospheric CH_4/H_2 molar ratio of ~1–2% is 14 to 28 times larger than the solar C/H_2 ratio. Water vapor is not observed at Neptune's cloud tops because the atmosphere is too cold. However, the CO/H_2 molar ratio of ~0.6×10^{-6} requires a water abundance several hundred times the solar O/H_2 ratio to produce the observed CO via the net reaction:

$$CH_4 + H_2O = 3 H_2 + CO \tag{1}$$

deep in Neptune's interior. The observed HCN/H_2 molar ratio of 0.3×10^{-9} requires an ammonia abundance greater than the solar N/H_2 ratio to produce sufficient amounts of N_2 (the precursor to HCN) via the net reaction

$$2 NH_3 = 3 H_2 + N_2 \tag{2}$$

deep in Neptune's interior. Vertical mixing transports the N_2 into Neptune's upper atmosphere where it reacts to form the observed HCN.

Neptune emits about 2.6 times as much heat as received from the sun. *Voyager 2* observations show that the upper troposphere has an adiabatic temperature profile (i.e., it is convective). The magnetic rotation period measured by *Voyager 2* is 15.8 hours; however, atmospheric rotation periods vary from ~18 hours near the equator to ~12 hours near the poles. Interior structure models suggest that Neptune is water-rich. Dynamo action in a conductive, water-rich interior is also apparently required to produce the observed magnetic field, which is offset 0.55 R_N from Neptune's center and tilted 47° to the rotational axis. In addition, the bulk D/H ratio of Neptune suggests a large abundance of hydrides (e.g., CH_4, NH_3, and H_2O) inside the planet. The bulk D/H ratio, which is derived from observations of CH_3D and the temperature dependent fractionation of D between HD and CH_4 is about 1.9×10^{-4}, 4–10 times larger than the bulk D/H ratios of Jupiter and Saturn. The enhanced D/H ratio is generally ascribed to a larger abundance of D-rich hydrides (i.e., a heavy element enrichment) on Neptune relative to Jupiter and Saturn.

Neptune has eight known satellites: six small satellites discovered by *Voyager 2*, with nearly circular orbits in Neptune's equatorial plane; Triton, with a retrograde, inclined, circular orbit; and Nereid, discovered by Kuiper in 1949, which has an inclined, highly elliptical orbit. Triton is the only Neptunian satellite whose mass and density are known. Hence, the compositions of the other satellites are unknown. The albedos and spectra indicate that the surfaces of Proteus (N1), Larissa (N2), Despina (N3), and Galatea (N4) are possibly carbonaceous, but the surface compositions of Thalassa (N5) and Naiad (N6) are unknown. Nereid's surface is possibly dirty ice or rock. Little is known about surface geology except for Triton.

Neptune's known ring system consists of six rings orbiting the planet from 1.69 R_N to 2.54 R_N. *Voyager 2* imaged the ring system in August 1989. Prior to this, Earth-based observations of stellar occultations by Neptune had indicated the presence of ring arcs around Neptune. The arcs are actually denser regions in the Adams ring. Neptune's rings range in width from ~15 km or less (Adams, Arago, LeVerrier rings) to 2000–4000 km (Galle and Lassell rings). The ring particles range in size from submicron to ~10 meters, and the different rings have different particle sizes and size ranges. The ring particles are dark and have a reddish color; they may be "dirty" ice (with or without silicates) or organic-rich material. Three small satellites, Despina, Thalassa, and Naiad, orbit Neptune between the Galle ring (1.69 R_N) and the LeVerrier ring (2.15 R_N).

Sources and further reading: Bergstralh, J. T., (ed.), 1984, *Uranus and Neptune,* NASA CP-2330. Cruikshank, D. P. (ed.), 1995, *Neptune and Triton*, Univ. of Arizona Press, Tucson, pp. 1249. Grosser, M., 1962, *The Discovery of Neptune*, Harvard Univ. Press, Cambridge, pp. 172. Lodders, K., & Fegley, B., Jr., 1994, *Icarus* 112, 368–375.

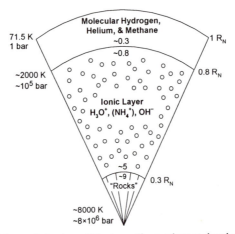

Figure 11.1 Internal structure of Neptune. The numbers at the phase boundaries are densities in g cm^{-3}. Structure models of Neptune are uncertain and nonunique.

Table 11.1 Some Physical Properties of Neptune

Property	Value	Property	Value
Equatorial radius at 1 bar (km)	24764±20	Rotational period (hours)	16.11±0.05
Polar radius at 1 bar (km)	24341±30	GM (10^{15} m^{3}s^{-2})	6.83473
Oblateness (R_{eq} − R_{pol})/R_{eq}	0.0171±0.0014	Equatorial gravity (m s^{-2})	11.00
Mass (10^{24} kg)	102.43	Polar gravity (m s^{-2})	11.41
Mass (Earth masses)	17.15	J_2 10^6	3538±9
Ice & rock mass (% of total)	82–96	J_4 10^6	−38±10
Mean density (g cm^{-3})	1.638	C/MR^2	0.24
Temperature at 1 bar (K)	71.5±2	Escape velocity (km s^{-1})	23.49
Effective temperature (K)	59.3±0.8	Solar constant (Wm^{-2})	1.47
Internal energy flux (W m^{-2})	0.433±0.046	Mean molec. wt. (g mol^{-1})	2.5–2.7
Intern. power/mass (10^{-11}J s^{-1}kg^{-1})	3.22±0.34	Scale height (km)	19.8–21.1
Thermal emis./absorbed solar energy	2.61±0.28	Magnetic axis offset	47.0°
Magnetic dipole moment (Tesla R$_{Nep}^3$)	0.133×10^{-4}	Solar wind–magnetopause boundary (R$_{Nep}$)	...

Sources: Conrath, B., Flasar, F. M., Hanel, R., Kunde, V., Maguire, W., Pearl, J., Pirraglia, J., Samuelson, R., Gierasch, P., Weir, A., Bézard, B., Gautier, D., Cruikshank, D., Horn, L., Springer, R., & Schaffer, W., 1989, *Science* 246, 1454–1459. Hubbard. W. B., Podolak, J. B., & Stevenson, D. J., 1995, in *Neptune and Triton* (Cruikshank, D. P., ed.), Univ. of Arizona Press, Tucson, pp. 109–138. Lindal, G. F., 1992, *Astron. J.* 103, 967–982. Lindal, G. F., Lyons, J. R., Sweetnam, D. N., Eshleman, V. R., Hinson, D. P., & Tyler, G. L., 1990, *Geophys. Res. Lett.* 17, 1733–1736. Stevenson, D. J., 1982, *Annu. Rev. Earth Planet. Sci.* 10, 257–295.

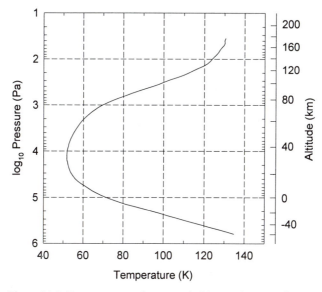

Figure 11.2 Temperature and pressure in Neptune's atmosphere

Table 11.2 Temperature, Pressure, and Density in Neptune's Atmosphere

Altitude (km)	Temperature (K)	Pressure (Pa)	Density (kg m⁻³)
0	71.5	100,000	0.438
20	54.4	31,333.	0.180
40	52.4	8,072	4.82(-2)
60	59.2	2,234	1.18(-2)
80	75.9	764	3.15(-3)
100	98.8	338	1.07(-3)
120	115	175	4.75(-4)
140	124	96.8	2.44(-4)
160	129	55.2	1.34(-4)
180	131	32.0	7.64(-5)
200	133	18.5	4.35(-5)

Exponents in parentheses.

Sources: Lindal, G. F., 1992, *Astron. J.* 103, 967–982. Lindal, G. F., Lyons, J. R., Sweetnam, D. N., Eshleman, V. R., Hinson, D. P., & Tyler, G. L., 1990, *Geophys. Res. Lett.* 17, 1733–1736.

Table 11.3 Chemical Composition of the Atmosphere of Neptune

Gas	Abundance	Comments	Sources
H_2	$\sim 80 \pm 3.2\%$	*Voyager* IRIS & radio occultation, by difference from sum of He + CH_4, many studies of the pressure-induced dipole & quadrupole lines and the ortho-para ratio	BB84, CS83, CGL91, FL79, Smi78
He	$19.0 \pm 3.2\%$	*Voyager* IRIS & radio occultation	CGL91
CH_4	$\sim 1-2\%$	$CH_4 \sim 14-28 \times$ solar C/H_2 ratio [GLS91]; abundances from *Voyager* radio occultation data on lapse rate, abundance from vis/IR spectroscopy is 1–10 %.	LLS90, LOC76, MGS78
HD	~ 192 ppm	based on D/H $\sim 1.2 \times 10^{-4}$, no reliable observations of HD lines [SSS89]	CS83, SSS89
CH_3D	~ 12 ppm	based on $CH_3D/CH_4 = 6^{+6}_{-4} \times 10^{-4}$ [DLO86] and 2% CH_4	BMM91, DLO86
C_2H_6	$1.5^{+2.5}_{-0.5}$ ppm	due to CH_4 photolysis, abundance varies with height & latitude	BRC91, KER90, OAS87, OBC90
CO	0.65 ± 0.35 ppm	present in troposphere & stratosphere	MGO93, RLR92
C_2H_2	60^{+140}_{-40} ppb	due to CH_4 photolysis, abundance varies with height & latitude	BRC91, CWF88, Mac80, OAS87, OBC90
H_2O	1.5–3.5 ppb	ISO observations	FLD97
CO_2	0.5 ppb	ISO observations	FLD97
HCN	0.3 ± 0.15 ppb	in stratosphere	MGO93, RLR92
H_2S *	<3 ppm	upper limits of 100 cm amagat	FL79
NH_3 *	<600 ppb	*Voyager* radio occultation upper limit at 6 bar level on Neptune	DR89, FL79, GD84, LLS90
CH_3CN	<5 ppb	in stratosphere	RLR92
HC_3N	<0.4 ppb	in stratosphere	RLR92

*Converted to a mixing ratio using an H_2 column abundance of 400 km amagat.

Sources: **[BB84]** Bergstralh, J. T., & Baines, K. H., 1984, in *Uranus and Neptune* (J. T. Bergstralh, ed.), NASA CP-2330, pp. 179–206. **[BMM91]** Bergstralh, J. T., Miner, E. D., & Matthews, M. S. (eds.), 1991, *Uranus*, Univ. of Arizona Press, Tucson, pp. 1076. **[BRC91]** Bézard, B., Romani, P. N., Conrath, B. J., & Maguire, W. C., 1991, *J. Geophys. Res.* 96, 18961–18975. **[CGL91]** Conrath, B. J., Gautier, D., Lindal, G. F., Samuelson, R. F., & Shaffer, W. A., 1991, *J. Geophys. Res.* 96, 18907–18919.

continued

Table 11.3 *(continued)*

[CS83] Cochran, W. D., & Smith, W. H., 1983, *ApJ.* 219, 756–762. **[CWF88]** Caldwell, J. T., Wagener, R., & Fricke, K. H., 1988, *Icarus* 74, 133–140. **[DLO86]** DeBergh, C., Lutz, B. L., Owen, T., Brault, J., & Chauville, J., 1986, *ApJ.* 311, 501–510. **[DR89]** DePater, I., & Richmond, M., 1989, *Icarus* 80, 1–13. **[FL79]** Fink, U., & Larson, H. P., 1979, *ApJ.* 233, 1021–1040. **[FLD97]** Feuchtgruber, H., Lellouch, E., de Graauw, T., Bézard, B., Encrenaz, T., & Griffin, M., 1997, *Nature* 389, 159–162. **[GD84]** Gulkis, S., & DePater, I., 1984, in *Uranus and Neptune* (Bergstralh, J. T., ed.) NASA CP-2330, pp. 225–262. **[KER90]** Kostiuk, T., Espenak, F., Romani, P., Zipoy, D., & Goldstein, J., 1990, *Icarus* 88, 87–96. **[LLS90]** Lindal, G. F., Lyons, J. R., Sweetnam, D. N., Eshleman, V. R., Hinson, D. P., & Tyler, G. L., 1990, *Geophys. Res. Lett.* 17, 1733–1736. **[LOC76]** Lutz, B. L., Owen, T., & Cess, R. D., 1976, *ApJ.* 203, 541–551. **[Mac80]** Macy, W., 1980, *Icarus* 41, 153–158. **[MGO93]** Marten, A., Gautier, D., Owen, T., Sanders, D., Tilanus, R. T., Matthews, H., Atreya, S. K., Tilanus, R. P. J., & Deane, J. R., 1993, *ApJ.* 406, 285–297. **[MGS78]** Macy, W., Gelfand, J., & Smith, W. H., 1978, *Icarus* 26, 428–436. **[OAS87]** Ortin, G. S., Aitken, D. K., Smith, C., Roche, P. F., Caldwell, J., & Snyder, R., 1987, *Icarus* 70, 1–12. **[OBC90]** Orton, G. S., Baines, K. H., Caldwell, J., Romani, P., Tokunaga, A. T., & West, R. A., 1990, *Icarus* 85, 257–265. **[RLR92]** Rosenqvist, J., Lellouch, E., Romani, P. N., Paubert, G., & Encrenaz, T., 1992, *ApJ.* 392, L99–L102. **[Smi78]** Smith, W. H., 1978, *Icarus* 33, 210–216. **[SSS89]** Smith, W. H., Schempp, W. V., Simon, J., & Baines, K. H., 1989, *ApJ.* 336, 962–966.

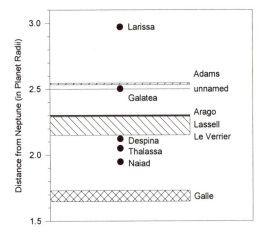

Figure 11.3   Neptune's satellites and rings. Not shown
are the satellites Triton, Nereid, and Proteus.

Table 11.4 Neptune's Rings

IAU-Designation	Ring Name	*a* (km)	*a*/R$_{Nep.}$	Ring Extension in Planet Radii	Ring Width (km)
1989N3R	Galle	41900	1.69	1.65–1.73	~2000
1989N2R	LeVerrier	53200	2.15	—	very narrow
1989N4R	Lassell	55300	2.2	2.15–2.3	~4000
	Arago	57600	2.3	—	very narrow
	unnamed	62000	2.5	—	very faint
1989N1R	Adams *	62933	2.54	2.15–2.4	15

IAU: International Astronomical Union.

a: semimajor axis

* The narrowest ring structures seen are the arcs within the Adam's ring, which are called
Courage, Liberté, Egalité, and Fraternité.

Source: Porco, C. C., Nicholson, P. D., Cuzzi, J. N., Lissauer, J. J., & Esposito, L. W.,
1995, in *Neptune and Triton* (Cruikshank, D. P., ed.), Univ. of Arizona Press, Tucson, pp.
703–804.

11.2 Triton

Triton, Neptune's largest satellite, was discovered by William Lassell in October 1846, less than three weeks after the discovery of Neptune (Grosser, 1962). Triton's nearly circular retrograde orbit and synchronous rotation are due to tidal torques imposed by Neptune. Triton is probably a captured satellite and its bulk density of 2.05 g cm^{-3}, predicted for icy planetesimals formed in the solar nebula, supports this concept. The orbital parameters of Neptune and Triton lead to complex variations in the subsolar point, which varies between 52° North and 52° South latitude. The resulting seasonal changes are extreme and drive large-scale atmospheric circulation patterns. The *Voyager 2* flyby of Neptune in August 1989 provided essentially all of our knowledge about Triton's mass, radius, density, surface geology, and atmospheric composition. Triton's average surface temperature of 38 K, is the lowest measured in the solar system, but Triton has a N_2-rich atmosphere with a surface pressure of 16±3 microbar. Methane is also observed in the atmosphere at ~0.01% of the N_2 abundance. Nitrogen, CH_4, CO, and CO_2 ices, but not water ice, are observed on Triton's surface. The *Voyager 2* images cover most of Triton south of 20° latitude and showed the presence of active plumes that are variously explained as active cryovolcanism, large dust devils, or geysers driven by solar energy.

Sources and further reading: Cruikshank, D. P. (ed.), 1995, *Neptune and Triton*, Univ. of Arizona Press, Tucson, pp. 1249. Davies, M. E., Rogers, P. G., & Colvin, T. R., 1991, *J. Geophys. Res.* 96, 15675–15681. Grosser, M., 1962, *The Discovery of Neptune*, Harvard Univ. Press, Cambridge, pp. 172. Ingersoll, A. P., 1990, *Nature* 344, 315–317. Jacobson, R. A., Riedel, J. E., & Taylor, A. H., 1991, *Astron. & Astrophys.* 247, 565–575. McCord, T. B., 1966, *Astron. J.* 71, 585–590. Owen, W. M., Vaughan, R. M., & Synnott, S. J., 1991, *Astron. J.,* 101, 1511–1515. Tyler, G. L., Sweetnam, D. N., Anderson, J. D., *and 14 additional authors*, 1989, *Science* 246, 1466–1473.

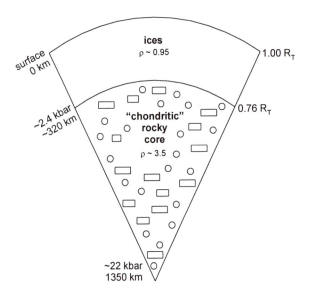

Figure 11.4 The internal structure of Triton

Table 11.5 Some Physical Properties of Neptune's Moon Triton

Property	Value	Property	Value
Radius (km)	1352.6±2.4	Semimajor axis (km)	3.5476×10^5 (14.33 R_N)
Mass (kg)	$(2.147\pm0.007)\times10^{22}$	Eccentricity of orbit	1.6×10^{-5}
Obs. density (g cm^{-3})	2.054±0.032	Inclination of orbit	157.345°
GM (m^{3}s^{-2})	1.433×10^{12}	Orbital period (days)	5.87687
GM/R^2 (m s^{-2})	0.783	Geometric albedo	0.70
$v_{esc.}$ (km s^{-1})	1.455	$T_{surface}$ (K)	38
C/MR^2	...	$P_{surface}$ (μbar)	~16
Ice (mass%)	30–45	Pressure scale height (km)	~15

Sources: Cruikshank, D. P. (ed.), 1995, *Neptune and Triton*, Univ. of Arizona Press, Tucson, pp. 1249. Davies, M. E., Rogers, P. G., & Colvin, T. R., 1991, *J. Geophys. Res.* 96, 15675–15681. Ingersoll, A. P., 1990, *Nature* 344, 315–317. Jacobson, R. A., Riedel, J. E., & Taylor, A. H., 1991, *Astron. & Astrophys.* 247, 565–575. McCord, T. B., 1966, *Astron. J.* 71, 585–590. Owen, W. M., Vaughan, R. M., & Synnott, S. J., 1991, *Astron. J.,* 101, 1511–1515. Tyler, G. L., Sweetnam, D. N., Anderson, J. D., *and 14 additional authors*, 1989, *Science* 246, 1466–1473.

12

PLUTO AND CHARON

Subsequent to Neptune's discovery in 1846, several astronomers, notably Percival Lowell, calculated that the remaining discrepancies between observed and predicted motions of Uranus and Neptune were due to the presence of a trans-Neptunian planet. Various searches were made for the predicted planet without success, until 18 February 1930 when Pluto was discovered by Clyde Tombaugh. After the discovery of Pluto, closer examination of photographic plates taken by Humason at Mount Wilson in 1919, during an unsuccessful search for the trans-Neptunian planet, also revealed the presence of Pluto (Tombaugh, 1961). Although both Tombaugh's and Humason's searches were motivated by perturbations of the orbits of Uranus and Neptune by the trans-Neptunian planet, it is now known that the perturbations were simply due to uncertainties in measurements and calculations of the observed and predicted planetary motions.

Pluto's orbit has a semimajor axis of 39.53 AU, but because of the high eccentricity of ~0.25, the aphelion and perihelion are 49.3 AU and 29.6 AU, respectively. Pluto comes inside Neptune's orbit (e.g., from 21 Jan. 1979 to 14 Mar. 1999), which has a semimajor axis of 30.07 AU. However, the two planets are never closer than ~17 AU because Pluto is never near perihelion when it is in conjunction with Neptune. This mutual avoidance is due to the 3:2 orbital resonance of Pluto and Neptune. In fact, Pluto comes closer to Uranus (~11 AU) than it comes to Neptune. Pluto's orbit also is highly inclined (~17°). In the 1930s, the unusual orbital parameters led to the well-known, but probably incorrect, idea that Pluto is an escaped satellite of Neptune. Although Triton and Pluto have common characteristics, it is more plausible that both bodies formed in the solar nebula and are primitive icy planetesimals (i.e., Kuiper belt objects).

Nearly 50 years after Pluto's discovery, James W. Christy of the U.S. Naval Observatory discovered Pluto's satellite, Charon, while attempting to refine Pluto's orbital parameters. The discovery of Charon was serendipitous because shortly thereafter Pluto and Charon started a five-year series of mutual eclipses (or events) that take place only once every ~124 years (the next series starts in 2109). The mutual events (transits of Pluto by Charon and occultations of Charon by Pluto) occur once about every 3.2 days, half of Charon's ~6.4 day orbital period. The mutual events

Pluto

occur because Pluto's rotational axis is tilted ~122° to its orbital plane; that is, Pluto, like Uranus, is tipped over on its side, and Charon orbits Pluto with an inclination of ~97°. During the past two decades, studies of the Pluto-Charon system during the mutual events, a stellar occultation by Pluto on 9 June 1988, recent Hubble Space Telescope images of Pluto, and spectroscopic observations of Pluto and Charon have provided much information. However, several uncertainties still remain.

The Pluto-Charon system is like a double planet system because of the similar sizes and masses of the two bodies. Pluto's radius is ~1150–1200 km and is uncertain because of the presence of an atmosphere. Charon's radius is ~590–640 km about 52% that of Pluto. The derived mass of the Pluto-Charon system is about 1.48×10^{22} kg, and the Charon/Pluto mass ratio is ~0.11–0.16. In comparison, the moon's radius is ~27% that of Earth, and the Moon/Earth mass ratio is ~0.012. Charon and Pluto orbit their center of mass, which is between the two bodies, with a Pluto-Charon distance of ~19,540 km. The Pluto-Charon system is also unusual because both bodies have apparently tidally despun each other (Charon's orbital period and Pluto's rotational period are both ~6.4 days).

Pluto's surface is dominantly N_2 ice with smaller amounts of CH_4 and CO; the surface temperature of 40 K implies a surface pressure of ~58 μbar N_2. Recently, CH_4 gas was detected in Pluto's atmosphere; the column abundance is $1.2^{+3.15}_{-0.87}$ cm amagat, consistent with CH_4 being only a minor atmospheric gas. Pluto's large orbital eccentricity causes dramatic seasonal changes in the atmosphere, which may condense onto the surface when Pluto is at aphelion. In contrast, Charon's surface is mainly H_2O ice; no CH_4, N_2, or CO ices have been detected. At present, no atmosphere has been detected on Charon; if present, any atmosphere must consist mainly of gases other than H_2O because of the low water vapor pressure over ice at Charon's surface temperature of ~40 K.

Sources and further reading: Lewis, J. S., & Prinn, R. G., 1984, *Planets and their atmospheres*, Academic Press, New York, pp. 470. McKinnon, W. B., Simonelli, D. P., & Schubert, G., 1997 in *Pluto and Charon* (Stern, S. A., & Tholen, D. J., eds.), Univ. of Arizona Press, Tucson, pp. 295–343. Stern, S. A., 1992, *Annu. Rev. Astron. Astrophys.* 30, 185–233. Stern, S. A., & Tholen, D. J. (eds.), *Pluto and Charon*, Univ. of Arizona Press, Tucson, pp. 718. Tombaugh, C. W., 1961, in *Planets and satellites*, (Kuiper, G. P., & Middlehurst, B. M., eds.), Univ. of Chicago Press, Chicago, pp. 12–30. Tryka, K. A., Brown, R. H., Cruikshank, D. P., Owen, T. C., & DeBergh, C., 1994, *Icarus* 112, 513–527. Young, L. A., Elliot, J. L., Tokunaga, A., DeBergh, C., & Owen, T., 1997, *Icarus* 127, 258–262.

Table 12.1 Some Physical Properties of the Pluto-Charon System

Property	Value	Property	Value
System mass (10^{22} kg)		Mass ratio (Charon/Pluto)	
[NO96] (selected value)	1.476±0.018	[NO96] (selected value)	0.124±0.008
[TB97]	1.471±0.002	[TB97]	$0.110^{+0.063}_{-0.056}$
[YOE94]	1.432±0.013	[YOE94]	0.1566
GM_{sys} (km^3 s^{-2})	985±12		±0.0035
Pluto's mass (10^{22} kg)		Charon's mass (10^{21} kg)	
[NO96] (selected value)	1.314±0.018	[NO96] (selected value)	1.62±0.09
[TB97]	1.325±0.071	[TB97]	1.46±0.07
[YOE94]	1.238±0.012	[YOE94]	1.94±0.04
Pluto's radius (km)		Charon's radius (km)	
[EY92]	1206±11	[YOE94]	642±11
[RBF94]*	1152±7	[RBF94]*	592±5
[TB90]	1151±6	[TB90]	593±13
[YB94]*	1180±24	[YB94]*	629±21
Pluto's density (g cm^{-3})	1.79–2.06	Charon's density (g cm^{-3})	1.46–1.86
Pluto's orbital parameters		Charon's semimajor axis (km)	
Semimajor axis (AU)	39.533	[BGT89, NO96]	19640±320
Sidereal period (⊕ years)	248.0	[TB97]	19636±8
Eccentricity of orbit	0.249	[YOE94]	19460±58
Inclination of orbit to ecliptic	17.146°	Charon's orbital period (days)[†]	6.3872
Sidereal rotation period (⊕ days), retrograde	6.3872	Charon's orbital eccentricity	0.00020 ±0.00021
Axial tilt to orbit	122.52°	Charon's orbital inclination	96.56°
Pluto's mean gravity (m s^{-2})	~0.645	Charon's mean gravity (m s^{-2})	~0.021

* Recalibrated to the semimajor axis of [NO96] by [McK97].
† Charon is in synchronous rotation.

Sources: **[BGT89]** Beletic, J. W., Goody, R. M., & Tholen, D. J., 1989, *Icarus* 79, 38–46. **[EY92]** Elliot, J. L., & Young, L. A., 1992, *Astron. J.* 103, 991–1015. **[McK97]** McKinnon, W. B., Simonelli, D. P., & Schubert, G., 1997, in *Pluto and Charon* (Stern, S. A., & Tholen, D. J., eds.), Univ. of Arizona Press, Tucson, pp. 245–343. **[NO96]** Null, G. W., & Owen, W. N., 1996, *Astron. J.* 111, 1368–1381. **[RBF94]** Reinsch, K., Burwitz, V., & Festou, M. C., 1994, *Icarus* 108, 209–218. **[St92]** Stern, A., 1992, *Annu. Rev. Astron. Astrophys.* 30, 185–233. **[TB90]** Tholen, D. J., & Buie, M. W., 1990, *Bull. Am. Astron. Soc.* 20, 807. **[TB97]** Tholen, D. J., & Buie, M. W., 1997, *Icarus* 125, 245–260. **[YB94]** Young, E. F., & Binzel, R. P., 1994 *Icarus* 108, 219–224. **[YOE94]** Young, E. F., Olkin, C. B., Elliot, J. L., Tholen, D. J., & Buie, M. W., 1994, *Icarus* 108, 186–199.

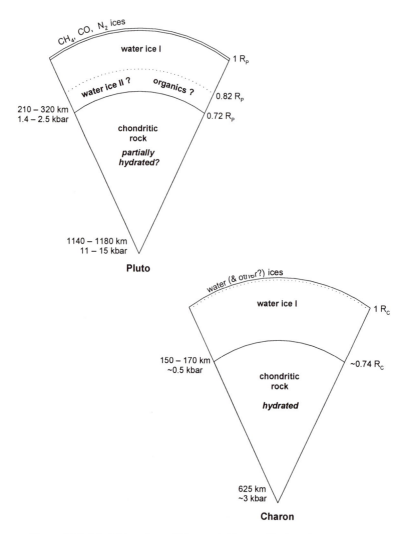

Figure 12.1 Model interiors of Pluto and Charon. The interior structures are not well constrained; other compositional models involving different ice modifications and/or ice-rock mixtures are also possible.

Sources: McKinnon, W. B., Simonelli, D. P., & Schubert, G., 1997, in *Pluto and Charon*, Univ. of Arizona Press, Tucson, pp. 245–343. Simonelli, D. P., & Reynolds, R. T., 1989, *Geophys. Res. Lett.* 16, 1209–1212.

13

THE ASTEROIDS

13.1 Introduction

In 1801, G. Piazzi discovered Ceres, the largest body in the asteroid belt located between the orbits of Mars and Jupiter. Since then, over 10 thousand asteroids (also called minor planets) have been detected. Ceres is about 914 km in diameter; only about 140 other asteroids known have diameters larger than 100 km. Ceres, with a mass of 1.2×10^{21} kg, contributes about a quarter of all the mass in the asteroid belt. Vesta (2.4×10^{20} kg) and Pallas (2.2×10^{20} kg) are the next two most massive asteroids.

General information about asteroids and listings of numbered asteroids (those with more or less well-defined orbital elements) can be found in *Asteroids II* [BGM89]. The Palomar-Leiden (PL) survey of faint asteroid extends the list of numbered asteroids [HHH70, WH87].

In 1996, the *Galileo* Mission brought more information about two smaller asteroids, 243 Ida and its moon Dactyl, and 951 Gaspra (see section 13.2). The *N*ear *E*arth *A*steroid *R*endezvous *(NEAR)* mission started on 17 February 1996 and the spacecraft encountered the slow-rotating C-asteroid 253 Mathilde on 27 June 1997. On 10 January 1999, another rendezvous is planned with 433 Eros, a near Earth asteroid [JGR97]

The statistical distribution of asteroids as a function of their semi-major axis (and orbital periods) shows preferred clusters and gaps, which are due to asteroid orbital resonances with the planets, especially Jupiter. Asteroid orbital resonances with Jupiter in rational proportions are the commensurabilities. For example, the Hilda group asteroids are in the 3:2 resonance since their orbital periods are 2/3 of Jupiter's orbital period. Hilda group asteroids can have high eccentricities resembling those of short-period comets. Asteroid 279 Thule is the lonely outpost at 6.3 AU in the 3:4 commensurability. The Trojan asteroids are in 1:1 commensurability with Jupiter and have the same orbital period as Jupiter but these asteroids move at the Lagrangian points 60° ahead (L4) and 60° behind (L5) Jupiter in its orbital path. The asteroid 5261 Eureka is in a similar situation with Mars and could be called a "Mars Trojan" [MIM93].

The mean motion resonances with Jupiter of 3:1, 5:2, 7:3, 2:1 are less favorably populated and are known as Kirkwood gaps, which are apparent in a plot of asteroid frequency versus asteroid semimajor axis (at 2.5, 2.82, 2.96, and 3.3 AU for the resonances listed here). Asteroids with orbits passing through the region of the terrestrial planets are the Amor, Apollo, and Aten groups. The latter two groups have Earth crossing orbits and are also known as near Earth asteroids (NEA).

Asteroid families are believed to form by catastrophic disruption of larger parent asteroids. The family members have statistically significant similar orbital proper elements (semimajor axis, eccentricity, inclination). The proper elements describe the invariant asteroid motion, thus, these parameters record the initial proximity of the orbits after the disruption of the parent asteroid. The identification of asteroid families was pioneered by Hirayama in 1918 [Hir18]. Recent studies of asteroid family identifications using different statistical methods are being done by, for example, [Wil92, Ben93, MK94, ZBC95]. Depending on the number of asteroids included and on the set of proper elements and statistical criteria used, family classification may vary among different authors; therefore the results are not directly comparable. Table 13.3 lists some asteroid families.

Members of a given asteroid family are expected to have similar albedos, color indices, or taxonomic types, when they form as fragments from a disruption of a parent body. The study of asteroid families can provide constraints on the evolution of the asteroid belt, as well as on the origin of meteorites. Impacts within the asteroid main belt leading to formation of families may be responsible for the injection of fragments into the important 3:1 mean motion resonance with Jupiter at 2.5 AU and the v_6 secular resonance at 2.18 AU. Fragments ejected into the inner solar system may be part of the NEA population or, in rare instances, may collide with the inner planets.

Depending on statistics of orbital proper elements, clusters and clans are recognized as subtypes of asteroid families. Clusters have sharp limits, and orbital proper elements of family members are clearly statistically resolved from other objects. Clans consist of larger associations and have less resolved boundaries in proper element statistics. Clans are often split into subgroups.

A well-documented asteroid family is the Vesta family, where the presence of a crater on Vesta, a number of NEA with basaltic compositions similar to Vesta, and the presence of a matching meteorite group (eucrites) support the evolutionary scenario of asteroid families [BX93]. Other known asteroid families include the Eos, Flora, Koronis, Maria and

Themis families. For more information, see [BGM89, Wil92, Ben93, MK94, ZBC95].

The criteria for the spectral classification of asteroids are listed in Table 13.2. Results of a large reflection spectroscopy survey of asteroids in the eight-color system are given by [ZTT85], from which the compositional zoning of the asteroid belt (type S to C to D) with increasing heliocentric distance is evident. For a review on meteorite and asteroid reflectance spectroscopy see [PM94].

Asteroid diameters, D, can be calculated from the absolute magnitude, H, and geometric albedo P from the relationship [BHD89]:

$$\log D \text{ (km)} = 0.5 \cdot (6.259 - 0.4 H - \log P)$$

Infrared spectroscopy and orbital considerations suggest an asteroid-meteorite link of certain meteorite classes to particular asteroids and their families. Some examples are listed in Table 13.1.

Table 13.1 Asteroids and Possibly Related Meteorite Groups

Asteroid	Meteorite Class	Sources
4 Vesta	Basaltic achondrites (eucrites)	BX93
2078 Nanking	Ordinary chondrites (H)	XBB95
3103 1982 BB	Enstatite achondrites (aubrites)	GRK92
3628 Boznemcova	Ordinary chondrites (L6, LL6?)	BXB93

Sources and further reading: [**Ben93**] Bendjoya, P., 1993, *Astron. & Astrophys. Suppl.* 102, 25–55. [**BGM89**] Binzel, R. P., Gehrels, T., Matthews, M. S. (eds.) 1989, *Asteroids II*, Univ. Arizona Press, Tucson, AZ., pp. 1258. [**BHD89**] Bowell, E., Hapke, B., Domingue, D., Lumme, K., Peltoniemi, J., & Harris, A. W., 1989, in [BGM89], pp. 524–556. [**BX93**] Binzel, R. P. & Xu, S. 1993, *Science* 260, 186–191. [**BXB93**] Binzel, R. P., Xu., S., Bus, S. J., Skrutskie, M. F., Meyer, M. R., Knezek, P., & Barker, E. S., 1993, *Science* 262, 1541–1543. [**GRK92**] Gaffey, M. J., Reed, K. L., & Kelley, M. S., 1992, *Icarus*, 100, 95–109. [**HHH70**] van Houten, C. J., van Houten-Groeneveld, I., Herget, P., & Gehrels, T., 1970, *Astron. & Astrophys. Suppl.* 2, 339–448. [**Hir18**] Hirayama, K., 1918, *Astron. J.* 31, 185–188. [**JGR97**] Special section about the NEAR mission in *J. Geophys. Res.* 102, (No. E10, 25 October 1997 issue), pp. 23,695–23,780. [**MIM93**] Mikkola, S., Innanen, K. A., Muinonen, K., & Bowell, E., 1993, *Celest. Mech. Dyn. Astron.* 58, 53–64. [**MK94**] Milani, A., & Knezevic, Z., 1994, *Icarus* 107, 219–254. [**PM94**] Pieters, C. M., & McFadden, L. A., 1994, *Annu. Rev. Earth Planet. Sci.* 22, 457–497. [**Wil92**] Williams, J. G., 1992, *Icarus* 96, 251–280. [**WH87**] Williams, J. G., & Hierath, J. E., 1987, *Icarus* 72, 276–303. [**XBB95**] Xu, S., Binzel, R. P., Burbine, T. H., & Bus, S. J., 1995, *Icarus* 115, 1–35. [**ZBC95**] Zappalà, V., Bendjoya, Ph., Cellino, A., Farinella, P., & Froeschlé, C., 1995, *Icarus* 116, 291–314. [**ZTT85**] Zellner, B., Tholen, D. J., & Tedesco, E. F., 1985, *Icarus* 61, 355–416.

Table 13.2 Asteroid Taxonomic Classes and Compositional Interpretations

Bell Superclass	Tholen Class	IRAS Albedo	Possible Mineralogy	Meteorite Analogs
Primitive	C	0.04–0.06	hydrated silicates, phyllosilicates, clays, organics	CM-, CI(?)-chondrites
	D	0.04–0.07	hydrated silicates? organics (kerogens)? ices?	...
	K		olivine, pyroxene, carbon	CV-, CO-chondrites
	P	<0.04	anhydrous silicates, organics (C-rich), ices	...
	Q	0.16–0.21	olivine, low-Ca pyroxene, metal (gray)?	ordinary chondrites
Metamorphic	B	0.04–0.08 subclass of C	hydrated silicates, clays, opaque materials	highly altered carbonaceous chondrites?
	F	0.04–0.08 subclass of C	hydrated silicates	highly altered carbonaceous chondrites?
	G	0.09 subclass of C	hydrated silicates, phyllosilicates	highly altered carbonaceous chondrites?
	T	<0.10	troilite? metal?	...
Igneous	A	0.12–0.39	olivine, metal?	olivine achondrites? pallasites?
	E	0.38	enstatite? metal?	aubrites, enstatite chondrites
	M	0.12–0.22	FeNi? enstatite?	iron meteorites, enstatite meteorites?
	R	0.34	low-Ca-pyroxene, olivine?	olivine-rich achondrites?
	S	0.14–0.17	olivine, low-Ca pyroxene, FeNi (red)? spinel?	primitive achondrites, pallasites? irons? ureilites?
	V	0.38	low-Ca pyroxene, plagioclase, olivine?	basaltic achondrites (eucrites)
	J	0.38 subclass of V	low-Ca pyroxene, plagioclase	basaltic achondrites (diogenites)

Additional descriptions

	U	added to classification letter if object has *unusual* spectrum
	I	*inconsistent* data for classification
	X	spectrum is of class E, M, or P, but good albedo data are needed

Sources and further information: Bell, J. F., 1986, *LPSC* XVII, 985–986. Binzel, R. P., Gehrels, T., & Matthews, M. S. (eds.), 1989, *Asteroids II*, Univ. Arizona Press, Tucson, AZ., pp. 1258. Binzel, R. P., & Xu, S., 1993, *Science* 260, 186–191. Gradie, J., & Tedesco, E., 1982, *Science* 216, 1405–1407. Pieters, C. M., & McFadden, L. A., 1994, *Annu. Rev. Earth Planet. Sci.* 22, 457–497. Xu, S., Binzel, R. P., Burbine, T. H., & Bus, S. J., 1995, *Icarus* 115, 1–35.

Table 13.3 Asteroid Zones, Groups, and Some Asteroid Families

Zone or Family	Limits for a, Q, q (AU)	Eccentricity e	Inclination i	Notes [a]
AAA Atens	$a < 1$, $Q > 0.983$			
Apollos	$a > 1.0$, $q < 1.017$			
Amors	$a > 1.0$, $1.017 < q < 1.3$			
MC Mars crossers	$a > 1$, $q \leq 1.666$			
HUN Hungaria zone	$1.78 \leq a \leq 2.00$	$e < 0.18$	$16° \leq i \leq 34°$	
Hungaria family	$1.909 \leq a \leq 1.949$	$0.054 \leq e \leq 0.073$	$21° \leq i \leq 22°$	
Fl Flora family	$2.169 \leq a \leq 2.211$	$0.127 \leq e \leq 0.153$	$4.36° \leq i \leq 5.91°$	limited by MC
I Main belt	$2.30 < a \leq 2.50$		$i < 18°$	
Vesta family	$2.356 \leq a \leq 2.390$	$0.093 \leq e \leq 0.100$	$6.43° \leq i \leq 7.87°$	W 169
Ausonia family	$2.358 \leq a \leq 2.395$	$0.112 \leq e \leq 0.120$	$6.20° \leq i \leq 6.89°$	W 165
Ny Nysa family	$2.419 \leq a \leq 2.477$	$0.146 \leq e \leq 0.177$	$2.92° \leq i \leq 3.67°$	W 24
Fortuna family	$2.430 \leq a \leq 2.449$	$0.119 \leq e \leq 0.138$	$1.66° \leq i \leq 2.52°$	W 158
PHO Phocaea group	$2.25 \leq a \leq 2.50$	$e \leq 0.10$	$18° \leq i \leq 32°$	
IIa Main belt	$2.500 < a \leq 2.706$		$i < 33°$	
Maria family	$2.530 \leq a \leq 2.586$	$0.071 \leq e \leq 0.104$	$14.7° \leq i \leq 15.4°$	W 4
Aurelia family	$2.570 \leq a \leq 2.620$	$0.238 \leq e \leq 0.258$	$4.30° \leq i \leq 5.51°$	W 146
Eunomia family	$2.612 \leq a \leq 2.735$	$0.127 \leq e \leq 0.170$	$12.9° \leq i \leq 14.0°$	W 140
IIb Main belt	$2.706 < a \leq 2.82$		$i \leq 33°$	
Eugenia family	$2.720 \leq a \leq 2.739$	$0.115 \leq e \leq 0.135$	$6.14° \leq i \leq 7.87°$	W 133
Pal Pallas zone	$2.500 \leq a \leq 2.82$			
Pallas family	$2.767 \leq a \leq 2.784$	$0.154 \leq e \leq 0.180$	$35.4° \leq i \leq 35.9°$	W 129
Ceres family	$2.767 \leq a \leq 2.812$	$0.086 \leq e \leq 0.105$	$9.26° \leq i \leq 10.4°$	W 67
IIIa Main belt	$2.82 < a \leq 3.03$	$e \leq 0.35$	$i \leq 30°$	
KOR Koronis zone	$2.833 \leq a \leq 2.919$	$0.039 \leq e \leq 0.067$	$1.83° \leq i \leq 2.41°$	W 3
EOS Eos zone	$2.997 \leq a \leq 3.026$	$0.052 \leq e \leq 0.089$	$9.56° \leq i \leq 10.8°$	W 2
IIIb Main belt	$3.03 < a \leq 3.27$	$e < 0.35$	$i \leq 30°$	
THE Themis family/zone	$3.077 \leq a \leq 3.225$	$0.134 \leq e \leq 0.172$	$0.74° \leq i \leq 2.12°$	W 1 (1A= fam. core)

continued

Table 13.3 *(continued)*

Zone or Family		Limits for a, Q, q (AU)	Eccentricity e	Inclination i	Notes [a]
	Hygiea family	$3.135 \leq a \leq 3.149$	$0.132 \leq e \leq 0.153$	$4.88° \leq i \leq 5.97°$	W 110
	Ursula family	$3.127 \leq a \leq 3.146$	$0.063 \leq e \leq 0.077$	$14.6° \leq i \leq 16.5°$	W 111
GR	Griqua group	$3.10 \leq a \leq 3.27$	$e \leq 0.35$	2:1 commensurability with Jupiter	
CYB	Cybele group	$3.27 < a \leq 3.70$	$e \leq 0.30$	$i \leq 25°$	
HIL	Hilda group	$3.70 < a \leq 4.20$	$e \leq 0.30$	$i \leq 20°$	
			3:2 commensurability with Jupiter		
T	Trojan group	$5.05 \leq a \leq 5.40$	1:1 commensurability with Jupiter		

a: semimajor axis; Q: aphelion; q: perihelion

[a] The numbers in the last column refer to the family numbers assigned by Williams, 1992.

Sources: Williams, J. G., 1992, *Icarus* 96, 251–280. Zellner, B., Thirunagari, A., & Bender, D., 1985, *Icarus* 62, 505–511.

13.2 Gaspra

On 29 October 1991, the *Galileo* spacecraft flew by 951 Gaspra, the first asteroid to be imaged by a spacecraft. The 57 images show Gaspra to be an irregularly shaped body with craters and grooves on its surface. The results of this flyby are described in several papers published in the January 1994 (Vol. 107, No. 1) issue of *Icarus*. Some orbital and physical properties of Gaspra are summarized in Table 13.4. Gaspra's mass and density are unknown and were not determined by the *Galileo* flyby.

Table 13.4 Some Physical Properties of 951 Gaspra

Property	Value	Property	Value
Mean radius (km)	6.1±0.4	Semimajor axis (AU)	2.21
Best fit ellipsoid (km)	18.2×10.4×9.4	Eccentricity of orbit	0.17
Area (km²)	525±50	Inclination of orbit	4.1°
Volume (km³)	954±200	Orbital period (yr)	3.28
Mass (kg)	...	Orbital velocity (km s⁻¹)	16.9–23.8
Visual geometric albedo	0.22±0.06	Rotation period (h)	7.042
Taxonomic class	S		prograde
$I/(MR^2)$	A = 0.28±0.03		
	B = 0.60±0.03		
	C = 0.63±0.03		

Sources: Results for Gaspra from the *Galileo* flyby are collected in: *Icarus* 107 (No. 1, Jan. 1994). Veverka, J., Belton, M., Klaasen, K., & Chapman, C., 1994, *Icarus* 107, 2–17.

13.3 Ida and Dactyl

The *Galileo* spacecraft flew by 243 Ida (a Koronis family asteroid) in August 1993 and the solid state imager (SSI) obtained 47 images showing Ida and its smaller satellite Dactyl, the first confirmed asteroid-satellite pair. The results of this flyby are described in several papers published in the March 1996 (Vol. 120, No. 1) issue of *Icarus*. Ida is an irregularly shaped asteroid with a mean diameter of 31.4 km, and Dactyl is a smaller elongated object with a mean diameter of 1.4 km. Dactyl's rotation rate is slow and may be longer than 8 hours. Both Ida and Dactyl are S-type asteroids. The existence of the Ida-Dactyl pair is intriguing because crater counting suggests that Ida is at least 2 billion years old, but Dactyl has a shorter collisional disruption lifetime.

Table 13.5 Some Physical Properties of 243 Ida and (243) 1 Dactyl

Property	243 Ida	1993 (243) 1 Dactyl
Mean radius (km)	15.7	0.7
Best fit ellipsoid (km)	59.8×25.4×18.6	1.6×1.4×1.2
Volume (km^3)	16100±1900	~1.4
Mass (kg)	4.2×10^{16}	~ 3.7×10^{12}
Density (g cm^{-3})	2.6±0.5	~ 2.6
Gravity (cm s^{-2})	0.31–1.1	~ 0.05
Rotation period (h)	4.633632±0.000007 retrograde	>8, possibly synchronous
Visual geometric albedo	0.21$^{+0.03}_{-0.01}$	0.20
Taxonomic class	S	S

Orbital parameters	Ida - Sun	Dactyl - Ida
Semimajor axis	2.864 AU	85 km
Eccentricity of orbit	0.04311	...
Inclination of orbit	1.371° (to ecliptic)	172° (to Ida's equator)
Orbital period	1770.1 days	> 24.7 h (retrograde)
Orbital velocity (km s^{-1})	16.9–18.4	0.006

Sources: Results from the *Galileo* flyby are collected in *Icarus* 107, (No. 1, March 1994). Belton, M. J. S., Chapman, C. R., Klaasen, K. P., Harch, A. P., Thomas, P. C., Veverka, J., McEwan, A. A., & Pappalardo, R. T., 1996, *Icarus* 120, 1–19.

13.4 Mathilde

Asteroid 253 Mathilde was discovered in 1885 by J. Palisa and became the first asteroid investigated by the *NEAR* spacecraft. The closest flyby of the *NEAR* spacecraft was on 27 June 1997, when it approached Mathilde to 1212 km. Mass and volume determinations from the spacecraft instruments yield a very low density of 1.3 g cm^{-3}, indicating that the asteroid may be very porous. The density is lower than that of carbonaceous chondrites, which are believed to be the meteoritic analogs of C-type asteroids (Table 13.1). However, Earth-based spectroscopy does not show hydrated minerals on Mathilde's surface, suggesting that carbonaceous chondrites may not be suitable meteoritic analogs. Imaging by the *NEAR* spacecraft also revealed that Mathilde's surface is cratered. Two very large craters with diameters of 33 and 26 km have dimensions comparable to Mathilde's mean radius.

Table 13.6 Some Physical Properties of 253 Mathilde

Property	Value	Property	Value
Mean radius (km)	26.5±1.3	Semimajor axis (AU)	2.647
Best fit ellipsoid (km)	66×48×46	Eccentricity of orbit	0.230
Volume (km^3)	$78,000^{+12,000}_{-11,000}$	Inclination of orbit	6.89°
Mass (kg)	$(1.033\pm0.044)10^{17}$	Orbital period (yr)	4.31
Density (g cm^{-3})	1.3±0.2	Orbital velocity (km s^{-1})	16.5–20.9
Taxonomic class	C	Visual geometric albedo	0.035–0.05, 0.036
Rotation period (h)	17.4		

Sources: Rivkin, A. S., Clark, B. E., Britt, D. T., & Lebofsky, L. A., 1997, *Icarus* 127, 225–257. Veverka, J., Thomas, P., Harch, A., Clark, B., Bell, J. F., Carcich, B., Joseph, J., Chapman C., Merline, W., Robinson, M., Malin, M., McFadden, L. A., Murchie, S., Hawkins, S. E., Farquhar, R., Izenberg, N., & Cheng, A., 1997, *Science* 2109–2114. Yeomans, D. K., Barriot, J. P., Dunham, D. W., Farquhar, R. W., Giorgini, J. D., Helfrich, C. E., Konopliv, A. S., McAdams, J. V., Miller, J. K., Owen, W. M., Scheeres, D. J., Synnott, S. P., & Williams, B. G., 1997, *Science* 278, 2106–2109.

13.5 Asteroid Data

Table 13.7 gives physical properties of some asteroids. The asteroids are listed in order of increasing semimajor axis for different asteroid zones, groups, or families. Within these groups, asteroids are listed by increasing asteroid number. For general sources and more data, see [BGM89]. The arrangement of the columns in Table 13.7 is as follows:

Column (1) Minor planet number and name.

Column (2) Taxonomic type, as defined in Table 13.2 [Tho89, TWM89, WMH97, XBB95].

Column (3)–(6) Proper orbital element data (a: semimajor axis of revolution; e: orbital eccentricity; q: perihelion distance $q = a (1 - e)$; and i: inclination to ecliptic) [Wil89, Wil92]; Amors, Atens, Apollos: [MTV89]; Hildas: [Sch82]; Trojans: [BS87].

Column (7) Rotation period [LHZ89, MDD95, WMH97].

Column (8)–(10) Diameter (in km), visual geometric albedo, P_{vis} from *IRAS* measurements, and absolute visual magnitude, H [Ted89, TWM89, WMH97, XBB95].

Column (11) Notes: Family membership numbers according to Williams [Wil92] are indicated by "W" + family number. Planet + "X" indicates planet crosser. Also indicated in the last column are alternate names, mass, or orbital commensurabilities, resonances, and Lagrangian points.

Entries listed in parentheses in columns (2)–(11) indicate that this property is not well known, except for provisional names of higher numbered asteroids.

Sources: **[BGM89]** Binzel, R. P., Gehrels, T., Matthews, M. S. (eds.), 1989, *Asteroids II*, Univ. Arizona Press, Tucson, pp. 1258. **[BS87]** Bien, R., & Schubart, J., 1987, *Astron. & Astrophys.* 175, 292–298. **[LHZ89]** Lagerkvist, C. I., Harris, A. W., & Zappalà, V., 1989, in *Asteroids II* (Binzel, R. P., Gehrels, T., Matthews, M. S., eds.), Univ. Arizona Press, Tucson, pp. 1162–1179. **[MDD95]** Mottola, S., de Angelis, G., di Martino, M., Erikson, A., Hahn, G., & Neukum, G., 1995, *Icarus* 117, 62–70. **[MTV89]** McFadden, L. A., Tholen, D. J., & Veeder, G. J., 1989, in *Asteroids II* (Binzel, R. P., Gehrels, T., Matthews, M. S., eds.), Univ. Arizona Press, Tucson, pp. 442–467. **[Sch82]** Schubart, J., 1982, *Astron. & Astrophys.* 114, 200–204. **[Ted89]** Tedesco, E. F., 1989, in *Asteroids II* (Binzel, R. P., Gehrels, T., Matthews, M. S., eds.), Univ. Arizona Press, Tucson, pp. 1090–1138. **[Tho89]** Tholen, D. J., 1989, *ibid.*, pp. 1139–1150. **[TWM89]** Tedesco, E. F., Williams, J. G., Matson, D. L., Veeder, G. J., Gradie, J. C., & Lebovsky, L. A., 1989, *ibid.*, pp. 1151–1161. **[Wil89]** Williams, J. G., 1989, *ibid.*, pp. 1034–1037. **[Wil92]** Williams, J. G., 1992, *Icarus* 96, 251–280. **[WMH97]** Wisniewski, W. Z., Michalowski, T. M., Harris, A. W., & McMillan, R. S., 1997, *Icarus* 126, 395–449. **[XBB95]** Xu, S., Binzel, R. P., Burbine, T. H., & Bus, S. J., 1995, *Icarus* 115, 1–35.

Table 13.7 Asteroid Data

Minor Planet	Type	a (AU)	e	q (AU)	i (deg)	$P_{Rotat.}$ (hrs)	Diam. (km)	Albedo P_{vis}	H (mag)	Notes
Aten Group Near Earth Asteroids $a < 1, Q > 0.983$										
2062 Aten	S	0.966	0.182	0.79	18.9	40.77	1.1	0.20	16.96	
2100 Ra-Shalom	C	0.83	0.436	0.47	15.8	19.79	2.4	0.09	16.12	Mercury X
2340 Hathor	CSU	0.84	0.450	0.46	5.9	...	(0.2)	...	20.2	Mercury X
3362 Khufu	...	0.99	0.469	0.53	9.9	...	0.7	0.16	18.1	Venus X
3554 Amun	M	0.97	0.280	0.70	23.4	...	2.0	0.17	15.82	Venus X
5381 Sekhmet	...	0.94	0.39	0.57	44.3	...	2	0.15	16.5	Venus X
5590 (1990 VA)	...	0.985	0.328	0.66	13.7	...	0.4	0.15	19.5	Venus X
Apollo Group Near Earth Asteroids $a > 1.0, q < 1.017$										
1566 Icarus	S	1.08	0.827	0.19	22.9	2.273	1.06	0.42	15.95	Me. X, (5×10^{12}kg)
1620 Geographos	S	1.24	0.336	0.83	13.3	5.227	2.0	0.19	14.97	
1685 Toro	S	1.37	0.436	0.77	9.4	10.196	12.2	0.03	13.96	
1862 Apollo	Q	1.47	0.560	0.65	6.3	3.065	1.5	0.21	16.23	Venus X
1863 Antinous	SU	2.26	0.607	0.89	18.4	4.02	1.8	0.18	15.81	
1864 Daedalus	SQ	1.46	0.615	0.56	22.2	8.57	(3.1)	...	15.02	Venus X, v_{16}
1865 Cerberus	S	1.08	0.467	0.58	16.1	6.80	1.0	0.26	16.91	Venus X
1866 Sisyphus	S	1.89	0.539	0.87	41.2	2.4	8.2	0.18	13.2	
1981 Midas	S	1.78	0.650	0.62	39.8	5.22	3.6	(0.18)	16.9	Venus X
2063 Bacchus	...	1.08	0.349	0.70	9.4	...	1	...	17.6	Venus X
2101 Adonis	...	1.87	0.764	0.44	1.4	...	1	...	18.2	Mercury X
2102 Tantalus	...	1.29	0.298	0.90	64	...	2	...	16.2	
2135 Aristaeus	...	1.60	0.5.03	0.80	23	...	1	...	17.94	
2201 Oljato	...	2.18	0.711	0.63	2.5	>24	1.4	0.42	15.41	Venus X
2212 Hephaistos	SG	2.16	0.835	0.36	11.9	...	6.5	(0.18)	13.41	Mercury X
2329 Orthos	...	2.40	0.658	0.82	24.4	...	3	...	15.1	Venus X
3103 Eger	E	1.41	0.355	0.91	20.9	5.71	1.4	0.63	14.7	(1982 BB)
3200 Phaethon	F	1.27	0.890	0.14	22.1	4.08	7.0	0.08	14.65	Me. X, (1983 TB)
3360 (1981 VA)	...	2.46	0.744	0.63	22.0	...	1.8	0.07	16.20	Mercury X
3361 Orpheus	...	1.21	0.323	0.82	2.7	...	0.8	...	19.03	(1982 HR)
3752 Camillo	...	1.41	0.303	0.99	55.5	...	2	...	15.5	(1985 PA)
3838 Epona	...	1.50	0.701	0.45	29.3	...	3	...	15.4	Me. X, (1986 WA)
4015 Wilson-Harrington	CF	2.64	0.623	0.995	2.8	...	5	...	15.99	(1979 VA)
4034 (1986 PA)	...	1.06	0.444	0.59	11.2	...	1	...	18.1	Venus X
4179 Toutatis	S	2.51	0.640	0.90	0.5	...	2.7	(0.18)	15.3	3:1, (1989 AC)
4183 Cuno	...	1.98	0.637	0.72	6.8	...	4	...	14.5	Venus X
4197 (1982 TA)	S	2.30	0.773	0.52	12.2	3.54	6.8	(0.18)	13.32	Venus X

continued

Table 13.7 *(continued)*

Minor Planet	Type	a (AU)	e	q (AU)	i (deg)	$P_{Rotat.}$ (hrs)	Diam. (km)	Albedo P_{vis}	H (mag)	Notes
4341 Poseidon	...	1.84	0.679	0.59	11.9	...	3	...	15.6	Ven. X, (1987 KF)
4450 Pan	...	1.44	0.587	0.55	5.5	...	1	...	17.1	Ven. X, (1987 SY)
4486 Mithra	...	2.20	0.661	0.75	3.0	...	...	...	15.6	(1987 SB)
4660 Nereus	...	1.49	0.360	0.95	1.4	...	1	...	18.3	(1982 DB)
5011 Ptah	...	1.68	0.523	0.80	7.9	...	1	...	17.0	(P-L 6743)
6063 Jason	S	2.22	0.764	0.52	4.8	...	...	...	...	Ven. X, (1984 KB)
(1986 JK)	C	2.80	0.680	0.90	2.1	...	...	...	...	
(1988 TA)	C	1.64	0.518	0.79	2.7	...	...	...	...	
Amor Group Asteroids $a > 1.0$, $1.017 < q < 1.3$										
433 Eros	S	1.46	0.223	1.13	10.8	5.27	22	0.18	11.24	$(5 \times 10^{15}$ kg)
719 Albert	...	2.58	0.545	1.17	11.2	...	...	...	15.6	
887 Alinda	S	2.49	0.559	1.10	9.3	73.97	4.2	0.23	13.83	3:1
1036 Ganymed	S	2.66	0.5.37	1.23	26.5	10.308	41	0.17	9.42	
1221 Amor	...	1.92	0.435	1.08	11.9	...	0.9	(0.18)	17.6	
1580 Betulia	C	2.19	0.490	1.12	52.1	6.13	7.4	0.03	14.55	near v_5
1627 Ivar	S	1.86	0.396	1.12	8.4	4.80	8.1	0.12	12.88	
1915 Quetzalcoatl	SMU	2.54	0.574	1.08	20.5	4.9	0.3	0.29	19.05	3:1
1916 Boreas	S	2.27	0.450	1.25	12.8	...	(3.1)	(0.18)	15.03	
1917 Cuyo	...	2.15	0.505	1.06	24.0	2.69	3.6	(0.18)	14.7	
1943 Anteros	S	1.43	0.256	1.06	8.7	...	1.8	0.22	15.83	
1951 Lick	A	1.39	0.062	1.30	39.1	4.42	4.98	0.026	16.1	
1980 Tezcatlipoca	SU	1.71	0.365	1.09	26.8	7.25	4.3	0.21	14.07	
2059 Baboquivari	...	2.65	0.526	1.26	11.0	...	31.7	0.092	14.7	
2061 Anza	TCG	2.26	0.538	1.05	3.7	11.50	(2.7)	0.05	16.7	
2202 Pele	...	2.29	0.512	1.12	8.8	...	2.0	0.06	17.2	
2608 Seneca	S	2.49	0.582	1.04	15.4	8.0	0.9	0.16	17.57	3:1
3102 Krok	QRS	2.15	0.449	1.19	8.4	148	...	...	16.04	(1981 QA)
3122 Florence	...	1.77	0.422	1.02	22.2	5	4	(0.18)	14.2	(1981 ET₃)
3199 Nefertiti	S	1.57	0.284	1.13	33.0	3.0	2.2	0.26	15.03	(1982 RA)
3288 Seleucus	S	2.03	0.458	1.10	5.9	75	2.8	0.17	15.34	(1982 DV)
3352 McAuliffe	...	1.88	0.369	1.19	4.8	...	...	...	15.8	(1983 SA)
3551 Verenia	V	2.09	0.488	1.07	9.5	4.930	0.8	0.40	16.75	(1983 RD)
3552 Don Quixote	D	4.23	0.713	1.21	30.8	3	38.8	0.02	13	(1983 SA)
3553 Mera	...	1.64	0.321	1.12	36.8	...	...	...	...	(1985 JA)
3757 (1982 XB)	S	1.84	0.446	1.02	3.9	9.012	0.5	0.15	18.95	
3908 (1980 PA)	V	1.93	0.459	1.04	2.2	...	1	0.40	17.4	

continued

Table 13.7 *(continued)*

Minor Planet	Type	*a* (AU)	*e*	*q* (AU)	*i* (deg)	$P_{Rotat.}$ (hrs)	Diam. (km)	Albedo P_{vis}	H (mag)	Notes
4055 (1985 DO₂)	V	1.82	0.326	1.23	23.2	...	...	...	...	
4688 (1980 WF)	QU	2.23	0.514	1.08	6.4	...	...	...	...	
5370 Taranis	C	3.35	0.631	1.23	19.0	...	4	0.05	16.0	2:1, (1986 RA)
5797 (1980 AA)	S	1.89	0.444	1.05	4.2	2.706	...	...	...	
6178 (1986 DA)	M	2.82	0.585	1.17	4.3	3.58	2.3	0.12	16.0	5:2
7236 (1987 PA)	C	2.74	0.557	1.21	16.1	...	...	...	...	
(1977 VA)	XC	1.86	0.394	1.13	3.0	...	...	...	...	
Mars Crosser Group *a* > 1, *q* ≤1.666										
132 Aethra	M	2.611	0.212	2.057	32.2	5.168	47.0	0.14	9.35	
475 Ocllo	X	2.596	0.251	1.944	19.5	...	31.0	0.033	11.86	
699 Hela	S	2.616	0.235	2.001	19.7	3.656	(13)	(0.18)	11.99	
1011 Laodamia	S	2.394	0.383	1.477	4.82	...	(8.5)	(0.18)	12.85	
1139 Atami	S	1.947	0.220	1.519	14.2	>15	8.4	(0.18)	12.55	
1198 Atlantis	S	2.249	0.281	1.617	4.24	...	4	(0.19)	14.6	
1747 Wright	AU	1.709	0.121	1.502	24.8	...	8.11	0.11	13.38	
2035 Stearns	E	1.884	0.142	1.616	25.6	...	5.3	(0.4)	12.78	
2099 Opik	S	2.304	0.325	1.555	28.8	...	(2.6)	(0.18)	15.44	
2368 Beltrovata	SQ	2.106	0.229	1.624	6.37	5.9	2.3	0.13	15.5	v_6
2744 Birgitta	S	2.299	0.274	1.669	8.34	9:02	(3.0)	(0.18)	15.09	(1975 RB)
Hungaria Zone 1.78 ≤ *a* ≤ 2.00, *e* < 0.18, 16° ≤ *i* ≤ 34°										
434 Hungaria	E	1.944	0.067	1.814	21.0	26.51	10.0	0.46	11.47	W190
1019 Strackea	S	1.912	0.065	1.788	7.76	...	9.55	0.15	12.73	W191
1025 Riema	E	1.979	0.055	1.870	24.8	...	5.5	(0.40)	12.87	
1103 Sequoia	E	1.934	0.073	1.793	21.1	3.049	6.1	0.48	12.49	W190
1235 Schorria	CX	1.910	0.075	1.767	25.8	...	13	(0.06)	12.96	W191
1355 Magoeba	X	1.853	0.070	1.723	23.8	...	...	...	13.18	W191
1453 Fennia	S	1.897	0.041	1.819	24.6	6	(7.3)	0.251	12.81	W191
1509 Esclangona	S	1.866	0.042	1.788	24.7	...	12.1	0.095	12.74	W191
1656 Suomi	S	1.878	0.087	1.715	24.2	2.42	9.31	0.11	13.1	W191
1727 Mette	S	1.854	0.071	1.722	8.11	2.63	(9.2)	(0.18)	12.7	
1750 Eckert	S	1.926	0.116	1.703	23.2	...	(6.3)	(0.18)	13.52	
1919 Clemence	X	1.936	0.068	1.804	21.4	...	...	...	13.77	W190
1920 Sarmiento	X	1.930	0.077	1.781	21.0	...	...	...	14.34	W190
2048 Dwonik	E	1.954	0.049	1.858	39.4	...	(3.5)	(0.45)	13.79	W190
2001 Einstein	X	1.933	0.106	1.728	24.8	...	...	...	12.96	
2131 Mayall	S	1.887	0.080	1.736	36.0	...	9.01	0.14	12.97	W190

continued

Table 13.7 *(continued)*

Minor Planet	Type	a (AU)	e	q (AU)	i (deg)	P_Rotat. (hrs)	Diam. (km)	Albedo P_vis	H (mag)	Notes
Flora Zone 2.10 ≤ a ≤ 2.3, i ≤ 11° (limited by Mars Crossers)										
8 Flora	S	2.201	0.141	1.891	5.57	12.79	141	0.22	6.48	W189
18 Melpomene	S	2.296	0.174	1.896	10.3	11.572	148	0.22	6.41	
40 Harmonia	S	2.267	0.019	2.224	3.67	9.136	111	0.20	7.14	
43 Ariadne	S	2.203	0.140	1.895	3.10	5.751	65.3	0.28	7.89	W185
80 Sappho	S	2.296	0.147	1.958	9.32	14.05	81.7	0.15	8.10	
228 Agathe	S	2.201	0.183	1.798	3.04	...	10.7	0.01	12.67	
281 Lucretia	SU	2.188	0.134	1.895	4.82	4.348	13.1	0.14	12.08	W189
296 Phaetusa	S	2.229	0.123	1.955	1.49	...	(9)	(0.18)	12.63	W187
317 Roxane	E	2.287	0.042	2.191	1.78	8.16	22.6	0.29	10.18	
336 Lacadiera	D	2.252	0.091	2.047	6.32	13.7	72.0	0.042	9.78	W174
341 California	S	2.199	0.129	1.915	5.28	...	16.6	0.26	10.96	W189
352 Gisela	S	2.194	0.130	1.909	4.01	6.7	22.5	0.31	10.11	W188
364 Isara	S	2.221	0.154	1.879	5.51	9.155	31.0	0.20	9.85	W180
376 Geometria	S	2.289	0.168	1.904	6.37	7.74	37.0	0.22	9.41	W175
443 Photographica	S	2.215	0.095	2.005	4.59	...	28.3	0.17	10.23	W195
453 Tea	S	2.183	0.136	1.886	5.22	6.4	24.2	0.14	10.81	W189
496 Gryhia	S	2.199	0.128	1.918	4.42	...	17.5	0.10	11.89	W189
512 Taurinensis	S	2.190	0.174	1.809	8.74	5.582	23.3	0.15	10.79	
540 Rosamunde	S	2.219	0.145	1.897	6.26	9.336	21.0	0.19	10.75	W183
548 Kressida	S	2.282	0.188	1.853	3.15	...	(16)	(0.18)	11.43	
851 Zeissia	S	2.228	0.14	1.916	2.29	9.9	14.2	0.17	11.75	W187
864 Aase	S	2.208	0.137	1.906	5.51	...	(8)	(0.18)	12.98	W189
901 Brunsia	S	2.224	0.164	1.859	4.42	4.872	14.9	(0.18)	11.61	W184
951 Gaspra	S	2.210	0.170	1.834	4.10	7.042	12.2	0.15	11.67	W189, Table 13.4
1047 Geisha	S	2.241	0.157	1.889	4.99	...	(13)	(0.18)	12.00	W184
1120 Crocus	...	2.216	0.121	1.948	4.19	737	...	...	12.2	W188
1451 Grano	S	2.203	0.146	1.892	5.45	...	(10)	(0.18)	12.6	W189
Main Belt, Zone I 2.30 < a ≤ 2.50, i ≤ 18°										
4 Vesta	V	2.362	0.097	2.133	6.43	5.342	520	0.38	3.16	W169, (2.6–3.0)×10²⁰kg
6 Hebe	S	2.425	0.146	2.071	15.0	7.275	185	0.25	5.70	(2×10¹⁹ kg)
7 Iris	S	2.386	0.210	1.885	6.60	7.139	203	0.21	5.76	(1.5×10¹⁹ kg)
9 Metis	S	2.386	0.125	2.088	4.76	5.078	190	0.15	6.32	W170
11 Parthenope	S	2.452	0.072	2.275	3.90	7.83	162	0.15	6.62	
12 Victoria	S	2.334	0.172	1.933	9.61	8.654	215	0.099	7.23	W171

continued

Table 13.7 *(continued)*

Minor Planet	Type	a (AU)	e	q (AU)	i (deg)	P_Rotat. (hrs)	Diam. (km)	Albedo P_vis	H (mag)	Notes
17 Thetis	S	2.469	0.141	2.121	5.05	12.275	93.2	0.15	7.77	
20 Massalia	S	2.408	0.162	2.018	1.49	8.098	151	0.19	6.52	W162
27 Euterpe	S	2.347	0.187	1.908	0.69	8.500	(140)	(0.18)	6.78	
30 Urania	S	2.360	0.103	2.117	2.87	13.686	104	0.13	7.74	
42 Isis	S	2.441	0.186	1.987	7.87	13.59	107	0.12	7.75	W157
51 Nemausa	G	2.366	0.111	2.103	10.2	7.785	153	0.086	7.36	$(9\times10^{17}\,kg)$
60 Echo	S	2.393	0.201	1.912	4.36	25.208	61.6	0.15	8.68	
63 Ausonia	S	2.395	0.119	2.110	6.32	9.298	108	0.17	7.35	W165
67 Asia	S	2.421	0.152	2.053	6.83	15.89	60.3	0.21	8.36	
79 Eurynome	S	2.444	0.175	2.016	5.16	5.979	68.8	0.27	7.83	W75
83 Beatrix	X	2.431	0.120	2.139	4.65	10.16	84.2	0.069	8.89	
113 Almathea	S	2.376	0.123	2.084	4.42	9.935	47.6	0.27	8.63	W170
115 Thyra	S	2.380	0.171	1.973	12.9	7.241	83.5	0.25	7.51	W163
118 Peitho	S	2.439	0.164	2.039	7.35	7.78	45.7	0.20	9.01	
131 Vala	CX	2.431	0.098	2.193	4.19	...	43.3	0.095	9.99	W161
161 Athor	M	2.379	0.095	2.153	8.92	7.288	45.7	0.12	9.55	
169 Zelia	S	2.358	0.093	2.139	5.51	...	36.5	0.19	9.60	W197
189 Phthia	S	2.450	0.011	2.423	5.62	...	38.5	0.18	9.51	
192 Prokne	S	2.403	0.207	1.906	7.47	13.622	107	0.21	7.13	
198 Ampella	S	2.458	0.175	2.028	11.1	...	58.7	0.19	8.55	
219 Thusnelda	S	2.354	0.164	1.968	11.4	29.76	43.6	0.15	9.43	
1290 Albertine	...	2.366	0.120	2.082	6.49	...	...	...	12.5	W165
1646 Rosseland	CX	2.361	0.099	2.127	7.87	69.2	26	0.04	12.05	W169
1906 Naef	V	2.373	0.098	2.140	6.55	...	6	0.42	12.7	W169
1929 Kollaa	V	2.363	0.112	2.098	7.12	...	8	0.37	12.2	W165
1933 Tinchen	V	2.353	0.090	2.141	6.89	...	6	0.24	13.3	W196
1959 Chandra	...	2.316	0.093	2.101	6.89	...	...	...	12.7	W196
2024 McLaughlin	S	2.325	0.096	2.102	6.60	...	10	0.09	13.3	W196
2590 Mourao	V	2.343	0.096	2.118	6.74	...	6	0.37	12.84	W196
3155 Lee	J	2.343	0.097	2.116	6.69	...	7	0.77	11.7	W196
3268 De Sanctis	V	2.347	0.100	2.112	7.02	...	5	0.38	13.2	W169
3657 Ermolova	J	2.313	0.088	2.109	6.58	...	7	0.34	12.6	W196, (1978 ST$_6$)
3944 Halliday	V	2.368	0.109	2.110	6.76	...	5	0.42	13.1	W165
3968 Koptelov	V	2.322	0.093	2.106	6.78	...	7	0.34	12.6	W196
4038 Kristina	V	2.366	0.099	2.132	6.33	...	4	0.45	13.5	W169
4147 Lennon	V	2.362	0.103	2.119	6.55	...	6	0.32	13.0	W169

continued

Table 13.7 *(continued)*

Minor Planet	Type	a (AU)	e	q (AU)	i (deg)	P$_{Rotat.}$ (hrs)	Diam. (km)	Albedo P$_{vis}$	H (mag)	Notes
4510 Shawna	S	2.360	0.098	2.129	6.83	...	7	0.23	13.0	W169
4546 Franck	V	2.356	0.091	2.142	6.52	...	4	0.38	13.7	W196
Nysa Zone 2.41 $\leq a \leq$ 2.5, 0.12 $\leq e \leq$ 0.21, 1.5° $\leq i \leq$ 4.3°										
19 Fortuna	C	2.442	0.131	2.122	2.24	7.445	200	0.07	7.09	W158
21 Lutetia	M	2.435	0.127	2.126	2.18	8.167	99.5	0.20	7.34	W158
44 Nysa	E	2.422	0.177	1.993	3.10	6.422	73.3	0.49	7.05	W24, hydrated
135 Hertha	M	2.427	0.174	2.005	2.75	8.40	82.0	0.13	8.21	W160
142 Polana	F	2.419	0.159	2.034	3.33	...	57.1	0.042	10.26	W24
650 Amalasuntha	...	2.458	0.161	2.062	3.21	...	...	...	13.03	W24
877 Walkure	F	2.486	0.132	2.158	3.50	17.49	39.6	0.047	10.94	
969 Leocadia	FXU	2.463	0.171	2.042	3.50	...	20.5	0.038	12.59	W24
Phocaea Group 2.25 $\leq a \leq$ 2.50, $e \leq$ 0.10, 18° $\leq i \leq$ 32°										
25 Phocaea	S	2.400	0.183	1.961	24.6	9.945	78.2	0.22	7.78	
105 Artemis	C	2.374	0.168	1.975	22.8	>24	123	0.032	8.89	
273 Atropos	SCTU	2.395	0.149	2.038	21.3	20	32.1	0.12	10.35	
326 Tamara	C	2.318	0.165	1.936	24.3	...	100	0.039	9.13	
391 Ingeborg	S	2.320	0.255	1.728	24.8	16	20.7	(0.18)	11.1	
502 Sigune	S	2.384	0.173	1.972	24.8	10.5	20.7	0.20	10.76	
654 Zelinda	C	2.297	0.192	1.856	19.4	31.9	132	0.043	8.43	
914 Palisana	CU	2.454	0.181	2.010	27.1	>14	79.0	0.084	8.82	
1108 Demeter	CX	2.428	0.163	2.032	27.9	...	41.0	0.063	11.88	
1170 Siva	S	2.326	0.212	1.833	24.1	...	12.3	0.11	12.52	
1310 Vlligera	S	2.393	0.236	1.828	25.1	...	(16)	(0.18)	11.55	v_4
1342 Brabantia	X	2.289	0.179	1.879	22.5	...	20.1	0.11	11.45	
1584 Fuji	S	2.376	0.195	1.913	27.3	10	24.7	0.13	10.81	
1657 Roemera	S	2.349	0.189	1.905	24.3	4.5	9.61	0.14	12.79	
1963 Bezovec	C	2.424	0.192	1.959	24.0	...	46.5	0.036	10.89	
2000 Herschel	S	2.381	0.226	1.843	25.1	...	(17)	(0.18)	11.36	
2050 Francis	S	2.326	0.255	1.733	26.2	...	(9)	(0.18)	12.79	
Main Belt, Zone IIa 2.500 < $a \leq$ 2.706, i < 33°										
3 Juno	S	2.670	0.218	2.088	14.2	7.210	234	0.22	5.31	(2×10^{19} kg)
5 Astraea	S	2.578	0.215	2.024	4.76	16.812	125	0.14	7.24	
13 Egeria	G	2.576	0.121	2.264	16.3	7.045	215	0.099	6.47	
15 Eunomia	S	2.644	0.143	2.266	13.4	6.083	272	0.19	5.22	W140, (4×10^{19}kg)
23 Thalia	S	2.626	0.249	1.972	10.4	12.308	111	0.21	7.07	
26 Proserpina	S	2.656	0.134	2.300	2.98	10.60	98.7	0.16	7.61	

continued

Table 13.7 *(continued)*

Minor Planet	Type	a (AU)	e	q (AU)	i (deg)	P$_{Rotat.}$ (hrs)	Diam. (km)	Albedo P$_{vis}$	H (mag)	Notes
29 Amphitrite	S	2.554	0.066	2.385	6.32	5.390	219	0.16	5.84	
32 Ponona	S	2.588	0.114	2.293	6.26	9.443	82.6	0.25	7.50	
34 Circe	C	2.687	0.153	2.276	5.74	>12	118	0.057	8.37	
37 Fides	S	2.642	0.165	2.206	3.50	7.332	112	0.17	7.28	W142, υ_{10}
46 Hestia	C	2.525	0.134	2.187	2.52	21.04	131	0.046	8.38	
50 Virginia	X	2.650	0.236	2.025	2.75	>24	...	...	9.20	
56 Melete	P	2.598	0.208	2.058	9.21	13.7	117	0.062	8.30	
58 Concordia	C	2.700	0.088	2.462	4.76	...	97.7	0.056	8.79	W132
64 Angelina	E	2.682	0.151	2.277	2.35	8.752	59.8	0.430	7.65	
66 Maja	C	2.646	0.171	2.194	3.38	...	78.3	0.050	9.39	W142, υ_{10}
75 Eurydike	M	2.671	0.267	1.958	5.22	8.92	58.3	0.12	9.02	
77 Frigga	(M)	2.668	0.109	2.377	2.75	9.012	71.0	0.13	8.57	W141
78 Diana	C	2.620	0.232	2.012	9.56	8	125	0.064	8.11	
85 Io	C	2.654	0.143	2.274	13.0	6.875	157	0.068	7.56	W140
89 Julia	S	2.552	0.089	2.325	17.2	11.387	159	0.16	6.57	
97 Klotho	M	2.668	0.228	2.060	12.8	35	87.1	0.19	7.70	
98 Ianthe	C	2.687	0.225	2.082	16.3	...	109	0.041	8.92	
99 Dike	C	2.664	0.215	2.091	13.3	>24	(79)	(0.05)	9.42	
101 Helena	S	2.584	0.104	2.315	10.5	23.16	68.3	0.15	8.45	W144
102 Miriam	1	2.661	0.234	2.038	6.03	...	86.0	0.049	9.23	
103 Hera	S	2.702	0.058	2.545	4.65	23.74	95.2	0.17	7.59	W134
109 Felicitas	C	2.696	0.277	1.949	0.167	26.3	91.6	0.060	8.87	
111 Ate	C	2.593	0.124	2.271	5.85	22.2	139	0.064	7.89	
114 Kassandra	T	2.676	0.181	2.192	4.76	10.76	103	0.084	8.24	
119 Althaea	S	2.581	0.049	2.455	6.20	...	60.7	0.17	8.61	
124 Alkeste	S	2.630	0.080	2.420	3.21	9.921	79.5	0.15	8.13	W141
134 Sophrosyne	C	2.565	0.105	2.296	12.2	...	122	0.041	8.67	
144 Vibilia	C	2.655	0.196	2.135	4.12	13.81	146	0.059	7.87	W136, υ_{10}
145 Adeona	C	2.673	0.160	2.245	12.0	8.1	155	0.044	8.05	W138
160 Una	...	2.728	0.052	2.586	3.957	...	85	0.059	9.04	W134
170 Maria	S	2.554	0.099	2.301	15.4	...	46.2	0.14	9.42	W4
194 Prokne	C	2.616	0.166	2.182	17.8	15.67	174	0.050	7.66	
201 Penelope	M	2.678	0.140	2.303	5.39	3.747	70.5	0.14	8.48	
204 Kallisto	S	2.671	0.177	2.198	8.97	...	50.8	0.17	9.00	
214 Aschera	E	2.611	0.057	2.462	3.96	6.835	23.7	0.52	9.45	W143
232 Russia	C	2.553	0.205	2.030	5.74	...	55.2	0.045	10.27	W152

continued

Table 13.7 *(continued)*

Minor Planet	Type	a (AU)	e	q (AU)	i (deg)	P_Rotat. (hrs)	Diam. (km)	Albedo P_vis	H (mag)	Notes
233 Asterope	T	2.660	0.064	2.490	8.45	19.70	108	0.073	8.30	
253 Mathilde	C	2.647	0.230	2.038	6.892	17.4	53	0.036	10.30	Table 13.6
324 Bamberga	C	2.683	0.285	1.918	13.3	29.43	242	0.057	6.82	
419 Aurelia	P, F	2.596	0.247	1.955	4.93	16.71	133	0.044	8.39	W146
Main Belt Zone IIb	$2.706 < a \leq 2.82$, $i \leq 33°$									
1 Ceres	G	2.767	0.097	2.499	9.73	9.075	913	0.10	3.32	W67, 1.17×10^{21} kg
2 Pallas	B	2.771	0.180	2.272	35.7	7.811	523	0.14	4.13	W129, 2.18×10^{20} kg
28 Bellona	S	2.776	0.176	2.287	8.801	15.695	126	0.15	7.17	
38 Leda	C	2.740	0.163	2.293	8.106	...	120	0.058	8.31	
41 Daphne	C	2.765	0.279	1.994	16.92	5.988	182	0.073	7.16	
45 Eugenia	C	2.721	0.115	2.408	6.14	5.699	214	0.048	7.27	W133
54 Alexandra	C	2.710	0.179	2.225	12.77	7.04	171	0.050	7.70	W138
55 Pandora	(E)	2.760	0.102	2.478	7.123	4.804	67.5	0.32	7.68	
59 Elpis	C	2.713	0.094	2.458	8.453	13.69	173	0.048	7.72	
68 Leto	S	2.782	0.144	2.381	7.585	14.848	127	0.20	6.84	W126
71 Niobe	S	2.755	0.117	2.433	25.72	11.21	87.3	0.28	7.26	
82 Alkmene	S	2.765	0.246	2.085	2.923	12.999	63.6	0.17	8.51	
93 Minerva	(C)	2.755	0.138	2.375	9.091	5.97	146	0.085	7.47	
88 Thisbe	C	2.768	0.143	2.372	6.370	6.042	232	0.051	7.05	
110 Lydia	M	2.733	0.047	2.605	5.164	10.927	89.1	0.17	7.79	
116 Sirona	S	2.768	0.176	2.281	2.866	12.028	75.5	0.22	7.86	
125 Liberatrix	M	2.743	0.086	2.507	4.531	3.969	47.5	0.18	9.06	
127 Johanna	CX	2.756	0.092	2.502	7.990	...	...	...	8.48	
128 Nemesis	C	2.750	0.088	2.508	5.221	39	194	0.045	7.55	W132
146 Lucina	C	2.719	0.086	2.185	12.18	18.54	137	0.052	8.15	
148 Gallia	GU	2.771	0.098	2.499	25.66	20.664	104	0.14	7.60	
156 Xanthippe	C	2.729	0.246	2.058	11.30	22.5	126	0.040	8.61	
173 Ino	C	2.743	0.160	2.304	14.06	5.93	159	0.053	7.79	
185 Eunike	C	2.739	0.099	2.468	22.89	10.83	165	0.053	7.73	
187 Lamberta	C	2.732	0.256	2.033	10.54	...	135	0.053	8.16	
188 Menippe	S	2.762	0.141	2.373	13.12	...	41.3	0.19	9.31	
200 Dynamene	C	2.737	0.084	2.507	7.816	19	132	0.053	8.20	
206 Hersilia	C	2.740	0.050	2.603	3.210	...	(110)	(0.053)	8.65	
210 Isabella	CF	2.722	0.095	2.463	4.876	...	90.0	0.041	9.32	
213 Lilaea	F	2.754	0.143	2.360	5.912	7.85	84.6	0.072	8.83	
216 Kleopatra	M	2.795	0.244	2.113	13.60	5.385	140	0.088	7.53	

continued

Table 13.7 *(continued)*

Minor Planet	Type	a (AU)	e	q (AU)	i (deg)	$P_{Rotat.}$ (hrs)	Diam. (km)	Albedo P_{vis}	H (mag)	Notes
288 Glauke	S	2.760	0.242	2.092	3.784	1150	37.5	0.11	10.08	
446 Aeternitas	A	2.788	0.093	2.529	10.30	...	43.0	0.35	8.57	W67
532 Herculina	S	2.772	0.184	2.262	16.60	9.405	231	0.16	5.78	
Main Belt Zone IIIa $2.82 < a \leq 3.03, e \leq 0.35, i \leq 30°$										
16 Psyche	M	2.922	0.100	2.630	2.58	4.196	264	0.10	5.99	$(4.0 \times 10^{19}$kg)
22 Kalliope	M	2.910	0.109	2.593	12.8	4.147	187	0.12	6.50	
33 Polyhymnia	S	2.865	0.300	2.006	2.235	18.601	(65)	(0.18)	8.43	
35 Leukothea	C	2.997	0.254	2.236	9.091	...	108	0.058	8.54	
47 Aglaja	C	2.881	0.111	2.561	5.279	13.0	133	0.072	7.86	
61 Danae	S	2.984	0.122	2.620	18.6	11.45	83.6	0.21	7.66	
69 Hesperia	M	2.979	0.174	2.461	9.21	5.655	143	0.12	7.10	
81 Terpsichore	C	2.854	0.179	2.343	8.569	...	124	0.046	8.49	
117 Lomia	C	2.991	0.028	2.907	15.31	...	154	0.040	8.18	
179 Klytaemnestra	S	2.972	0.070	2.764	9.033	11.173	81.0	0.14	8.20	
195 Eurykleia	C	2.879	0.068	2.683	7.065	...	89.7	0.053	9.05	
747 Winchester	PC	2.998	0.245	2.263	21.3	9.40	178	0.047	7.68	
773 Irmintraud	D	2.858	0.047	2.724	17.5	...	99.1	0.033	9.34	
Koronis Family (W3) $2.833 \leq a \leq 2.919, 0.039 \leq e \leq 0.067, 1.83° \leq i \leq 2.41°$										
158 Koronis	S	2.869	0.045	2.740	2.18	14.18	39.8	0.17	9.49	W3
167 Urda	S	2.854	0.043	2.731	2.12	16	42.2	0.21	9.16	W3
208 Lacrimosa	S	2.893	0.045	2.763	2.12	13.5	44.3	0.21	9.05	W3
243 Ida	S	2.864	0.043	2.741	1.371	4.63	31.4	0.21	10.02	W3, Table 13.5
277 Elvira	S	2.886	0.051	2.739	2.12	30	29.5	0.21	9.96	W3
462 Eriphyla	S	2.874	0.050	2.730	2.063	8.6	38.0	0.30	9.01	W3
Eos Family (W 2) $2.997 \leq a \leq 3.026, 0.052 \leq e \leq 0.089, 9.56° \leq i \leq 10.8°$										
221 Eos	K	3.012	0.079	2.774	9.88	10.436	110	0.12	7.69	W2
339 Dorothea	SK	3.012	0.067	2.810	9.788	...	43.7	0.16	9.34	W2
562 Salome	S	3.019	0.066	2.820	10.2	10.4	...	0.13	10.02	W2
Main Belt Zone IIIb $3.03 < a \leq 3.27, e < 0.35, i \leq 30°$										
10 Hygiea	C	3.144	0.136	2.716	5.28	27.659	429	0.075	5.27	W110, 9×10^{19}kg
31 Euphrosyne	C	3.156	0.099	2.844	28.0	5.531	248	0.070	6.53	
48 Doris	C	3.112	0.064	2.913	6.66	11.89	225	0.064	6.83	
49 Pales	C	3.090	0.193	2.494	4.876	10.42	154	0.051	7.91	
52 Europa	C	3.097	0.119	2.728	6.49	5.631	312	0.057	6.25	
57 Mnemosyne	S	3.153	0.095	2.853	15.72	...	116	0.21	6.95	
86 Semele	C	3.108	0.176	2.561	3.842	16.634	127	0.043	8.51	

continued

Table 13.7 *(continued)*

Minor Planet	Type	*a* (AU)	*e*	*q* (AU)	*i* (deg)	$P_{Rotat.}$ (hrs)	Diam. (km)	Albedo P_{vis}	H (mag)	Notes
94 Aurora	C	3.158	0.068	2.943	8.337	7.22	212	0.038	7.55	
95 Arethusa	C	3.068	0.112	2.724	13.95	8.688	145	0.062	7.84	
106 Dione	G	3.172	0.136	2.741	3.669	...	152	0.083	7.42	
108 Hecuba	S	3.218	0.123	2.822	4.934	...	67.2	0.19	8.27	
120 Lachesia	C	3.118	0.088	2.844	7.759	>20	178	0.045	7.73	
130 Elektra	G	3.119	...	...	...	5.225	189	0.089	6.86	υ_6
137 Meliboea	C	3.119	0.159	2.623	15.31	>20	150	0.048	8.04	W113
147 Protogeneia	C	3.137	0.011	3.102	3.038	...	137	0.029	8.76	
152 Atala	D	3.140	0.074	2.908	11.77	5.282	(130)	(0.04)	8.58	
159 Aemilia	C	3.106	0.117	2.743	5.221	...	131	0.061	8.07	
181 Eucharis	K	3.132	0.195	2.521	19.0	>7	107	0.12	7.77	
184 Dejopeja	M	3.183	0.113	2.823	2.178	6.7	68.2	0.18	8.29	
196 Philomela	S	3.114	0.039	2.993	6.20	8.333	146	0.18	6.64	
209 Dido	(C)	3.148	0.076	2.909	7.585	8	149	0.044	8.15	
211 Isolda	C	3.044	0.149	2.590	5.049	...	148	0.059	7.84	
375 Ursula	C	3.127	0.073	2.899	16.5	16.83	216	0.042	7.43	W111
423 Diotima	C	3.068	0.052	2.908	10.3	4.622	217	0.038	7.48	W200
451 Patientia	C	3.063	0.059	2.882	14.1	9.727	230	0.073	6.65	
511 Davida	C	3.178	0.171	2.635	14.7	5.13	337	0.053	6.17	$(3.0\times10^{19}kg)$
601 Nerthus	X	3.130	0.073	2.902	15.8	...	...	...	9.66	W111
702 Alauda	C	3.194	0.041	3.063	21.8	8.36	202	0.056	7.23	
704 Interamnia	F	3.062	0.081	2.814	18.9	8.727	333	0.064	6.00	
744 Aguntina	FX	3.173	0.153	2.688	7.00	...	62.0	0.039	10.19	
Themis Family 3.077 ≤ *a* ≤ 3.225, 0.134 ≤ *e* ≤ 0.172, 0.74° ≤ *i* ≤ 2.12°										
24 Themis	C	3.133	0.159	2.635	1.15	8.374	200	0.067	7.07	W 1A
62 Erato	BU	3.122	0.146	2.666	1.32	...	99.3	0.090	8.24	W 1A
90 Antiope	C	3.148	0.150	2.676	1.38	...	125	0.051	8.37	W 1A
104 Klymene	C	3.149	0.141	2.705	2.522	9	127	0.052	8.31	W 1
171 Ophelia	C	3.134	0.161	2.629	1.38	13.4	121	0.053	8.39	W 1A
222 Lucia	(C)	3.222	0.071	2.993	1.891	7	58.0	0.082	9.52	W 1
223 Rosa	CP	3.089	0.136	2.669	1.55	...	90.7	0.022	9.95	W 1
268 Adorea	FC	3.097	0.170	2.571	1.43	6.1	142	0.038	8.40	W 1
Griqua Group 3.10 ≤ *a* ≤ 3.27, *e* ≤ 0.35 (2:1 commensurability with Jupiter)										
1362 Griqua	CP	3.276	...	...	...	7	31.1	0.066	11.10	2:1
1921 Pala	...	3.276	...	...	...	...	...	...	14.5	2:1
1922 Zulu	...	3.276	...	...	...	...	...	...	11.8	2:1

continued

Table 13.7 *(continued)*

Minor Planet	Type	*a* (AU)	*e*	*q* (AU)	*i* (deg)	$P_{Rotat.}$ (hrs)	Diam. (km)	Albedo P_{vis}	H (mag)	Notes
Cybele Group 3.27 < *a* ≤ 3.70, *e* ≤ 0.30, *i* ≤ 25°										
65 Cybele	C	3.429	0.129	2.987	3.21	4.041	245	0.057	6.79	
76 Freia	P	3.390	0.186	2.759	3.03	9.98	190	0.029	8.08	
87 Sylvia	P	3.486	0.051	3.308	9.85	5.183	271	0.040	6.95	
107 Camilla	C	3.488	0.084	3.195	9.85	4.840	237	0.060	6.80	
121 Hermione	C	3.451	0.089	3.144	6.66	8.97	217	0.042	7.39	
168 Sibylla	C	3.379	0.025	3.295	5.11	...	154	0.050	7.93	
225 Henrietta	C	3.382	0.150	2.875	24.9	...	124	0.041	8.62	
229 Adelina	BCU	3.411	0.120	3.002	2.06	...	96.0	0.037	9.29	
260 Huberta	CX	3.445	0.084	3.156	6.03	...	101	0.034	9.26	
414 Liriope	C	3.503	0.070	3.258	8.34	...	75.2	0.047	9.55	
420 Bertholda	P	3.418	0.044	3.268	7.76	11.04	146	0.038	8.35	
466 Tisiphone	C	3.358	...	...	...	...	121	0.056	8.34	υ_6
483 Seppina	S	3.426	0.015	3.375	18.5	...	73.5	0.013	8.45	
522 Helga	X	3.629	0.039	3.487	3.27	...	113	0.027	9.28	
536 Merapi	X	3.500	0.038	3.367	18.4	...	158	0.042	8.08	
566 Stereoskopia	C	3.387	0.065	3.167	3.78	...	175	0.032	8.15	
570 Kythera	ST	3.429	0.068	3.196	2.69	...	106	0.052	8.70	
643 Scheherezade	P	3.352	0.080	3.084	15.1	...	76.1	0.036	9.83	
692 Hippodamia	S	3.369	0.092	3.059	26.6	...	47.7	0.180	9.08	
713 Luscinia	C	3.399	0.108	3.032	11.3	...	109	0.041	8.90	
721 Tabora	D	3.551	0.083	3.256	7.76	...	82.6	0.050	9.28	
733 Mocia	CF	3.398	0.061	3.191	20.5	...	92.0	0.049	9.07	
790 Pretoria	P	3.406	0.169	2.830	23.0	10.37	176	0.034	8.05	
940 Kordula	FC	3.379	0.116	2.987	5.57	...	...	...	9.33	
1004 Belopolskya	PC	3.397	0.060	3.193	2.46	...	76.6	0.035	9.82	
1028 Lydina	C	3.402	0.101	3.058	8.63	...	76.3	0.052	9.41	
1154 Atronomia	FXU	3.399	0.067	3.171	3.38	...	64.3	0.027	10.50	
1167 Dubiago	D	3.413	0.073	3.164	6.55	...	69.0	0.039	9.94	
1177 Gonnessia	XFU	3.350	0.037	3.226	16.1	...	95.5	0.039	9.25	
1266 Tone	P	3.363	0.038	3.235	18.2	...	76.0	0.060	9.27	
1280 Baillauda	X	3.413	0.018	3.352	7.70	...	55.3	0.044	10.30	
1328 Devota	X	3.496	0.109	3.115	6.49	...	59.6	0.036	10.35	
1390 Abastumani	P	3.435	...	...	...	...	104	0.033	9.24	υ_6
1467 Mashona	GC	3.386	0.129	2.949	23.1	...	112	0.054	8.55	
1556 Wingolfia	XC	3.420	0.058	3.222	14.9	...	30.8	0.10	10.57	

continued

Table 13.7 *(continued)*

Minor Planet	Type	a (AU)	e	q (AU)	i (deg)	$P_{Rotat.}$ (hrs)	Diam. (km)	Albedo P_{vis}	H (mag)	Notes
1579 Herrick	F	3.424	0.132	2.972	9.38	...	48.5	0.040	10.69	
1796 Riga	XFCU	3.359	0.071	3.121	22.8	...	76.5	0.041	9.66	
2196 Ellicott	CFXU	3.435	0.099	3.095	10.47	...	62.2	0.036	10.24	(1965 BC)
2208 Pushkin	D	3.499	0.040	3.359	4.15	...	45.2	0.036	10.96	
2266 Tchaikovsky	D	3.384	0.184	2.761	13.47	...	53.6	0.029	10.81	
2311 El Leoncito	D	3.636	0.023	3.552	5.54	...	61.0	0.029	10.55	

Hilda Group $3.70 < a \leq 4.20$, $e \leq 0.30$, $i \leq 20°$ (3:2 commensurability with Jupiter)

Minor Planet	Type	a (AU)	e	q (AU)	i (deg)	$P_{Rotat.}$ (hrs)	Diam. (km)	Albedo P_{vis}	H (mag)	Notes
153 Hilda	C	(3.97)	0.172	...	8.9	(8.11)	175	0.060	7.46	
190 Ismene	P	(3.97)	0.168	...	5.9	...	(210)	(0.04)	7.56	
334 Chicago	C	(3.97)	0.049	...	3.67	9.19	170	0.064	7.46	
361 Bononia	DP	(3.97)	0.206	...	12.0	...	149	0.039	8.27	
499 Venusia	C	(3.97)	0.202	...	3.3	...	86.0	0.033	9.64	
748 Simeisa	P	(3.97)	0.168	...	3.5	...	107	0.039	8.99	
1038 Tuckia	DTU	(3.97)	0.163	...	8.2	...	(41)	(0.05)	10.82	
1144 Oda	TD	(3.97)	0.047	...	9.15	...	51	(0.07)	10.12	
1162 Larissa	P	(3.97)	0.142	...	1.6	...	56.8	0.080	9.58	
1180 Rita	P	(3.97)	0.168	...	6.2	...	(100)	(0.04)	9.15	
1212 Francette	P	(3.97)	0.230	...	7.2	>16	90.7	0.038	9.38	
1268 Libya	P	(3.97)	0.134	...	5.0	...	97.5	0.040	9.17	
1269 Rollandia	D	(3.97)	0.124	...	1.9	...	109	0.047	8.73	
1345 Potomac	X	(3.97)	0.203	...	10.9	...	79.3	0.036	9.74	
1439 Vogtia	XFU	(3.97)	0.175	...	3.8	...	60.1	0.027	10.65	
1512 Oulu	PD	(3.97)	0.194	...	6.5	...	90.0	0.032	9.59	
1529 Oterma	P	(3.97)	0.153	...	7.9	...	(66)	(0.04)	10.04	
1578 Kirkwood	D	(3.97)	0.202	...	0.6	...	57.0	0.040	10.33	
1746 Brouwer	D	(3.97)	0.141	...	9.2	...	(70)	(0.04)	9.91	
1748 Mauderli	D	(3.97)	0.176	...	2.3	...	(53)	(0.04)	10.52	
1754 Cunningham	P	(3.97)	0.192	...	11.4	...	82.6	0.033	9.74	
1902 Shaposhnikov	PC	(3.97)	0.188	...	11.4	...	101	0.028	9.49	
1911 Schubart	CP	(3.97)	0.190	...	2.9	...	83.0	0.023	10.11	
2067 Aksnes	P	(3.97)	0.176	...	2.5	...	50.3	0.044	10.49	
2246 Bowell	D	(3.97)	0.151	...	6.1	...	52.1	0.034	10.71	
2312 Duboshin	D	(3.97)	0.112	...	4.2	...	60.0	0.039	10.24	
2760 Kacha	P	(3.97)	...	...	...	...	62.6	0.043	10.04	(1980 TU$_6$)

Trojan Group $5.05 \leq a \leq 5.40$ (1:1 commensurability with Jupiter)

Minor Planet	Type	a (AU)	e	q (AU)	i (deg)	$P_{Rotat.}$ (hrs)	Diam. (km)	Albedo P_{vis}	H (mag)	Notes
588 Achilles	DU	(5.20)	0.103	...	11.4	...	147	0.030	8.59	L4, (3.22×10^{14} kg)

continued

Table 13.7 *(continued)*

Minor Planet	Type	a (AU)	e	q (AU)	i (deg)	P_Rotat. (hrs)	Diam. (km)	Albedo P_vis	H (mag)	Notes
617 Patroclus	P	(5.20)	0.100	...	21.5	(>40)	149	0.043	8.17	
624 Hektor	D	(5.20)	0.054	...	19.0	6.921	234	0.034	7.47	L4
659 Nestor	XC	(5.20)	0.130	...	5.0	...	115	0.040	8.80	L4
884 Priamus	D	(5.20)	0.087	...	10.1	...	(112)	(0.04)	8.89	
911 Agamemnon	D	(5.20)	0.021	...	22.7	7	175	0.041	7.88	L4
1143 Odysseus	D	(5.20)	0.052	...	4.0	...	135	0.041	8.43	L4
1172 Aneas	D	(5.20)	0.060	...	17.8	...	151	0.038	8.26	
1173 Anchises	P	(5.20)	0.094	...	8.2	11.60	135	0.026	8.91	
1208 Troilus	FCU	(5.20)	0.046	...	33.0	>24	111	0.036	9.00	
1437 Diomedes	DP	(5.20)	0.017	...	22.0	18	171	0.029	8.30	L4
1583 Antilchus	D	(5.20)	0.018	...	29.1	...	109	0.051	8.66	L4
1647 Menelaus	...	(5.20)	0.058	...	6.7	...	72.0	0.028	10.2	L4
1867 Deiphobus	D	(5.20)	0.029	...	28.2	>24	131	0.037	8.60	
2207 Antenor	D	(5.20)	0.058	...	6.2	...	92.6	0.058	8.87	
2223 Sarpedon	DU	(5.20)	0.038	...	16.8	...	105	0.027	9.41	
2241 Alcathous	D	(5.20)	0.101	...	17.9	...	123	0.040	8.66	(1979 WM)
2260 Neoptolemus	DTU	(5.20)	0.019	...	16.4	...	85.0	0.064	8.95	L4
2357 Phereclos	D	(5.20)	0.048	...	2.7	...	103	0.042	8.99	
2363 Cebriones	D	(5.20)	0.033	...	32.7	...	91.7	0.066	8.7	
2674 Pandarus	D	(5.20)	0.088	...	2.1	8.480	102	0.041	9.05	
2893 Peiroos	D	(5.20)	0.047	...	13.2	...	92.8	0.055	8.92	
Other										
279 Thule	D	4.294	0.01	4.25	2.33	7.44	135	0.030	8.77	4:3
944 Hidalgo	D	5.764	0.656	1.96	42.5	10.064	43.5	(0.06)	10.48	Jupiter X
1373 Cincinnati	...	3.409	0.323	2.308	...	...	...	...	13.1	
1474 Beira	FX	2.735	0.154	2.313	36.9	...	...	...	12.61	
2060 Chiron *	B	13.6	0.38	8.43	6.9	5.918	182	0.14	5.82	Saturn X

* See also Table 14.10.

14

CENTAUR OBJECTS AND KUIPER BELT OBJECTS

14.1 Centaur Objects

By the end of 1996, six small bodies orbiting the Sun with semimajor axes between those of Jupiter and Neptune became known as Centaur Objects. The first Centaur object discovered was (2060) Chiron. Initially classified as an asteroid with an unusual orbit, Chiron later was designated a comet, after it exhibited cometary activity. The subsequent discovery of other small objects with similar unusual orbital characteristics led to the identification of a new dynamical group of bodies in the solar system.

The differences in photometric colors among the Centaurs suggest two different types of surface materials. Chiron and 1995 GO with near solar colors may possess surfaces composed of material similar to carbonaceous chondrites plus ices, whereas the other Centaurs showing very red colors may have surfaces covered with complex organic substances.

The Centaur objects most likely did not form in their present locations in the solar nebula since dynamical studies indicate that their orbits are unstable over a timescale of about 10^6 years.

Table 14.1 Centaur Objects

Object	a (AU)	e	i (deg.)	H_R (Mag)	Dia. (km)	P_{rot} (hr)	Sources
2060 Chiron (1977 UB)	13.6	0.38	6.9	5.82	182	5.92	CTO94, LJ96
5145 Pholus (1992 AD)	20.2	0.57	24.7	6.74	185±16	9.98	BB92, DTB96, LJ96
7066 Nessus (1993 HA₂)	24.5	0.52	15.7	9.78	~75 *	...	LJ96
1995 DW₂	24.9	0.24	4.2	8.90	~110 *	...	LJ96
1994 TA	16.8	0.30	5.4	11.5	~30 *	...	JH97
8405 1995 GO	18.0	0.62	17.6	9.0	~110 *	8.87	BL97, JH97
1997 CU₂₆	15.7	0.17	23	6.0	302±30	...	JK98

* assuming an albedo of 0.04. The geometric albedos of Chiron, Pholus, and 1997 CU_{26} are 0.14, 0.044, and 0.045, respectively.
Orbital elements were provided by B. G. Marsden and G. V. Williams,
Harvard-Smithsonian Center for Astrophysics.

continued

Table 14.1 (continued)

Sources: **[BB92]** Buie, M. W., & Bus, S. J., 1992, *Icarus* 100, 288–294. **[BL97]** Brown, W. R., & Luu, J. X., 1997, *Icarus* 126, 218–224. **[CTO94]** Campins, H., Tedesco, C. M., Osip, D. J., Rieke, G. H., Rieke, M. J., Schulz, B., *Astron. J.*, 108, 2318–2322. **[DSC93]** Davies, J. K., Sykes, M. V., & Cruikshank, D. P., 1993, *Icarus* 102, 166–169. **[DTB96]** Davies, J. K., Tholen, D. J., & Ballantune, D. R., 1996, in *Completing the inventory of the solar system*, ASP Conf. Series (Rettig, T. W., & Hahn, J. M., eds.), Vol. 107, pp. 97–105. **[HTM90]** Hartmann, W. K., Tholen, D. J., Meech, K. J., & Cruikshank, D. P., 1990, *Icarus* 83, 1–15. **[JH97]** Jedicke, R., & Herron, J. D., 1997, *Icarus* 127, 494–507. [JK98] Jewitt, D., & Kalas, P., 1998, *ApJ.* 499, L103–L106. **[LJ96]** Luu, J., & Jewitt, D., 1996, *Astron. J.* 111, 499–503, *ibid.*, 112, 2310–2318. **[MTH92]** Mueller, B., E. A., Tholen, D. J., Hartmann, W. K., & Cruikshank, D. P., 1992, *Icarus* 97, 150–154. **[RTL97]** Romanishin, W., Tegler, S. C., Levine, J., & Butler, N., 1997, *Astron. J.,* 113, 1893–1898. **[WTR97]** Weintraub, D. A., Tegler, S. C., & Romanishin, W., 1997, *Icarus* 128, 456–463.

Table 14.2 Centaur Objects: Colors

Object	B–V	V–R	R–I	J–H	H–K	V–J	V–K	Sources
2060 Chiron	0.70	0.37	0.31	0.29	0.07	1.13	1.49	HTM90, LJ96
	0.66	0.35	0.54					
5145 Pholus	1.19	0.75	0.76	0.45	–0.04	2.53	2.94	DSC93, DTB96, LJ96,
	1.35	0.81		0.30	0.04	2.67	3.01	MTH92, RTL97, WTR97
7066 Nessus	0.88	0.72	0.69	...	...	2.2	2.4	DTB96, LJ96, RTL97
		0.77						
1995 DW₂	0.64	0.51	...	...	...	...	...	LJ96
8405 1995 GO	0.78	0.73	...	0.43	0.24	1.72	2.39	BL97, RTL97, WTR97
	0.75	0.47						
Solar colors	0.67	0.36	0.33	...	...	...	...	HTM90, LJ96

Note: see Table 14.1 for sources.

14.2　Kuiper Belt Objects

Evolutionary and dynamical considerations [Edg49, Kui51] require a source region of small bodies beyond Neptune's orbit to provide the short period comets. Observational proof of this region at heliocentric distances between 30 and 50 AU came in 1993, when Jewitt and Luu [JL93] reported the discovery of 1992 QB$_1$. By the end of 1997, about 60 objects were found. These objects are now known as Kuiper belt objects (KBOs), Edgeworth-Kuiper objects (EKOs), and trans-Neptunian objects (TNOs).

Because all observations of the TNOs are very recent, their orbital parameters and diameters are still relatively uncertain. Most objects appear to be ≥100 km in diameter; the total estimated population is 35,000 to 70,000 objects of ≥100 km between 30 and 50 AU [JL95, LJ96].

BVRI photometry indicates a diversity of surface compositions for the TNOs [LJ96, JL98]. The near infrared spectrum of object 1993 SC shares similarities with spectra of Pluto and Neptune's moon Triton (a captured TNO?). The spectrum of object 1993 SC suggests the presence of simple hydrocarbon ices, such as CH_4, C_2H_6, etc., and more complex organic compounds on the surface [BCP97].

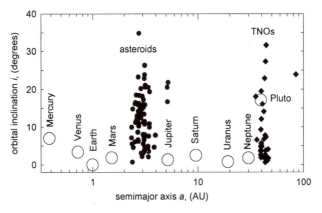

Figure 14.1 A comparison of the orbital inclinations of the planets, asteroids (with diameters ≥150 km), and the TNOs. Note the highly inclined orbit of Pluto.

About 40% of all discovered objects are in the 3:2 mean motion resonance with Neptune. Interestingly, Pluto is also in the 3:2 resonance. The assembly of bodies in this resonance and the unusually high inclination of Pluto's orbit (see Figure 14.1) raise the more philosophical question of whether Pluto is a planet or simply the largest TNO known at present.

Sources and further reading: [BCP97] Brown, R. H., Cruikshank, D. P., Pendleton, Y., & Veeder, G. J., 1997, *Science* 276, 937–939. [DL97] Duncan, M. J., & Levison, H. F., 1997, *Science* 276, 1670 –1672. [DMG97] Davies, J. K., McBride, N., & Green, S. F., 1997, *Icarus* 125, 61–66. [Edg49] Edgeworth, K. E., 1949, *Mon. Not. R. Astron. Soc.* 109, 600–609. [JL93] Jewitt, D., & Luu, J., 1993, *Nature* 362, 730–732. [JL95] Jewitt , D. C., & Luu, J. X., 1995, *Astron. J.,* 109, 1867–1876. [JL98] Jewitt, D., & Luu, J., 1998, Astron. J. 115, 1667–1670. [JLC96] Jewitt, D., Luu, J., & Chen, J., 1996, *Astron. J.* 112, 1225–1238. [Kui51] Kuiper, G. P., 1951, in *Astrophysics: A topical symposium* (Hynek, J. A., ed.), McGraw-Hill, New York, pp. 357–424. [L94] Luu, J. 1994, in *Asteroids, Comets, Meteors 1993* (Milani, A., ed.), IAU, Kluwer, Dortrecht, pp. 31–44. [LJ96] Luu, J. & Jewitt, D., 1996, *Astron. J.,* 112, 2310–2318. [LMJ97] Luu, J., Marsden, B. G., Jewitt, D., Trujuillo, C. A., Hergenrother, C. W., Chen, J., & Offutt, W. B., 1997, *Nature* 387, 573–575. [TR97] Tegler, S. C., & Romanishin, W., 1997, *Icarus* 126, 212–217. [WOF95] Williams, I. P., O'Ceallaigh, D., Fizimmons, A., & Marsden, B. G., 1995, *Icarus* 116, 180–185.

Table 14.3 Properties of Some Kuiper Belt Objects

Object	a (AU)	e	i (deg.)	H_R (Mag)	Dia. [a] (km)	Sources
1992 QB$_1$	43.34	0.078	2.2	6.62	283	JL93, JL95, LJ96, L94
1993 FW	43.48	0.044	7.8	6.58	286	JL95, LJ96
1993 RO	39.64	0.206	3.7	8.38	139	JL95, LJ96
1993 RP	39.33	0.114	2.6	8.96	96	JL95, LJ96
1993 SB	39.69	0.322	1.9	7.51	188	JL95, WOF95
1993 SC	39.91	0.191	5.1	6.36	319	JL95, LJ96, WOF95
1994 ES$_2$	45.52	0.114	1.1	7.87	159	JL95, LJ96
1994 EV$_3$	42.71	0.048	1.7	6.99	267	JL95, LJ96
1994 GV$_9$	43.47	0.058	0.6	6.77	264	JL95
1994 JQ$_1$	43.89	0.048	3.8	5.97	382	JL95
1994 JR$_1$	39.36	0.117	3.8	7.00	238	JL95
1994 JS	42.21	0.22	14.1	6.78	256	JL95, JLC96, LJ96
1994 JV	35.25	0.0 *	18.1	6.86	237	JL95, JLC96, LJ96
1994 TB	38.98	0.322	12.1	6.45	299	JL95, JLC96, LJ96
1994 TG	42.25	0.0 *	6.8	7.0	261	JL95, JLC96
1994 TG$_2$	42.45	0.0 *	2.2	7.75	168	JL95

continued

Table 14.3 *(continued)*

Object	*a* (AU)	*e*	*i* (deg.)	H_R (Mag)	Dia. [a] (km)	Sources
1994 TH	40.94	0.0 *	16.1	6.59	245	JL95, JLC96
1995 DA_2	36.21	0.070	6.6	8.0	180	JLC96, LJ96
1995 DB_2	46.33	0.135	4.1	7.9	320	JLC96, LJ96
1995 DC_2	43.85	0.070	2.3	6.9	378	JLC96, LJ96
1995 GA_7	39.46	0.119	3.5	7.2	213	JLC96
1995 GJ	42.91	0.091	22.9	6.5	299	JLC96
1995 HM_5	39.32	0.250	4.8	7.9	158	JLC96
1995 KJ_1	43.47	0.0 *	2.7	6.1	365	JLC96
1995 KK_1	39.47	0.190	9.3	7.8	165	JLC96
1995 QY_9	40.13	0.272	4.8	7.6	209	LJ96, JLC96
1995 QZ_9	39.80	0.153	19.5	7.0	238	JLC96
1995 WY_2	46.51	0.126	1.7	6.67	310	LJ96
1996 RQ_{20}	44.33	0.115	31.6	7.0	270	DL97
1996 TL_{66}	84.82	0.587	23.9	5.32	490	DL97, JL98, LMJ97
1996 TO_{66}	43.73	0.128	27.3	4.52	840	JL98
1996 TP_{66}	39.77	0.336	5.7	6.97	270	JL98
1996 TS_{66}	44.18	0.128	7.3	6.11	400	JL98

Note: see section 14.2 for source listing. Orbital elements were provided by B. G. Marsden and G. V. Williams, Harvard-Smithsonian Center for Astrophysics.
a: approximate diameters adopting an albedo of 0.04.
* A circular orbit is assumed.

Table 14.4 Kuiper Belt Objects: Colors

Object	B–V	V–R	R–I
1992 QB$_1$	0.65	0.77	0.74
1993 RO	1.07	0.57	0.65
1993 FW	0.89	0.62	0.50
1993 SC	0.92, 0.94, 1.27	0.54, 0.57, 0.68, 0.70	0.43, 0.68, 0.86
1994 ES$_2$	0.71	0.94	0.97
1994 EV$_3$	1.50	0.54	...
1994 TB	0.88, 1.10	0.85, 0.58	0.65
1994 JV	...	0.78	0.59
1994 JS	...	0.85	...
1995 WY$_2$	0.99	0.68	0.43
1995 QY$_9$	0.68	0.46	0.40
1996 TO$_{66}$	0.59	0.32	0.36
1996 TP$_{66}$	0.80	0.65	0.71
1996 TS$_{66}$	0.93	0.43	0.67
1996 TL$_{66}$	0.58	0.13	0.54
Solar colors	0.67	0.36	0.33

Sources: DMG97, JL98, LJ96, TR97, WOF95; see section 14.2 for source listing

15

COMETS

Comets (from the Greek *aster kometes,* which means long-haired star) are members of the population of small bodies in the outskirts of our solar system. During their motion around the sun and movement through the inner solar system, comets often yield spectacular celestial displays that have attracted human attention since ancient times.

The five principal parts of a comet are the cometary nucleus, the coma, the dust tail, the ion tail, and the corona. The last four components develop once a comet approaches the sun. The irregularly shaped, kilometer-size cometary nucleus is the major solid component of a comet and consists of a conglomerate of ices and rocky dust particles. The potato-shaped nucleus of comet P/Halley is $16\times8\times7$ km in size and possess a very dark surface. The cometary coma becomes visible when a comet approaches perihelion and icy material evaporates, producing an atmosphere of gases (H_2O, CO, CO_2, N_2, NH_3, CH_4) and releasing dust. Closer to the sun, subsurface ices in the nucleus can be heated and create gas jets. If these gas jets are strong enough, the cometary orbit may be changed by these nongravitational forces. Cometary nuclei are relatively stable against solar heating; for example, comet Ikeya-Seki 1965 approached the sun's surface within 470,000 km without major damage, whereas comet West 1976 broke up into at least four fragments after its perihelion passage.

The dust trail along a comet's orbit consists of dust particles that are detached from the coma by solar radiation pressure. The radiation pressure cannot cause the ion trail, which points straight away from the sun. Instead, ions produced in the coma are dragged by the interplanetary magnetic field carried by the solar wind. Once a comet is in the inner solar system, hydrogen gas produced by photodissociation of OH from the coma forms a corona of several millions of km in radius. The corona is only detectable in the UV by spacecraft.

Periodically appearing comets with orbital periods <200 years are called short period (SP) comets (e.g., P/Encke, P/Halley); those with orbital periods >200 years are termed long period (LP) comets. Comets have elliptical to nearly parabolic orbits, but their orbits are typically unstable because of gravitational perturbations from the planets (mainly Jupiter);

orbits are also affected by mass loss during outgassing. The evolution and instability of cometary orbits eventually lead to comet loss (most spectacular losses being cometary collisions with Jupiter or the sun), which requires that comets are replenished from a reservoir far out in the solar system. In 1950, Jan Oort investigated the orbital evolution of observed comets and deduced that many original cometary orbits (i.e., orbits prior to planetary perturbations) must have had aphelion distances greater than 20,000 AU. Cometary orbits may be stable up to 200,000 AU, giving the distances of the reservoir for long period comets known as the Oort cloud. Comets may escape from this cloud because of perturbations from nearby passing stars. The Oort cloud may contain about $10^{12} - 10^{13}$ comets, with a total mass of ~30 Earth masses.

The reservoir of short period comets is the Kuiper belt at closer distances of 30–1,000 AU, populated by 10^8–10^9 cometary objects with a total mass of ~0.1 Earth masses (see section 14.2). The lower orbital

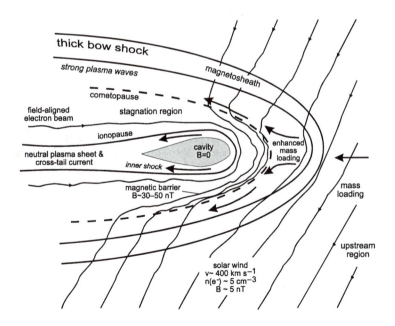

Figure 15.1 Comet – solar wind interactions (schematic and not to scale)

inclination of the SP comets suggest that the Kuiper belt is a flat, disk-shaped region, whereas the highly inclined orbits of LP comets suggest that the Oort cloud is a nearly spherical reservoir around the solar system. Comets thought to be on their first pass through the solar system from their source region are called new comets, whereas those returning are called old comets.

Our knowledge of the composition of cometary volatiles was greatly increased by the recent observations of the long period comets C/1995 O (Hale-Bopp) and C/1996 B2 (Hyakutake). As noted in Table 15.2, the number of parent molecules identified in comets approximately doubled as a result of observations of Hale-Bopp and Hyakutake. In particular, many nitrogen-bearing (e.g., HNC, HNCO, NH_3, CH_3CN, HC_3N, NH_2CHO) and sulfur-bearing (e.g., CS_2, SO, SO_2, OCS, H_2CS) species were observed in these two comets.

In the next few years, these advances in Earth-based observations of comets will be complemented by several planned spacecraft missions that are designed to return samples of cometary and interstellar dust (STAR-DUST mission), to rendezvous with and study several comets (Deep Space 1 and CONTOUR missions), to land on and study a comet nucleus (Rosetta mission), and to return samples of a cometary nucleus (Deep Space 4 mission). Some details of past, continuing, and planned comet spacecraft missions are summarized in Table 15.1.

Sources and further reading: Festou, M. C., Rickman, H., & West, R. M., *Astron. Astrophys. Rev.* 4, 363–447, & 5, 37–163. Newburn, R. L., Neugebauer, M., & Rahe, J., 1991, *Comets in the post-Halley era*, Vol. 1 & 2, pp. 1350. Mumma, M. J., Weissman, P. R., & Stern, S. A., 1993, in *Protostars and planets* (Levy, E. H., & Lunine, J. I., eds.) Univ. of Arizona Press, Tucson, pp. 1177–1252. Oort, J. H., 1950, *Bull. Astron. Inst. Neth.* 11, 91–110. Wilkening, L. L. (ed.), *Comets,* Univ. Arizona Press, Tucson, pp. 766.

Table 15.1   Space Missions to Comets

Mission	Launch Date	Mission Target	Mission Objectives
ISEE-3/ICE (USA)	12 Aug. 1978	Giacobini-Zinner (in 1985)	flyby; studied ion solar-wind/comet interactions; planned probe recovery in 2014
Sakigake (Japan)	7 Jan. 1985	P/Halley (in 1986)	flyby; magnetic field measurements
Giotto (ESA)*	2 July 1985	P/Halley (in 1986) Grigg-Skjellerup (in 1992)	flyby; first close imaging of cometary nucleus; dust and gas composition measurements; solar-wind/comet interactions
Suisei (Japan)	18 Aug. 1985	P/Halley (in 1986)	UV imaging, studied solar wind/comet plasma interactions
Vega 1 & 2 (USSR)	15 & 21 Dec. 1984	P/Halley (in 1986)	flyby; nucleus imaging, tempera-ture & rotation rate determinations
Galileo (USA)	18 Oct. 1989	multi-target mission	orbiter; direct imaging of Shoemaker-Levy 9 impact on Jupiter in 1994
Deep Space 1 (USA)	1 July 1998	planned asteroid/comet flybys in 1999 & 2000	
STARDUST (USA)	Feb. 1999	81P/Wild 2	planned imaging; dust collection of cometary and interstellar dust; dust composition measurements
CONTOUR (USA)	July 2002	planned flybys to three comets	
Rosetta (ESA)	23 Jan. 2003	planned asteroid flybys (in 2007, 2008); lander on comet Wirtanen (in 2012)	
Deep Space 4 (USA)	12 April 2003	planned lander on comet Temple 1 and sample return	

*ESA: European Space Agency

Table 15.2 Species Observed in Comets

Species	Characteristic Band or Line (μm)	Observed in [a]
H I	0.10257 (Ly-β), 0.12157 (Ly-α), 0.6563 (H-α)	Austin, Kohoutek, Levy
H_2	0.1608	
C I	0.1561, 0.1657, 0.1931	Kohoutek, West
C^+	0.13353	West
C_2	0.2313, 0.5165, 0.7715, 1.21, 1.45, 1.78	Halley, West
$^{12}C^{13}C$	0.4745, 0.5120–0.5170	Kohoutek, West
C_3	0.0.3880–0.4100, 0.4020–0.4045	many
CH	0.3889, 0.4315, 3.35	many, Halley, Hale-Bopp
CH^+	0.4225	Halley
CH_4	3.31	Halley, Hale-Bopp, Hyakutake
C_2H_2	3.03	Hale-Bopp, Hyakutake
C_2H_6	3.35	Hale-Bopp, Hyakutake
CN	0.3555–0.3595, 0.3883, 0.7873, 1.10, 1.46, 4.90	Halley, Hale-Bopp, West
^{13}CN	0.3870	Bennett
CN^+	0.2181:, 0.3185	
HCN	3.02, 88.6 GHz, 265.9 GHz, 354.5 GHz	Halley, Hale-Bopp, Hyakutake
$H^{13}CN$		Hyakutake
HNC	90.7 GHz, 272.0 GHz, 362.6 GHz	Hale-Bopp, Hyakutake
DCN	362.0 GHz	Hale-Bopp, Hyakutake
CH_3CN	92 GHz, 110.7 GHz, 147 GHz, 221 GHz	Hale-Bopp
HNCO		Hyakutake
CO	0.1510, 0.2000–0.2400, 4.67, 115.3 GHz, 230.5 GHz, 345.8 GHz, 461.0 GHz, 691 GHz	many comets, Halley, Hale-Bopp, Hyakutake
CO^+	0.2190, 0.3954, 0.4273	many, West
CO_2	0.1193, 4.25	Halley, Hale-Bopp
CO_2^+	0.2890, 0.3509, 0.3674	Bradfield, West
H_2CO	3.52, 3.59, 4.83 GHz, 140.8 GHz, 145.6 GHz, 218 GHz, 225.7 GHz, 351.8 GHz	many comets, Austin, Halley, Hale-Bopp, Hyakutake, Levy
CH_3OH	3.33, 3.37, 3.52, 97 GHz, 145.1 GHz, 157 GHz, 242 GHz, 252 GHz, 461.7 GHz, 464.8 GHz	many comets, Austin, Halley, Hale-Bopp, Hyakutake
CS	0.2576, 98.0 GHz, 147.0 GHz, 244.9 GHz, 342.9 GHz	Hale-Bopp, West
CS_2		Halley, Hale-Bopp

continued

Table 15.2 *(continued)*

Species	Characteristic Band or Line (μm)	Observed in [a]
OCS	4.86	Halley, Hale-Bopp, Hyakutake
N_2^+	0.3914	Daniel, Morehouse
NH	0.3360	Cunningham
NH_2	0.4900–0.6900, 0.5700, 3.23	Cunningham, Hyakutake
NH_3	3.00	Halley, Hale-Bopp
NO	0.2150	
O I	0.10260, 0.13035, 0.13556, 0.29723, 0.63002, 0.63639	many, Austin, Bradfield, Halley, Kohoutek, Mrkos
OH	0.2810–0.2850, 0.3085, 0.3125–0.3180, 2.87, 3.04, 119.45, 1.66540 GHz, 1.66736 GHz	many, Hale-Bopp, Halley, Hyakutake, Kohoutek
OH^+	0.3565, 0.3885–0.4030	Bradfield, Cunningham, Halley
H_2O	1.38, 1.94, 2.44, 2.65, 2.8, 4.63, 4.85, 6.3, 174.6, 179.5	many, d'Arrest, Hale-Bopp, Halley, Hyakutake, Wilson
H_2O ice	1.5, 2.04, 3.0, (45), (65)	Halley, Bowell, Cernis, Hale-Bopp
HDO	464.9 GHz, 490.6 GHz	Hyakutake
H_2O^+	0.6198	Hale-Bopp, Kohoutek
Na I	0.3303, 0.5890, 0.58959	many
S I	0.1425, 0.1474, 0.1813	West, Wilson
S_2	0.2820–0.3060	Hyakutake
H_2S	168.8 GHz, 216.7 GHz	many, Hale-Bopp, Hyakutake
SO_2		Halley
K I	0.40442, 0.76649, 0.76990	Ikeya-Seki
Ca II	0.3934, 0.42267	Ikeya-Seki
Ca^+	0.39337, 0.39685	Ikeya-Seki
Cr I	0.3579–0.5208	Ikeya-Seki
Mn I	0.4031	Ikeya-Seki
Co I	0.3220–0.4150	Ikeya-Seki
Ni I	0.3221–0.5476, 0.3381, 0.3446, 0.3458, 0.3462, 0.3525, 0.3566, 0.3619	Cruls-Tebbutt, Ikeya-Seki
Fe I	0.3214–0.5506, 0.3441, 0.3570, 0.3581, 0.3720, 0.3749, 0.3816, 0.3820, 0.3860, 0.4046	Cruls-Tebbutt
Cu I	0.348, 0.3208–0.5105	Ikeya-Seki

continued

Table 15.2 *(continued)*

Note: Gases with uncertain spectral identification include HCH, HCO, NH_4^+, SH, SH^+, SO, Mg, Al, Ti, and V.

[a] Full comet designations: 6P/d'Arrest, C/1989 X1 Austin, C/1969 Y1 Bennett, C/1980 E1 Bowell, C/1979 Y1 Bradfield, C/1983 O1 Cernis, C/1882 R1 Cruls-Tebbutt, C/1940 R2 Cunningham, C/1907 L2 Daniel, C/1995 O1 Hale-Bopp, 1P/Halley, C/1996 B2, Hyakutake, C/1965 S1 Ikeya-Seki, C/1973 E1 Kohoutek, C/1990 K1 Levy, C/1908 R1 Morehouse, C/1957 P1 Mrkos, C/1975 V1 West, C/1986 P1 Wilson

Sources: Crovisier, J., 1998, *Earth, Moon, & Planets*, in press. Crovisier, J., Leech, K., Bockelée-Morvan, D., Brooke, T. Y., Hanner, M. A., Altierii, B., Keller, H. U., & Lellouch, E., 1997, *Science* 275, 1904–1907. Delsemme, A. H., 1982, in *Comets* (Wilkening, L. L., ed.), Univ. of Arizona Press, Tucson, pp. 85–130. Eberhardt, P., 1996, in *Proceedings of the 1996 Asteroids, comets, meteors conference*, Versailles, COSPAR Colloquia Series, in press. Festou, M. C., Rickman, H., & West, R. M., 1993, *Astron. Astrophys. Reviews* 4, 363–447, & 5, 37–163. Huebner, W. F., Boice, D., C., Schmidt, H. U., & Wegmann, R., 1991, in *Comets in the post-Halley era*, (Newburn, R. L., Neugebauer, M., & Rahe, J., eds.), Kluwer Acad. Publ., Dortrecht, The Netherlands, Vol. 2, pp. 907–936. Lis, D. C., Keene, J., Young, Y., Phillips, T. G., Bockelée-Morvan, D., Crovisier, J., Schilke, P., Goldsmith, P. F., & Bergin, E. A., 1997, *Icarus* 130, 355–372. Meier, R., Owen, T. C., Jewitt, D. C., Matthews, H. E., Senay, M., Biver, N., Bockelée-Morvan, D., Crovisier, J., & Gautier, D., 1998, *Science* 279, 1707–1710. Notesco, G., Laufer, D., & Bar-Nun, A., 1997, *Icarus* 125, 471–473. Wagner, R. M., & Schleicher, D. G., 1997, *Science* 275, 1918–1920. Weaver, H. A., Feldman, P. D., A'Hearn, M. F., Arpigny, C., Brandt, J. C., Festou, M. C., Haken, M., McPhate, B., Stern, S. A., & Tozzi, G. P., 1997, Science 275, 1900–1904. Wyckoff, S., 1982, in *Comets* (Wilkening, L. L., ed.), Univ. Arizona Press, Tucson, pp. 3–55.

Table 15.3 Elemental Abundances in Comet Halley, CI-Chondrites, and the Solar Photosphere [a]

Element	Comet P/Halley		CI-Chondrites	Solar Photosphere
	Dust	Dust & Ice		
H	2025	4062	520	2.63×10^6
C	814	1010	74	933
N	42	95	5.9	245
O	890	2040	748	1950
Na	10	10	5.61	5.62
Mg	100	100	100	100
Al	6.8	6.8	8.32	7.76
Si	185	185	97.7	93.3
S	72	72	43.7	42.7
K	0.2	0.2	0.363	0.347
Ca	6.3	6.3	6.31	6.03
Ti	0.4	0.4	0.234	0.288
Cr	0.9	0.9	1.32	1.23
Mn	0.5	0.5	0.912	0.646
Fe	52	52	83.2	85.1
Co	0.3	0.3	0.224	0.219
Ni	4.1	4.1	4.90	4.68

[a] atoms/100 Mg

Note: see also Table 3.5 for solar photospheric abundances and Tables 2.1 and 16.9 for abundances on CI-chondrites

Sources: Jessberger, E. K., & Kissel, J., 1991, in *Comets in the post-Halley era* (Newburn, R., Neugebauer, M., & Rahe, J., eds.), Kluwer Acad. Publ., Dortrecht, The Netherlands, Vol. 2, pp. 1075–1092. Mumma, M. J., Weissman, P. R., & Stern, S. A., in *Protostars & planets III* (Levy, E. H., & Lunine, J. I., eds.) Univ. of Arizona Press, Tucson, 1177–1252.

Table 15.4 Relative Abundances in P/Halley (by Number)

Molecule	Abundance	Molecule	Abundance	Molecule	Abundance
H_2O	100	H_2CO	0–5	N_2	~0.02
CH_4	0–2	CH_3OH	~1	NH_3	1–2
CO	7–8	OCS	<7	HCN	≤0.1
CO_2	3	CS_2	1	SO_2	<0.002

Note: see Table 15.3 for sources

Table 15.5 Orbital Elements of Some Short-period Comets

Designation	Name	a (AU)	e	q (AU)	i (deg.)	$P_{orbital}$ (yrs)	Apps.	H_o (mag)	Notes
2P	Encke	2.266	0.850	0.340	11.9	3.28	56	11.0	wide varia. in H_o during apparitions
D/1766 G1	P/Helfenzrieder	2.671	0.848	0.406	7.9	4.35	1	6.8	
D/1819 W1	P/Blanpain	2.963	0.699	0.892	9.1	5.10	1	8.5	
26P	P/Grigg-Skjellerup	2.961	0.664	0.995	21.1	5.10	17	12.5	stable since 4th perihelion passage
79P	P/du Toit-Hartley	3.005	0.601	1.199	2.9	5.21	3	...	2 components seen in 1982
45P	P/Honda-Mrkos-Pajdusáková	3.023	0.824	0.532	4.2	5.27	9	10.6	no fading since discovery
79P	P/du Toit-Neujmin-Delporte	3.005	0.601	1.199	2.9	5.21	3	14.0	
73P	P/Schwassmann-Wachmann 3	3.059	0.695	0.933	11.4	5.34	4	11.5	no fading since discovery
D/1884 O1	P/Barnard 1	3.067	0.583	1.279	5.5	5.38	1	8.9	
D/1886 K1	P/Brooks 1	3.089	0.571	1.325	12.7	5.44	1	8.9	
25D	P/Neujmin 2	3.090	0.567	1.338	10.6	5.43	2	11.3	
41P	P/Tuttle-Giacobini-Kresák	3.096	0.656	1.065	9.2	5.46	8	13.0	
5D	P/Brorsen	3.105	0.810	0.590	29.4	5.46	5	9.2	
10P	P/Tempel 2	3.105	0.522	1.484	12.0	5.48	19	10.0	
46P	P/Wirtanen	3.112	0.652	1.083	11.7	5.50	7	16.3	
9P	P/Tempel 1	3.115	0.520	1.494	10.6	5.50	9	9.4	
71P	P/Clark	3.118	0.502	1.553	9.5	5.50	5	12.0	
88P	P/Howell	3.145	0.552	1.409	4.4	5.58	4	...	
D/1770 L1	P/Lexell	3.150	0.786	0.674	1.6	5.60	1	7.7	
11D	P/Tempel-Swift	3.185	0.638	1.153	5.4	5.68	4	12.8	
D/1783 W1	P/Pigott	3.257	0.552	1.459	45.1	5.89	1	6.9	
D/1978 R1	P/Honda-Campos	3.287	0.665	1.101	5.9	5.97	1	12.6	
83P	P/Russell 1	3.337	0.517	1.612	22.7	6.10	2	15.0	
37P	P/Forbes	3.350	0.568	1.447	7.2	6.13	8	10.0	no fading since discovery
54P	P/de Vico-Swift	3.412	0.524	1.624	3.6	6.31	3	14.5	
D/1978 C2	P/Tritton	3.424	0.580	1.438	7.0	6.35	1	16.5	
D/1952 B1	P/Harrington-Wilson	3.431	0.515	1.664	16.3	6.36	1	12.1	
7P	P/Pons-Winnecke	3.432	0.634	1.256	22.3	6.37	21	14.5	
104P	P/Kowal 2	3.440	0.564	1.500	15.8	6.39	2	14.5	
81P	P/Wild 2	3.441	0.540	1.583	3.2	6.39	4	6.5	

continued

Table 15.5 *(continued)*

Designation	Name	*a* (AU)	*e*	*q* (AU)	*i* (deg.)	P$_{orbital}$ (yrs)	Apps.	H$_o$ (mag)	Notes
31P	P/Schwassmann-Wachmann 2	3.444	0.399	2.07	3.8	6.39	11	11.4	no fading since discovery
57P	P/du Toit-Neujmin-Delporte	3.447	0.501	1.720	2.8	6.39	5	...	
76P	P/West-Kohoutek-Ikemura	3.451	0.543	1.577	30.5	6.41	4	9.6	
22P	P/Kopff	3.465	0.544	1.58	4.7	6.45	14	13.4	
6P	P/d'Arrest	3.487	0.614	1.346	19.5	6.51	17	6.5	brightness increase in 1976
43P	P/Wolf-Harrington	3.488	0.539	1.608	18.5	6.51	8	13.0	
D/1892 T1	P/Barnard 3	3.493	0.590	1.432	31.3	6.52	1	9.8	
87P	P/Bus	3.493	0.375	2.183	2.6	6.52	3	...	
94P	P/Russell 4	3.510	0.365	2.229	6.2	6.58	3	...	
67P	P/Churyumov-Gerasimenko	3.514	0.630	1.300	7.1	6.59	5	10.0	
21P	P/Giacobini-Zinner	3.517	0.706	1.034	31.8	6.61	12	10.0	
3D-A	P/Biela	3.529	0.756	0.861	12.5	6.62	6	8.1	Sp: 1840/3.59 (2)
44P	P/Reinmuth 2	3.532	0.464	1.893	7.0	6.64	8	10.5	
D/1896 R2	P/Giacobini	3.532	0.588	1.455	11.4	6.65	1	9.9	Sp: 1986/2.36 (2)
62P	P/Tsuchinshan 1	3.533	0.576	1.498	10.5	6.65	5	14.0	
75P	P/Kohoutek	3.536	0.498	1.775	5.9	6.65	3	...	
D/1918 W1	P/Schorr	3.548	0.469	1.884	5.6	6.67	1	11.0	
18P	P/Perrine-Mrkos	3.563	0.643	1.272	17.8	6.72	5	18.5	
15P	P/Finlay	3.572	0.710	1.036	3.7	6.76	12	13.0	
51P	P/Harrington	3.581	0.561	1.572	8.7	6.78	5	14.8	
60P	P/Tsuchinshan 2	3.593	0.504	1.782	6.7	6.82	5	14.0	
49P	P/Arend-Rigaux	3.595	0.600	1.438	17.9	6.82	7	8.9	
65P	P/Gunn	3.599	0.316	2.462	10.4	6.83	6	10.0	
19P	P/Borrelly	3.621	0.623	1.365	30.3	6.88	12	13.0	
16P	P/Brooks 2	3.621	0.491	1.843	5.5	6.89	14	13.5	Sp: 1886/5.38 (4) 1886/4.25 (2)
86P	P/Wild 3	3.626	0.366	2.299	15.5	6.91	3	...	
84P	P/Giclas	3.643	0.493	1.847	7.3	6.96	4	13.5	
48P	P/Johnson	3.648	0.366	2.313	13.7	6.97	7	10.0	
77P	P/Longmore	3.651	0.343	2.399	24.4	6.98	4	...	
69P	P/Taylor	3.652	0.466	1.950	20.6	6.97	4	12.0	Sp: 1915/1.65 (2)
33P	P/Daniel	3.681	0.552	1.649	20.1	7.06	8	11.5	

continued

Table 15.5 *(continued)*

Designation	Name	*a* (AU)	*e*	*q* (AU)	*i* (deg.)	$P_{orbital}$ (yrs)	Apps.	H_o (mag)	Notes
17P	P/Holmes	3.690	0.410	2.177	19.2	7.09	8	13.5	
113P	P/Spitaler	3.690	0.422	2.133	5.8	7.10	2	9.0	
D/1895 Q1	P/Swift	3.730	0.652	1.298	3.0	7.20	1	11.4	
98P	P/Takamizawa	3.741	0.575	1.590	9.5	7.22	2	...	
102P	P/Shoemaker 1	3.747	0.470	1.986	26.2	7.26	2	...	
106P	P/Schuster	3.754	0.590	1.539	20.1	7.26	2	13.2	
30P	P/Reinmuth 1	3.763	0.502	1.874	8.1	7.31	9	14.0	
D/1984 H1	P/Kowal-Mrkos	3.774	0.483	1.951	3.0	7.32	1	...	
4P	P/Faye	3.775	0.578	1.593	9.1	7.34	19	8.4	stable since 10th perihelion passage
89P	P/Russell 2	3.793	0.400	2.276	12.0	7.38	3	...	
D/1894 F1	P/Denning	3.798	0.698	1.147	5.5	7.40	1	10.4	
47P	P/Ashbrook-Jackson	3.828	0.395	2.316	12.5	7.49	7	7.1	no fading since discovery
91P	P/Russell 3	3.831	0.343	2.517	14.1	7.50	2	...	
61P	P/Shajn-Schaldach	3.832	0.388	2.345	6.1	7.49	5	12.0	
52P	P/Harrington-Abell	3.857	0.540	1.774	10.2	7.59	6	15.0	
97P	P/Metcalf-Brewington	3.919	0.594	1.591	13.0	7.76	2	...	
D/1984 W1	P/Shoemaker 2	3.952	0.666	1.32	21.6	7.84	1	...	
70P	P/Kojima	3.952	0.393	2.399	0.9	7.85	4	...	
39P	P/Oterma	3.958	0.144	3.388	4.0	7.88	3	9.5	
78P	P/Gehrels 2	3.980	0.410	2.348	6.7	7.94	3	...	
50P	P/Arend	3.996	0.537	1.850	19.9	7.99	6	14.5	
82P	P/Gehrels 3	4.037	0.151	3.427	1.1	8.11	3	9.5	
80P	P/Peters-Hartley	4.045	0.598	1.626	29.8	8.13	3	8.0	
58P	P/Jackson-Neujmin	4.074	0.661	1.381	13.5	8.24	5	16.7	
24P	P/Schaumasse	4.075	0.705	1.202	11.8	8.22	9	11.0	
14P	P/Wolf	4.088	0.406	2.428	27.5	8.25	14	13.0	
36P	P/Whipple	4.175	0.259	3.094	9.9	8.53	10	...	
74P	P/Smirnova-Chernykh	4.188	0.147	3.572	6.6	8.57	4	8.3	
32P	P/Comas Solá	4.273	0.568	1.846	12.9	8.83	9	8.5	no fading since discovery
59P	P/Kearns-Kwee	4.318	0.487	2.215	9	8.96	4	11.2	
72P	P/Denning-Fujikawa	4.333	0.820	0.78	8.6	9.01	2	11.5	
93P	P/Lovas 1	4.352	0.614	1.68	12.2	9.09	2	...	
64P	P/Swift-Gehrels	4.399	0.692	1.355	9.3	9.21	4	15.0	

continued

Table 15.5 (continued)

Designation	Name	a (AU)	e	q (AU)	i (deg.)	P$_{orbital}$ (yrs)	Apps.	H$_o$ (mag)	Notes
42P	P/Neujmin 3	4.833	0.586	2.001	4.0	10.6	4	14.5	related to P/Van Biesbroeck?
40P	P/Väisälä 1	4.885	0.635	1.783	11.6	10.8	6	13.5	
68P	P/Klemola	4.925	0.640	1.773	10.9	10.9	3	9.7	
34P	P/Gale	4.950	0.761	1.183	11.7	11.0	2	10.5	
85P	P/Boethin	5.018	0.778	1.114	5.8	11.2	2	13.6	
56P	P/Slaughter-Burnham	5.127	0.504	2.543	8.2	11.6	4	13.6	
53P	P/Van Biesbroeck	5.371	0.553	2.401	6.6	12.4	4	7.5	related to P/Neujmin 3?
92P	P/Sanguin	5.383	0.663	1.814	18.7	12.5	2	13.5	
P/1983 M1	P/IRAS	5.582	0.696	1.697	46.2	13.2	1	...	
63P	P/Wild 1	5.609	0.647	1.980	19.9	13.3	2	14.0	
8P	P/Tuttle	5.670	0.824	0.998	54.7	13.5	11	8.0	very little fading since discovery
29P	P/Schwassmann-Wachmann 1	6.044	0.045	5.772	9.4	14.9	6	5.0	
66P	P/du Toit	6.075	0.787	1.294	18.7	15.0	2	16.0	
99P	P/Kowal 1	6.085	0.232	4.673	4.4	15.0	2	9.0	
90P	P/Gehrels 1	6.100	0.510	2.989	9.6	15.1	2	11.5	
D/1960 S1	P/van Houten	6.251	0.367	3.957	6.7	15.6	1	8.0	
P/1983 C1	P/Bowell-Skiff	6.254	0.689	1.945	3.8	15.7	1	...	
P/1983 J3	P/Kowal-Vávrová	6.333	0.588	2.609	4.3	15.9	1	...	
101P	P/Chernykh	6.803	0.594	2.356	5.1	14.0	2	6.5	
D/1993 F2	Shoemaker-Levy 9	6.814	0.210	5.380	5.88	17.8	1	...	see chapter 8
28P	P/Neujmin 1	6.933	0.776	1.553	14.2	18.2	5	10.2	
P/1983 V1	P/Hartley-IRAS	7.723	0.834	1.282	95.7	21.5	1	...	
27P	P/Crommelin	9.074	0.919	0.735	29.1	27.4	5	10.7	
55P	P/Tempel-Tuttle	10.23	0.904	0.982	162.7	32.9	4	13.5	
38P	P/Stephan-Oterma	11.24	0.860	1.574	18.0	37.7	3	5.3	
D/1827 M1	P/Pons-Gambart	14.94	0.946	0.807	136.5	57.5	1	7.0	
20D	P/Westphal	15.68	0.920	1.254	40.9	61.9	2	9.3	
D/1921 H1	P/Dubiago	15.70	0.929	1.115	22.3	62.3	1	10.5	
13P	P/Olbers	16.83	0.930	1.178	44.6	69.6	3	5.5	
23P	P/Brorsen-Metcalf	17.11	0.972	0.479	19.3	70.5	3	8.7	
12P	P/Pons-Brooks	17.20	0.955	0.774	74.2	70.9	3	5.9	
1P	P/Halley	17.79	0.967	0.587	162.2	76.0	30	4.6	

continued

Table 15.5 *(continued)*

Designation	Name	*a* (AU)	*e*	*q* (AU)	*i* (deg.)	$P_{orbital}$ (yrs)	Apps.	H_o (mag)	Notes
122P	P/de Vico	17.81	0.963	0.659	85.4	74.4	2	7.2	
D/1942 EA	P/Väisälä 2	19.50	0.934	1.287	38.0	85.4	1	13.2	
109P	P/Swift-Tuttle	26.61	0.964	0.958	113.4	135	5	4.0	
D/1917 F1	P/Mellish	27.14	0.993	0.190	32.7	145	1	7.4	
D/1889 M1	P/Barnard 2	27.63	0.960	1.105	31.2	145	1	9.0	
D/1984 A1	P/Bradfield 1	28.27	0.952	1.357	51.8	151	1	...	
35P	P/Herschel-Rigollet	28.77	0.974	0.748	64.2	155	2	8.5	
D/1937 D1	P/Wilk	32.58	0.981	0.619	26.0	187	1	...	

a: semimajor axis; *e:* eccentricity; *q:* perihelion distance; *i:* inclination to ecliptic. Apps: apparitions = number of observed perihelion passages prior to 1 January 1995. Notes: Sp: Split comet: year of observed break-up/heliocentric distance (AU), number of fragments in parentheses.

Sources: Marsden, B. G., & Williams, G. V., 1996, *Catalogue of cometary orbits 1996*, 11th ed., IAU, Central Bureau for Astronomical Telegrams, Minor Planet Center, Harvard-Smithsonian Center for Astrophysics, (a good source where more detailed information about orbital elements can be found). Sekanina, Z., 1982, in *Comets* (Wilkening, L. L., ed.) Univ. of Arizona Press, Tucson, pp. 251–287. Vsekhsviatskii, S. K., 1964, *Physical characteristics of comets*, Jerusalem, Israel Program for Scientific Translations, NASA TT F-80, pp. 596.

Table 15.6 Orbital Elements of Some Long-period Comets

Designation	Name	a	e	q	i	$P_{orb.}$	Notes
		(AU)		(AU)	(deg.)	(yrs)	
C/1857 O1	Peters	38.13	0.980	0.747	32.8	235	
C/1932 Y1	Dodwell-Forbes	40.94	0.972	1.131	24.5	262	
C/1840 U1	Bremiker	43.89	0.966	1.480	57.9	286	
C/1932 P1	Peltier-Whipple	43.45	0.976	1.037	71.7	291	
C/1979 Y1	Bradfield	43.96	0.988	0.545	148.6	291	
C/1932 G1	Houghton-Ensor	45.00	0.972	1.254	74.3	302	
C/1874 Q1	Coggia	45.40	0.963	1.688	34.1	306	
C/1941 B1	Friend-Reese-Honda	50.16	0.981	0.942	26.3	355	
C/1955 L1	Mrkos	50.23	0.989	0.534	86.5	356	
C/1931 O1	Nagata	50.34	0.979	1.047	42.3	357	
C/1979 S1	Meier	53.43	0.973	1.432	67.1	391	
C/1964 N1	Ikeya	53.51	0.985	0.822	171.9	391	
C/1861 J1	Great Comet	55.08	0.985	0.822	85.4	409	
C/1861 G1	Thatcher	55.68	0.983	0.921	79.8	415	
C/1898 F1	Perrine	56.03	0.980	1.095	72.5	419	
C/1940 O1	Whipple-Paraskevopoulos	56.49	0.981	1.082	54.7	425	
C/1975 T2	Suzuki-Saigusa-Mori	58.41	0.986	0.838	118.2	446	
C/1930 F1	Wilk	61.72	0.992	0.482	67.1	485	
C/1855 G1	Schweizer	63.00	0.965	2.194	128.6	500	
C/1843 D1	Great March Comet	64.05	1.000	0.006	144.4	513	
C/1974 O1	Cesco	67.23	0.980	1.373	173.2	551	
C/1846 J1	Brorsen	66.11	0.990	0.634	150.7	538	
C/1906 V1	Thilee	69.79	0.983	1.213	56.4	583	
C/1937 P1	Hubble	71.04	0.972	1.954	11.6	599	
C/1952 H1	Mrkos	74.64	0.983	1.283	112.0	645	
C/1811 W1	Pons	82.88	0.981	1.582	31.3	755	
C/1882 R1-B	Great September Comet	83.17	1.000	0.008	142.0	759	Sp: 1882/0.017 (2)
C/1961 T1	Seki	83.19	0.992	0.681	155.7	759	
C/1886 H1	Brooks	83.87	0.997	0.270	87.7	768	
C/1853 G1	Schweizer	84.82	0.989	0.909	122.2	781	
C/1973 H1	Huchra	88.23	0.973	2.384	48.3	829	
C/1965 S1-A	Ikeya-Seki	91.82	1.000	0.008	141.9	880	Sp: 1965/0.008 (2)
C/1936 O1	Kaho-Kozik-Lis	92.40	0.994	0.518	121.9	888	
C/1935 A1	Johnson	93.27	0.991	0.811	65.4	901	
C/1963 R1	Pereyra	93.41	1.000	0.005	144.6	903	
C/1963 A1	Ikeya	95.45	0.993	0.632	160.6	932	

continued

Table 15.6 *(continued)*

Designation	Name	*a* (AU)	*e*	*q* (AU)	*i* (deg.)	$P_{orb.}$ (yrs)	Notes
C/1894 G1	Gale	97.17	0.990	0.983	87.0	958	
C/1887 B2	Brooks	99.97	0.984	1.630	104.3	1000	
C/1960 Y1	Candy	105.1	0.990	1.062	151.0	1080	
C/1931 P1	Ryves	111.1	0.999	0.075	169.3	1170	
C/1854 R1	Klinkerfues	118.3	0.993	0.799	40.9	1290	
C/1785 E1	Méchain	120.7	0.996	0.427	92.6	1330	
C/1964 L1	Tomita-Gerber-Honda	123.0	0.996	0.500	161.8	1370	
C/1922 B1	Reid	125.0	0.987	1.629	32.4	1400	
C/1936 K1	Peltier	133.7	0.992	1.100	78.5	1550	
C/1976 D1	Bradfield	137.0	0.994	0.848	46.8	1600	
C/1969 Y1	Bennett	141.2	0.996	0.538	90.0	1680	
C/1807 R1	Great Comet	143.2	0.995	0.646	63.2	1710	
C/1939 B1	Kozik-Peltier	146.3	0.995	0.716	63.5	1770	
C/1922 W1	Skjellerup	147.4	0.994	0.924	23.4	1790	
C/1968 H1	Tago-Honda-Yamamoto	154.1	0.996	0.680	102.2	1910	
C/1858 L1	Donati	156.1	0.996	0.578	117.0	1950	
C/1854 Y1	Winnecke-Dien	156.4	0.991	1.359	14.2	1960	
C/1909 L1	Borrelly-Daniel	160.9	0.995	0.843	52.1	2040	
C/1871 V1	Tempel	161.3	0.996	0.691	98.3	2050	
C/1911 O1	Brooks	163.5	0.997	0.489	33.8	2090	
C/1769 P1	Messier	163.5	0.999	0.123	40.7	2090	
C/1888 D1	Sawerthal	169.4	0.996	0.699	42.2	2200	Sp: 1888/0.76 (2)
C/1942 X1	Whipple-Fedtke-Tevzadze	173.5	0.992	1.354	19.7	2290	Sp: 1943/1.43 (2)
C/1926 B1	Blathwayt	176.7	0.992	1.345	128.3	2350	
C/1888 U1	Barnard	179.5	0.991	1.528	56.3	2410	
C/1840 B1	Galle	180.8	0.993	1.220	120.8	2430	
C/1881 K1	Great Comet (Tebbutt)	181.2	0.996	0.735	63.4	2440	
C/1857 Q1	Klinkerfeus	182.4	0.997	0.563	124.0	2460	
C/1911 N1	Kiess	184.6	0.996	0.684	148.4	2510	
C/1995 O1	Hale-Bopp	185.6	0.995	0.914	89.42	2530	
C/1920 X1	Skjellerup	193.9	0.994	1.148	22.0	2700	
C/1846 B1	de Vico	194.9	0.992	1.481	47.4	2720	
C/1881 W1	Swift	195.8	0.990	1.925	144.8	2740	
C/1961 R1	Humason	204.5	0.990	2.133	153.3	2930	
C/1811 F1	Great Comet (Flaugergues)	212.4	0.995	1.035	106.9	3100	
C/1947 F1	Rondanina-Bester	217.6	0.997	0.560	39.3	3210	

continued

Table 15.6 *(continued)*

Designation	Name	*a* (AU)	*e*	*q* (AU)	*i* (deg.)	P_orb. (yrs)	Notes
C/1873 Q1	Borrelly	225.7	0.996	0.794	96.0	3390	
C/1978 T1	Seargebt	228.5	0.998	0.370	67.8	3450	
C/1893 U1	Brooks	231.3	0.996	0.812	129.8	3520	
C/1947 X1-A	Southern Comet	243.4	1.000	0.110	138.5	3800	Sp: 1947/0.15 (2)
C/1955 N1	Bakharev-Macfarlane-Krienke	244.7	0.994	1.427	50.0	3830	
C/1825 K1	Gambart	246.6	0.996	0.889	123.3	3870	
C/1864 N1	Tempel	249.2	0.996	0.909	178.1	3930	
C/1825 N1	Pons	271.6	0.995	1.241	146.4	4480	
C/1980 V1	Meier	284.4	0.995	1.520	101.0	4800	
C/1871 G1	Winnecke	299.3	0.998	0.654	87.6	5180	
C/1913 J1	Schaumasse	309.2	0.995	1.457	152.4	5440	
C/1822 N1	Pons	310.8	0.996	1.145	127.3	5480	
C/1925 F2	Reid	334.4	0.995	1.633	27.0	6120	
C/1939 V1	Friend	336.6	0.997	0.945	93.0	6180	
C/1826 P1	Pons	340.0	0.997	0.853	25.9	6270	
C/1939 H1	Jurlof-Achmarof-Hassel	346.7	0.998	0.528	138.1	6460	
C/1887 J1	Barnard	356.8	0.996	1.394	17.5	6740	
C/1844 Y1	Great Comet (Wilmot)	359.1	0.999	0.251	45.6	6800	
C/1964 P1	Everhard	361.1	0.997	1.259	68.0	6860	
C/1944 H1	Väisälä	370.9	0.994	2.411	17.3	7140	
C/1953 G1	Mrkos-Honda	391.8	0.997	1.022	93.9	7760	
C/1952 Q1	Harrington	407.0	0.996	1.665	59.1	8210	
C/1907 L2	Daniel	424.8	0.999	0.512	9.0	8760	
C/1911 S2	Quénisset	429.7	0.998	0.788	108.1	8910	
C/1889 O1	Davidson	435.2	0.998	1.040	66.0	9080	Sp: 1889/1.06 (2)
C/1847 C1	Hind	471.3	1.000	0.043	48.7	10200	
C/1877 G2	Swift	485.7	0.998	1.009	77.2	10700	
C/1972 E1	Bradfield	494.6	0.998	0.927	123.7	11000	
C/1890 V1	Zona	495.8	0.996	2.047	154.3	11000	
C/1948 W1	Bester	509.2	0.997	1.273	87.6	11500	
C/1914 S1	Campbell	534.2	0.999	0.713	77.8	12400	Sp: 1914/0.82 (2)
C/1913 R1	Metcalf	555.9	0.998	1.356	143.4	13100	
C/1957 P1	Mrkos	559	0.999	0.355	93.9	13200	
C/1849 G1	Schweizer	568.8	0.998	0.894	67.0	13600	
C/1874 H1	Coggia	572.7	0.999	0.676	66.3	13700	

continued

Table 15.6 *(continued)*

Designation	Name	*a* (AU)	*e*	*q* (AU)	*i* (deg.)	P$_{orb.}$ (yrs)	Notes
C/1930 D1	Peltier-Schwassmann-Wachmann	581.7	0.998	1.087	99.9	14000	
C/1975 T1	Mori-Sato-Fujikawa	628.9	0.997	1.604	91.6	15800	
C/1863 V1	Tempel	630.1	0.999	0.706	78.1	15800	
C/1863 G2	Respighi	682.6	0.999	0.629	85.5	17800	
C/1929 Y1	Wilk	691.1	0.999	0.672	124.5	18200	
C/1968 Y1	Thomas	705.7	0.995	3.316	45.2	18800	
C/1877 G1	Winnecke	731.0	0.999	0.950	121.2	19800	
C/1850 J1	Petersen	771.6	0.999	1.081	68.2	21400	
C/1963 F1	Alcock	792.4	0.998	1.537	86.2	22300	
C/1892 E1	Swift	809.1	0.999	1.027	38.7	23000	
C/1874 O1	Borrelly	840.3	0.999	0.983	41.8	24400	
C/1941 B2	de Kock-Paraskevopoulos	880.3	0.999	0.790	168.2	26100	
C/1980 Y1	Bradfield	945.2	1.000	0.260	138.6	29100	
C/1972 F1	Gehrels	1072	0.997	3.277	175.6	35100	
C/1961 O1	Wilson-Hubbard	1064	1.000	0.040	24.2	34700	
C/1973 D1	Kohoutek	1082	0.999	1.382	121.6	35600	
C/1927 X1	Skjellerup-Maristany	1100	1.000	0.176	85.1	36500	
C/1966 P2	Barbon	1111	0.998	2.019	28.7	37000	
C/1903 A1	Giacobini	1244	1.000	0.411	30.9	43900	
C/1893 N1	Rodame-Quénisset	1250	0.999	0.675	160.0	44200	
C/1847 N1	Mauvais	1252	0.999	1.766	96.6	44300	
C/1924 F1	Reid	1253	0.999	1.756	72.3	44400	
C/1863 G1	Klinkerfues	1271	0.999	1.068	112.6	45300	
C/1873 Q2	Henry	1422	1.000	0.385	121.5	53600	
C/1864 O1	Donati-Toussaint	1449	0.999	0.931	109.7	55200	
C/1949 N1	Bappu-Bok-Newkirk	1517	0.999	2.058	105.8	59100	
C/1890 O2	Denning	1550	0.999	1.260	98.9	61100	
C/1980 Y2	Panther	1616	0.999	1.657	82.6	64900	
C/1948 L1	Honda-Bernasconi	1656	1.000	0.208	23.1	67400	
C/1974 C1	Bradfield	1661	1.000	0.503	61.3	67700	
C/1969 O1-A	Kohoutek	1965	0.999	1.719	86.3	87100	Sp: 1970/1.79 (2)
C/1967 Y1	Ikeya-Seki	2000	0.999	1.697	129.3	89400	
C/1927 E1	Stearns	2024	0.998	3.684	87.7	91100	
C/1948 V1	Eclipse Comet	2083	1.000	0.135	23.1	95100	
C/1977 R1	Kohler	2183	1.000	0.991	48.7	102000	

continued

Table 15.6 *(continued)*

Designation	Name	*a* (AU)	*e*	*q* (AU)	*i* (deg.)	P$_{orb.}$ (yrs)	Notes
C/1916 G1	Wolf	2833	0.999	1.686	25.7	151000	
C/1951 P1	Wilson-Harrington	2833	1.000	0.740	152.5	151000	
C/1844 N1	Mauvais	3521	1.000	0.855	131.4	209000	
C/1966 P1	Kilston	3817	0.999	2.385	40.3	236000	
C/1981 M1	Gonzáles	3861	0.999	2.334	107.1	240000	
C/1948 N1	Wirtanen	3891	0.999	2.517	130.3	243000	
C/1952 M1	Peltier	4608	1.000	1.202	45.6	313000	
C/1898 V1	Chase	4630	1.000	2.285	22.5	315000	
C/1959 X1	Mrkos	4975	1.000	1.253	19.6	351000	
C/1976 J1	Harlan	5155	1.000	1.569	38.8	370000	
C/1969 T1	Tago-Sato-Kosaka	5882	1.000	0.473	75.8	451000	Sp: 1970/1.2 (2)
C/1975 V1-A	West	6849	1.000	0.197	43.1	567000	Sp: 1976/0.22–0.41 (4)
C/1974 F1	Lovas	7576	1.000	3.011	50.6	659000	
C/1910 P1	Metcalf	9615	1.000	1.948	121.1	943000	
C/1888 P1	Brooks	9804	1.000	0.902	74.2	971000	
C/1977 V1	Tsuchinshan	9804	1.000	3.603	168.5	971000	
C/1882 F1	Wells	10870	1.000	0.061	73.8	1.1×10^6	
C/1958 R1	Burnham-Slaughter	12200	1.000	1.628	61.3	1.3×10^6	
C/1889 G1	Barnard	12350	1.000	2.256	163.9	1.4×10^6	
C/1902 R1	Perrine	12660	1.000	0.401	156.4	1.4×10^6	
C/1958 D1	Burnham	23260	1.000	1.323	15.8	3.6×10^6	
C/1910 A1	Great January Comet	25000	1.000	0.129	138.8	4.0×10^6	
C/1972 X1	Araya	52630	1.000	4.861	113.1	12×10^6	
C/1937 N1	Finsler	58820	1.000	0.863	146.4	14×10^6	

a: semimajor axis; *e*: eccentricity; *q:* perihelion distance; *i*: inclination to ecliptic. Notes: Sp: Split comet: year of observed break-up/heliocentric distance (AU), number of fragments in parentheses.

Sources: Marsden, B. G., & Williams, G. V., 1996, *Catalogue of cometary orbits 1996*, 11th ed., IAU, Central Bureau for Astronomical Telegrams, Minor Planet Center, Harvard-Smithsonian Center for Astrophysics, (a good source where more detailed information about orbital elements can be found). Sekanina, Z., 1982, in *Comets* (Wilkening, L. L., ed.) Univ. of Arizona Press, Tucson, pp. 251–287. Vsekhsviatskii, S. K., 1964, *Physical characteristics of comets*, Jerusalem, Israel Program for Scientific Translations, NASA TT F-80, pp. 596.

Table 15.7 Some Meteor Streams

Meteor Stream	Possible Associated Comet or Asteroid	Maximum Date	Radiant R.A.	Radiant Dec.	ZHR
Quadrantids	...	3–4 Jan.	229°	+49°	145
Virginids	1620 Geographos	12 Apr.	177°	+6°	20
April Lyrids	Thatcher, 1861 I	22 Apr.	271.4°	+33.6°	13±1
η Aquarids *	Halley, 1835 III	3–5 May	335.6°	–1.9°	67±5
τ Herculids	Schwassmann-Wachmann 3, 1930 VI	3 June	228°	+39°	...
χ Scorpiids	1862 Apollo	5 June	247°	–13°	...
Daytime ζ Perseids ***	...	7–8 June	62°	+23°	40 radar
Daytime Arietids	1566 Icarus	8 June	45°	+23°	54±12
Sagittariids	2102 Adonis	11 June	304°	–35°	20
June Boötids	Pons-Winnecke, 1915 III	28 June	219°	+49°	133±16
Daytime β Taurids #	Encke, 1971 II	29–30 June	79°	+21°	30 radar
δ Aquarids-South	...	2 July	333.1°	–16.5°	11.4±1.2
o Draconids	Metcalf, 1919 V	16 July	271°	+58°	...
Perseids	Swift-Tuttle, 1862 III	12–13 Aug.	46.2°	+57.4°	˙84±5
δ Aquarids-North	...	13 Aug.	339°	–5°	1.0±0.2
Aurigids	Kiess, 1911 II	1 Sept.	84.6°	+42°	...
Annual Andromedids	Biela, 1852 III	3 Oct.	26°	+37°	5
October Draconids	Giacobini-Zinner 1946 V	9 Oct.	262.1°	+54.1°	...
Northern Piscids ***	...	12 Oct.	26°	+14°	...
ε Geminids	Ikeya, 1964 VIII	19 Oct.	103°	+28°	2.9±0.6
Orionids *	Halley, 1835 III	21 Oct.	94.5°	+15.8°	25±4
Taurids-North #	Encke, 1971 II	4–7 Nov.	58.3°	+22.3°	25
Taurids-South #	Encke, 1971 II	4–7 Nov.	50.5°	+13.6°	45
Leonids	Temple-Tuttle, 1965 IV	17 Nov.	152.3°	+22.2°	23±6
December Phoenicids	Blanpain, 1819 IV	4–5 Dec.	15°	–55°	2.8±0.8
Monocerotids	Mellish, 1917 I	10 Dec.	99.8°	+14.0°	2.0±0.4
χ Orionids-North **	2201 1947 XC	10 Dec.	84°	+26°	...
χ Orionids-South **	2201 1947 XC	11 Dec.	85°	+16°	...
Geminids	3200 Phaeton	13–14 Dec.	112.3°	+32.5°	88±4
Ursids	Tuttle, 1939 X	22 Dec.	223°	+78°	(12±3)

Notes: Radiant (1950.0) positions for time of maximum. Radiants increase ~1° per day. ZHR is zenithal hourly rate of meteors. The *, **,***, and # indicate twin showers. See Table 15.8 for sources.

Table 15.8   Some Meteor Streams: Orbital Elements

Meteor Stream	a (AU)	e	q (AU)	i (deg.)	ω (deg.)	Ω (deg.)
Geminids	1.36	0.896	0.142	23.6	324.3	261.0
Daytime ζ Perseids	1.492	0.755	0.365	6.5	60.5	80.8
Daytime Arietids	1.6	0.94	0.09	21	29	77
Taurids-South	1.93	0.806	0.375	5.2	113.2	40.0
Northern Piscids	2.06	0.80	0.40	3	291	199
Daytime β Taurids	2.2	0.85	0.34	6	246	276.4
Taurids-North	2.59	0.861	0.359	2.4	292.3	230.0
δ Aquarids-North	2.62	0.97	0.069	20	332	139
Virginids	2.63	0.90	0.26	3	304	350
τ Herculids	2.70	0.63	0.97	19	204	72
δ Aquarids-South	2.86	0.976	0.069	27.2	152.8	305.0
December Phoenicids	2.96	0.68	0.98	16	0	73
Quadrantids	3.064	0.682	0.974	70.3	168.1	282.3
Annual Andromedids	3.22	0.82	0.58	4	267	190
June Boötids	3.27	0.69	1.02	18	180	98
October Draconids	3.51	0.717	0.996	30.7	171.8	196.3
Ursids	5.70	0.85	0.9389	53.6	205.85	270.66
Leonids	11.5	0.915	0.985	162.6	172.5	234.5
η Aquarids	13	0.958	0.560	163.5	95.2	42.4
Orionids	15.1	0.962	0.571	163.9	82.5	28.0
ε Geminids	26.77	0.97	0.77	173	237	209
Lyrids	28	0.968	0.919	79.0	214.3	31.7
Perseids	28	0.965	0.953	113.8	151.5	139.0
Monocerotids	42	0.997	0.14	24.8	135.8	77.6
Aurigids	∞	1.000	0.802	146.4	121.5	157.9
o Draconids	∞	1.00	1.01	43	190	113
Sagittariids	∞	1.00	0.10	99	142	260

Note: meteor streams are listed in order of increasing semimajor axis *a*.
orbital elements for equinox 1950.0

Sources: Cook, A. F., 1973, in *Evolution and physical properties of meteoroids*, NASA-SP 319, pp. 183–191. Drummond, J. D., 1981, *Icarus* 45, 545–553, & *Icarus* 47, 500–517. Jenniskens, P., 1994, *Astron. & Astrophys.* 287, 990–1013. Kronk, G. W., 1988, *Meteor showers: a descriptive catalogue,* Enslow Publ., Hillside, NJ, pp. 291. Sekanina, Z., 1976, *Icarus* 27, 265–321.

METEORITES

16.1 Introduction to Meteorites

Meteorites have interested humans since ancient times. About 6000 years ago, before iron ore processing was discovered, iron meteorites provided the only available source for iron metal. More recently, the study of meteorites has helped unravel the formation processes of the early solar system, which may be recorded in meteorite composition and mineralogy, because meteorites are the oldest remnants of early, relatively unprocessed matter we have available.

The origin of meteorites as extraterrestrial objects was put forward in the book "Ueber den Ursprung der von Pallas gefundenen und anderer ihr ähnlicher Eisenmassen, und über einige damit in Verbindung stehende Naturerscheinungen" ("On the origin of the mass of iron found by Pallas and other similar iron masses, and on some related natural phenomena") in 1794 by Ernst F. F. Chladni.

The number of recognized meteorites has increased dramatically since Chladni's time. The number of meteorite specimens in museums and private collections may easily reach about 40,000 pieces. Nevertheless, it is difficult to arrive at the total number of known meteorites because (1) pieces of the same meteorite are kept in several collections and, (2) more importantly, meteorites are now recovered more frequently, and systematic meteorite searches in Antarctica or desert regions significantly increase the number of meteorites (see section 16.2 and 16.3).

The infall rate of extraterrestrial material is about 10,000 tons per year, but meteorites constitute only a small fraction of this material, which is dominantly interplanetary dust particles (IDPs) from comets, asteroids, and other sources. Small samples of interplanetary dust particles are collected in the earth's stratosphere using a modified version of the U2 spy plane. Interplanetary dust particles are also collected by melting ice from the Greenland ice sheet and Antarctic polar cap and by dragging magnetic rakes over the ocean floor.

In meteoritics, a meteorite fall indicates that the fall was observed and the meteorite was subsequently recovered shortly after the fall. A meteorite find means that the meteorite was found and that, in general, no observations of its fall are recorded. Meteorites are named after the location where they are recovered. This is generally the nearest community or post office, although different systems, described in section 16.2 and 16.3, are used to name the Antarctic and desert meteorites.

Table 16.1 lists (alphabetically by country) museums and other institutions where larger meteorite collections are stored. The two largest collections are at the National Institute of Polar Research, Tokyo, Japan, with ~8900 specimens (in 1994) and at the NASA Johnson Space Center in Houston, Texas, USA, housing ~7650 specimens (in 1995). Both of these collections predominantly consist of Antarctic meteorites.

Table 16.1 Meteorite Collections

Location	City	Country
Naturhistorisches Museum	Vienna	Austria
Museum d'Histoire Naturelle	Paris	France
Museum für Naturkunde, Humboldt Universität	Berlin	Germany
Max-Planck-Institut für Chemie	Mainz	Germany
Geological Survey of India	Calcutta	India
Vatican Observatory Collection	Rome	Italy
National Institute of Polar Research (NIPR)	Tokyo	Japan
Academy of Sciences	Moscow	Russia
The Open University	Milton Keynes	UK
The Natural History Museum	London	UK
U.S. National Museum, Washington	Washington, D.C.	USA
Center for Meteorite Studies	Tempe, AZ	USA
NASA Johnson Space Center	Houston, TX	USA
Field Museum of Natural History	Chicago, IL	USA
American Museum of Natural History	New York, NY	USA

Systematic meteorite searches are carried out in Antarctica and desert regions where meteorites are more easily spotted. These meteorite hunting expeditions have greatly increased our collection of extraterrestrial material during the past three decades. All these meteorites are finds and have experienced some degree of terrestrial weathering. Nevertheless, more members of rare meteorites have become available for research, and new types of meteorite groups were discovered. Listings of new meteorites are reported in the Meteoritical Bulletin, published in the journal *Meteoritics and Planetary Science*.

16.2 Antarctic Meteorites

The collection of meteorites from the ice fields in Antarctica has vastly enlarged the number of meteorites available for research. More than 15,000 meteorite specimens have been found and are mainly stored at the National Institute of Polar Research, Tokyo, Japan, and at the NASA Johnson Space Center, Houston, Texas, USA. The Antarctic meteorites receive their names from the find location, the year recovered and a specimen number. Listings of Antarctic meteorites in the Japanese collection can be found in Catalog of the Antarctic meteorites, (Hirasawa, T., editor-in-chief), NIPR, Tokyo, pp. 230. Meteorites of the U. S. Antarctic Meteorite program (ANSMET) are listed in the *Meteoritical Bulletin 76* (*Meteoritics* 29, pp. 100–143) and *Meteoritical Bulletin 79*, (*Meteoritics & Planet. Sci.* 31, A161–A174).

Table 16.2 lists the geographic location, the three-letter code for the meteorite name, and the latitude and longitude for Antarctica meteorite fields. Most of the Antarctic meteorites have been collected by Japanese- or American-sponsored expeditions; more recently, the European Community has also sponsored meteorite hunting expeditions in the Antarctic.

Table 16.2 Find Locations of Antarctic Meteorites

Geographic Name	Abbreviation	Latitude	Longitude
Allan Hills	ALH	76°43'S	159°40'E
Asuka (Sor Rondane Mountains)	A	~72°S	~26°E
Belgica Mountains	B	72°35'S	31°15'E
Bates Nunataks	BTN	80°15'S	153°30'E
Bowden Neve	BOW	83°30'S	165°00'E
David Glacier	DAV	75°19'S	162°00'E
Derrick Peak	DRP	80°04'S	156°23'E
Dominion Range	DOM	85°20'S	166°30'E
Elephant Moraine	EET	76°15'S	157°30'E
Frontier Mountain	FRO	72°15'S	162°20'E
Geologists Range	GEO	82°30'S	155°30'E
Grosvenor Mountains	GRO	85°40'S	175°00'E
Inland Forts	ILD	77°38'S	161°00'E
Lewis Cliff	LEW	84°17'S	161°05'E
			continued

Table 16.2 *(continued)*

Geographic Name	Abbreviation	Latitude	Longitude
Lonewolf Nunataks	LON	81°20'S	152°50'E
MacAlpine Hills	MAC	84°13'S	160°30'E
MacKay Glacier	MCY	76°58'S	162°00'E
Meteorite Hills	MET	79°41'S	155°45'E
Miller Range	MIL	83°15'S	157°00'E
Mount Baldr	MBR	77°35'S	160°34'E
Mount Howe	HOW	87°22'S	149°30'W
Outpost Nunatak	OTT	75°50'S	158°12'E
Patuxent Range	PAT	84°43'S	64°30'W
Pecora Escarpment	PCA	85°38'S	68°42'W
Purgatory Peak	PGP	77°20'S	162°18'E
Queen Alexandra Range	QUE	84°00'S	168°00'E
Reckling Peak	RKP	76°16'S	159°15'E
Taylor Glacier	TYR	77°44'S	162°10'E
Thiel Mountains	TIL	85°15'S	91°00'W
Wisconsin Range	WIS	84°45'S	125°00'W
Yamato Mountains	Y	71°30'S	35°40'E

16.3 Meteorites Recovered from Deserts

The hot deserts of Nullarbor (Australia), Roosevelt County (New Mexico), and the Sahara have provided a plentiful source of meteorites. By 1995, the number of specimens recovered from the Sahara was about 470; additional expeditions keep increasing the number of meteorites from desert locations. Meteorite names consist of the find location and a specimen number. Listings of meteorites recovered from desert regions are given in the Meteoritical Bulletins published annually in Meteoritics and Planetary Science. A description of Saharan meteorites is given by Bischoff, A., & Geiger, T., *Meteoritics* 30, 1995, 113–124. Table 16.3 lists the latitude and longitude of some Saharan find locations.

Table 16.3 Meteorite Find Locations in the Sahara

Geographic Name	Latitude	Longitude
Acfer (Algeria)	27.6°N	4–4.5°E
Adrar (Algeria)	27.9°N	0.5°E
Aguemour (Algeria)	27°N	4.5°E
El Atchane (Algeria)	29.7°N	4.0°E
El Djouf (Algeria)	27.4°N	1.5°W
Dar al Gani (Libya)	27.0°N	16.25°E
Daraj (Libya)	29.5°N	12.0°E
Hammadah al Hamra (Libya)	27.8°N	12.5°E
Ilafegh (Algeria)	21.6°N	1.3°E
Reggane (Algeria)	25.4°N	0.3°E
Tanezrouft (Algeria)	24.2°–25.2°N	0.3°–1.0°E

Note: Latitude and longitude are approximate.
Sources: Otto, J., 1992, *Chem. Erde* 52, 33–40. Bischoff, A., & Geiger, T., *Meteoritics* 30, 1995, 113–124.

16.4 Meteorite Literature

Several scientific journals publish articles describing research in meteorit ics; however, many articles about meteorites appear in three journals: *Me teoritics and Planetary Science*, *Geochimica et Cosmochimica Acta*, anc *Earth and Planetary Science Letters*. Valuable reference works dealing with meteoritics include *Meteorites: A Petrologic-Chemical Synthesis* by R. T. Dodd (Cambridge, 1981); *Meteorites* by B. Mason (Wiley, 1962); and *Meteorites and the Early Solar System* edited by J. F. Kerridge and M. S. Matthews (Univ. of Arizona Press, 1988). An excellent, brief introduction to all aspects of meteoritics is given in *Kleine Meteoritenkunde* by F. Heide and F. Wlotzka (Springer, 1988; translated in *Meteorites: Messengers from Space*, Springer, 1995).

Two indispensable resources for scientists studying meteorites are the British Museum *Catalogue of Meteorites* by A. L. Graham, A. W. R. Bevan, and R. Hutchison (British Museum of Natural History, London, 1985) and (the three-volume) *The Handbook of Iron Meteorites* by V. F. Buchwald (Univ. of California Press, 1975). Finally, meteorite literature from the late 15th century to the mid-20th century is indexed in *A*

Bibliography on Meteorites by H. Brown, G. Kullerud, and W. Nichiporuk (Univ. of Chicago Press, 1953).

16.5 Meteorite Classification and Composition Tables

The following tables summarize information about meteorite classification, chemical composition, isotopic composition, radiometric ages, mineralogy, and petrology. Of special interest are the carbonaceous chondrites of the Ivuna type (CI- or C1-chondrites), which are chemically very close to the composition of the sun (see Chapter 2). Several authors have evaluated the mean elemental composition of the CI-chondrites; data from six compilations together with a "selected" composition are summarized in Table 16.9.

The mean compositions of the other meteorite groups are obtained from a database containing more than 5000 individual analyses for different types of meteorites. If possible, only data from meteorite falls were considered in the computation of mean abundances in Tables 16.10, 16.11, 16.17, and 16.18, because changes in chemical composition may be caused by weathering of meteorites which are finds.

Table 16.4 Meteorite Classes

Class	Major Minerals	Examples, Observed Falls (Finds)
Carbonaceous chondrites		
CI	hydrated phyllosilicates (serpentine)	Ivuna, Alais, Orgueil, Tonk
CM	hydrated silicates, pyroxene, olivine	Mighei, Boriskino, Murchison
CV	olivine, sulfide	Vigarano, Allende, Bali, Grosnaja, Kaba
CO	olivine, sulfide, hydrated silicates	Ornans, Felix, Kainsaz, Warrenton
CK	olivine, pyroxene	Karoonda, Adelaide
CR	olivine, pyroxene, metal	Renazzo, Al Rais
CH (High iron)	metal, pyroxene, olivine	(Acfer 182, ALH85085, PCA 91328, RKP 92335)
Ordinary chondrites and related meteorites		
H (High iron) (Bronzite chondrites)	bronzite, olivine, iron-nickel	Dhajala, Lost City, Richardton, Sitathali
L (Low iron) (Hypersthene chondrites)	hypersthene, olivine, iron-nickel	Bruderheim, Leedey, Mirzapur
LL (Low iron, Low metal) (Amphoterite chondrites)	hypersthene, olivine, iron-nickel	Chainpur, Cherokee Springs, Saint Sèverin
R (Rumurutites)	olivine, feldspar, pyroxene	Rumuruti, Carlisle Lakes (ALH 85151, Y 75302)
K (Kakangari-type)	low Ca-pyroxene, olivine, sulfide, metal	Kakangari (Lea County 002, LEW 87232)
Acapulcoites	pyroxene, olivine, plagioclase, sulfide, metal	Acapulco (Monument Draw)
Lodranites	orthopyroxene, olivine, metal	Lodran (FRO90011, Gibson, MAC88177)
Enstatite Chondrites		
EH (High iron)	enstatite, iron-nickel, sulfides	Abee, Indarch, Parsa, Qingzhen
EL (Low iron)	enstatite, iron-nickel, sulfides	Daniel's Kuil, Hvittis, Khaipur, Pillistfer

continued

Table 16.4 *(continued)*

Class	Major Minerals	Examples, Observed Falls (Finds)
Achondrites		
EHD-meteorites		
Eucrites	pigeonite, plagioclase	Ibitira, Juvinas, Moore County, Pasamonte
Howardites	hypersthene, plagioclase	Bialystock, Binda, Frankfort, Kapoeta
Diogenites	hypersthene	Johnstown, Roda, Shalka
SNC-meteorites		
Shergottites	pigeonite, maskelynite (= glassy feldspar)	Shergotty, Zagami, (ALH77005, ALH84001, EETA79001, LEW88516, QUE94201)
Nakhlites	augite, olivine	Nakhla, Governador Valadares, Lafayette
Chassignites	olivine	Chassigny
Angrites	fassaite	Angra dos Reis (LEW 86010, LEW 87051)
Aubrites (enstatite achondrites)	enstatite	Aubres, Norton County, Peña Blanca Spring
Brachinaites	olivine	Brachina
Lunar	anorthite	(ALH81005, Calcalong Creek, MAC88105, QUE94281)
Ureilites	olivine, pigeonite, iron-nickel	Novo Urei, Goalpara, Kenna
Winonaites	olivine, orthopyroxene	Winona
Stony Irons		
Mesosiderites (related to EHD)	pyroxene, olivine, plagioclase, iron-nickel	Barea, Chinguetti, Esterville, Patwar
Pallasites	iron-nickel, olivine	Admire, Brenham
Irons		
Ataxites	taenite	Colfax, Dayton
Hexahedrites	kamacite	Coahuila, Hex River
Ocatahedrites	kamacite, taenite	Canyon Diablo, Toluca

Table 16.5 Petrological Classification of Chondrites

Petrographic Type	1	2	3	4	5	6	7
Texture	no chondrules	very clearly defined chondrules		well defined chondrules	chondrules can be recognized	poorly recognizable chondrules	relict chondrules
Matrix	fine grained opaque	chiefly fine, opaque	clastic and minor opaque	coarse grained transparent, recrystallized, coarsening from type 4 to 7			
Homogeneity of ol + px (Fe, Mg content)	—	>5% mean deviation of Fe		0–5%	homogeneous		
Low-Ca-pyroxene polymorph	—	mainly cpx, monoclinic		cpx abundant, monoclinic >20%	monoclinic <20%	orthorhombic; CaO < 1 wt%	CaO > 1 wt%
Feldspar	—	primary only; minor and calcic crystalline, secondary feldspar absent		secondary feldsp. very fine grained < 2 μm	fine grained, small secondary grains < 50 μm	grains clearly visible, coarsening from type 5 to 7, grains > 50 μm	
Glass in chondrules	—	clear and isotropic		turbid, devitrified	no glass		
Metal, maximum Ni content	—	taenite minor or absent, < 20 wt% Ni		kamacite and taenite (>20 wt% Ni) in exsolution			
Sulfides, mean Ni content	—	> 0.5 wt%		< 0.5 wt%			
H₂O content (wt%)	18–20	2–16	0.3–3	< 2			
Carbon content (wt%)	3–5	1.5–2.8	0.1–1.1	< 0.2			
	hydrous alteration ⇐⇐⇐⇐			not thermally equilibrated ⇐⇐⇐⇐ ⇒⇒⇒⇒ thermally equilibrated			
Metamorphic temperatures		400–600°C		600–700°C	700–750°C	750–950°C	> 950°C

Sources: Van Schmus, W. R., & Wood, J. A., 1967, *Geochim. Cosmochim. Acta* 31, 747–765. Dodd, R. T., 1981, *Meteorites, a petrologic-chemical synthesis,* Cambridge Univ. Press, Cambridge, pp. 368. Sears, D. W. G., & Dodd, R. T., 1988, in *Meteorites and the early solar system,* (Kerridge, J. F., & Matthews, M. S., eds.), Univ. of Arizona Press, Tucson, pp. 3–31.

Table 16.6 Shock Classification of Chondrites

Shock Stage	Peak Shock Pressure Effects		in plagioclase	Pressure (GPa)		
	in olivine			293 K[a]	293 K[b]	920 K[b]
S1 unshocked	sharp optical extinction, irregular fractures	angular variation of extinction position: a: low < 1° b: high 1–2°	sharp optical extinction, irregular fractures	<4–5		
S2 very weakly shocked	undulatory extinction, irregular fractures	angular variation of extinction position > 2°	undulatory extinction, irregular fractures	5–10		
S3 weakly shocked	planar fractures, undulatory extinction, irregular fractures	sets of planar fractures: a: low: max. of 2 b: high: ≥ 3	undulatory extinction	15–20	10–15	10–15
S4 moderately shocked	mosaicism (weak)	planar fractures & planar deformation features a: low: incipient mosaicism b: high: mosaicism	a: low: undulatory extinction b: high: partially isotropic, planar deformation features	30–35	25–30	20–25
S5 strongly shocked	mosaicism (strong), planar fractures and planar deformation features		maskelynite	45–55	45–60	35–45
S6 very strongly shocked	restricted to local regions in or near melt zones recrystallization, yellow-brown staining, melting, ringwoodite		shock melt (normal glass)	75–90		45–60
shock melted	whole rock melting (impact melt rocks and melt breccias)					

[a] experiments with nonporous single crystals and dunite.
[b] experiments with unshocked H6-type chondrite. Pressures for 920 K experiments may be lower than indicated.

Sources: Stöffler, D., Keil, K., & Scott, E. R. D., 1991, *Geochim. Cosmochim. Acta* 55, 3845–3867. Schmitt, R. T., & Stöffler D., 1995, *Meteoritics* 30, 574–575.

Table 16.7 Minerals in Meteorites

Mineral Name	Chemical Formula	Found in	Notes
Akaganéite	β-FeO(OH,Cl) (tetragonal)	I	w
Åkermanite (Ak)	$Ca_2MgSi_2O_7$	C, Inc	
Alabandite	(Mn,Fe)S	C, EC, EA, I	r
Albite (Ab)	$NaAlSi_3O_8$	EC, EA, I	
Almandine	$Fe_3Al_2(SiO_4)_3$		sh
Amakinite	$(Fe^{2+},Mg)(OH)_2$		w
Amesite (serpentine)	$Mg_2Al(Si,Al)O_5(OH)_4$	CM	
Anatase	TiO_2		sh
Andradite	$Ca_3Fe_2Si_3O_{12}$	CV	r
Anhydrite	$CaSO_4$	C, OC, SNC	
Ankerite	$(Ca,Fe,Mn)(CO_3)_2$	C	
Anorthite (An)	$CaAl_2Si_2O_8$	many, Inc	
Anthophyllite	$(Mg,Fe)_7Si_8O_{22}(OH)_2$	CV	
Antigorite (serpentine)	$Mg_3Si_2O_5(OH)_4$	CM	
Apatite	$Ca_3(PO_4)_3(F,Cl,OH)$	many, Inc	w
Aragonite	$CaCO_3$ (orthorhombic)	C, U	
Armalcolite	$FeMgTi_2O_5$	CV, Inc	
Arupite	$Ni_3(PO_4)_2 \cdot 8H_2O$	I	w
Astrakhanite, blödite	$Na_2Mg(SO_4)_2 \cdot 4H_2O$	CI	r
Augite	$Mg(Ca,Fe^{2+},Al)_2(Si,Al)_2O_6$	SNC	
Awaruite	$Ni_3Fe–Ni_3Fe$	CI, CV, I	r
Baddeleyite	ZrO_2	OC, C, SNC, Inc	r, sh
Barite	$BaSO_4$		w
Barringerite	$(Fe,Ni)_2P$	Pa	r
Barringtonite	$MgCO_3 \cdot 2H_2O$		w
Bassanite	$CaSO_4 \cdot \frac{1}{2}H_2O$	EA, SNC	w
Beckelite	$(Ce,Ca)_5(SiO_4)_3(OH,F)$	Inc	r
Berthierine	$(Fe,Mg)_{4-6}(Si,Al)_4O_{10}(OH)_6$	C	
Beusite	$(Mn,Fe,Ca,Mg)_3(PO_4)_2$	I	w
Biotite	$K(Mg,Fe)_3(Si_3Al)O_{10}(OH,F)_2$	SNC	
Bornite	Cu_5FeS_4	some	r, w
Brainerite	$(Fe,Mg)CO_3$	C	
Bravorite	$(Fe,Ni)S_2$	some	r, w
Breunnerite	$(Mg,Fe)CO_3$		r, w

continued

Table 16.7 *(continued)*

Mineral Name	Chemical Formula	Found in	Notes
Brezinaite	Cr_3S_4	I	m, r
Brianite	$Na_2CaMg(PO_4)_2$	I	m, r
Brucite	$Mg(OH)_2$	CI	r, w
Buchwaldite	$NaCaPO_4$	I	r
Bunsenite	NiO	I	r, w
Calcite	$CaCO_3$ (trigonal)	CI, CM, Inc, SNC	r, w
Calcium-armalcolite	$CaTi_2O_5$	Inc	
Calcium oxide	CaO	Inc	
Carbonate-fluorapatite	$Ca_5(PO_4,CO_3)_3F$		w
Carlsbergite	CrN	I	m, r
Cassidyite	$Ca_2(Ni,Mg)(PO_4)_2 \cdot 2H_2O$	I	r, w
Caswellsilverite	$NaCrS_2$	EC, EA	m, r
Celsian	$Ba(Al_2Si_2O_8)$	Angrites	r
Chalcocite	Cu_2S		w
Chalcopyrite	$CuFeS_2$	CV, I, SNC	r, w
Chamosite (chlorite)	$Fe_6Mg_3[(Si_4O_{10})(OH)_8]_2$	CI, CM	
Chaoite	C (hexagonal)	U	sh
Chengbolite	$PtTe_2$	CK	r
Chladniite	$Na_2CaMg_7(PO_4)_6$	I	
Chlorapatite	$Ca_5(PO_4)_3Cl$	many, SNC	
Chrysotile (serpentine)	$Mg_3Si_2O_5(OH)_4$	CM	
Chromite	$FeCr_2O_4$	many, Inc, SNC	
Chromium sulfide	CrS, Cr_2S_3	I	r
Cinnabar	HgS	CI, CV	r
Cliftonite	cubic graphite	I	
Clinochlore (chlorite)	$(Mg,Fe^{2+})_5Al(Si_3Al)O_{10}(OH)_8$	CI, CM	
Clinopyroxene	$(Ca,Mg,Fe)SiO_3$	many	
Clintonite	$Ca(Mg,Al)_3(Al,Si)_4O_{10}(OH,F)_2$	CV	
Cobaltite	CoAsS	OC	w
Cohenite	$(Fe,Ni)_3C$	many	r
Collinsite	$Ca_2(Mg,Ni)(PO_4)_2 \cdot 2H_2O$	I	w
Cooperite	PtS	CK	r
Copiapite	$MgFe_4(OH)_2(SO_4)_6 \cdot 18H_2O$	some	r, w
Copper	Cu	many	r, w

continued

Table 16.7 *(continued)*

Mineral Name	Chemical Formula	Found in	Notes
Cordierite	$(Mg,Fe)_2Al_3(AlSi_5O_{18})\cdot xH_2O$	CV, Inc	r
Corundum	Al_2O_3	Inc	pre
Coulsonite	FeV_2O_4	Inc	
Covellite	CuS	EC	r
Cristobalite	SiO_2 (tetragonal)	some	r, sh
Cronstedtite	$(Mg,Fe^{2+})_2(Al,Si,Fe^{3+})O_5(OH)_4$	CI	
Cubanite, isocubanite	$CuFe_2S_3$	CI	r
Cuprite	Cu_2O	OC	r, w
Daubréelite	$FeCr_2S_4$	C, EC, EA, I	m, r
Diamond	β-C (cubic)	C, OC, U, I	pre, sh
Diopside	$MgCaSi_2O_6$	many, Inc	
Djerfisherite	$K_3(Cu,Na)(Fe,Ni)_{12}(S,Cl)_{14}$	EC, EA, I	r
Dolomite	$CaMg(CO_3)_2$	CI	r
Erlichmanite	OsS_2	CK	r
Enstatite	$MgSiO_3$	EC, EA	
Epsomite	$MgSO_4\cdot 7H_2O$	CI	r, w
Eskolaite	Cr_2O_3	CI, U	r, w
Farringtonite	$Mg_3(PO_4)_2$	OC, Pa, I	m, r
Fassaite	$Ca(MgFe^{2+},Fe^{3+},Al,Ti)(Si,Al)_2O_6$	C, Inc	r
Fayalite (fa)	Fe_2SiO_4	many	
Ferrihydrite	$Fe_{4-5}(OH,O)_{12}$	CI, SNC	
Ferrosilite	$Fe_2(SiO_3)_2$	CV	
Fluorapatite	$Ca_5(PO_4)_3F$	I, SNC	w
Fluorrichterite	$Na_2CaMg_5Si_8O_{22}(F,OH)_2$	EC, EA, I	r
Forsterite (fo)	Mg_2SiO_4	many, Inc	
Galena	PbS		w
Galileiite	$NaFe_4(PO_4)_3$	I	
Gehlenite (Ge)	$Ca_2Al_2SiO_7$	C, Inc	
Geikilite	$MgTiO_3$	C, Inc	
Gentnerite	$Cu_8Fe_3Cr_{11}S_{18}$	I	m, r
Gersdorffite	$NiAsS$		w
Goethite	α-FeO(OH) (orthorhombic)	C, I, SNC	w
Gold alloys	(Au,Ag,Fe,Ni,Pt)	Inc	

continued

Table 16.7 *(continued)*

Mineral Name	Chemical Formula	Found in	Notes
Graftonite	$(Fe,Mn)_3(PO_4)_2$	C, I	r
Graphite	$\alpha\text{-}C$	C, EA, U, I	pre, r, w
Greenalite (serpentine)	$(Fe^{2+},Fe^{3+})_{2\text{-}3}Si_2O_5(OH)_4$	CM, CH	
Greigite	Fe_3S_4	EC	
Grossite	$CaAl_4O_7$	C, Inc	
Grossular	$Ca_3Al_2Si_3O_{12}$	CV, Inc	r
Gypsum	$CaSO_4 \cdot 2H_2O$	CI, CM, SNC	r, w
Haapalite	$4(Fe,Ni)S \cdot 3(Mg,Fe^{2+})(OH)_2$	C	w
Halite	$NaCl$	C, SNC, U	r, w
Haxonite	$(Fe,Ni)_{23}C_6$	I	m, r
Heazlewoodite	Ni_3S_2	I	r, w
Hedenbergite	$CaFeSi_2O_6$	CV, Inc	
Heideite	$(Fe,Cr)_{1+x}(Ti,Fe)_2S_4$	EA	r
Hematite	$\alpha\text{-}Fe_2O_3$ (trigonal)	CI	r, w
Hercynite	$(Fe,Mg)Al_2O_4$	CV, SNC, Inc	r
Hexahydrite	$MgSO_4 \cdot 6H_2O$	C	w
Hibbingite	$\gamma\text{-}Fe_2(OH)_3Cl$	I	w
Hibonite	$CaAl_{12}O_{19}$	C, OC, EC, Inc	r
Hollandite	$(Fe_{15}Ni)O_{12}(OH)_{22}Cl$	I	w
Honessite	$(Fe,Ni)_8SO_4(OH)_{16} \cdot xH_2O$	I	w
Hornblende	$(Ca,Na)_2(Mg,Al,Fe,Mn,Ti)_5\text{-}$ $(Si,Al,P)_8O_{22}(OH,F)_2$	C	r
Hydromagnesite	$Mg_5(CO_3)_4(OH)_2 \cdot 4H_2O$	C	w
Hydroxyapatite	$Ca_5(PO_4)_3OH$	SNC	w
Hysingerite	$Fe_4Si_4O_{10}(OH)_8 \cdot 4H_2O$	SNC	u
Idaite	Cu_5FeS_6	EC	
Illite	$(K,H_3O)A_{12}(Si_3Al)O_{10}(H_2O,OH)_2$	SNC	w
Ilmenite	$FeTiO_3$	many, SNC	r
Iridarsenite	$(Ir,Ru)As_2$	CK	r
Iron carbide	Fe_5C_2	I	u
Jadeite	$Na(Al,Fe)Si_2O_6$	C, OC	r
Jarosite	$KFe_3(OH)_6(SO_4)_2$	I	r, w
Jimthomsonite	$(Mg,Fe)_5Si_6O_{16}(OH)_2$	CV	r
Johnsomervilleite	$Na_2Ca(Fe,Mg,Mn)_7(PO_4)_6$	I	

continued

Table 16.7 *(continued)*

Mineral Name	Chemical Formula	Found in	Notes
Kaerusite	$Ca_2(Na,K)(Mg,Fe)_4TiSi_6Al_2O_{22}F_2$	SNC	
Kamacite	α-FeNi (cubic, Ni < 7.5 wt%)	many, Inc	
Kirschsteinite	$(Ca,Fe)_2SiO_4$	Angrites, CV, Inc	
Krinovite	$NaMg_2CrSi_3O_{10}$	I	m, r
Kutnohorite	$Ca(Mn,Mg,Fe^{2+})(CO_3)_2$	C	
Laurite	RuS_2	CK	r
Lawrencite	$(Fe,Ni)Cl_2$	EC, I	m
Lepidocrocite	γ-FeO(OH,Cl)	I	w
Limonite	$FeOOH \cdot xH_2O$	I	w
Lipscombite	$Fe_3(OH)_2(PO_4)_2$	I	r, w
Lizardite (serpentine)	$Mg_3Si_2O_5(OH)_4$	CI, CM	
Löllingite	$FeAs_2$	CK	r
Lonsdaleite	C-diamond (hexagonal)	U, I	m, pre, r, sh
Mackinawite	$(Fe,Ni)_9S_8$ (tetragonal)	I, Ach	r
Magarite	$CaAl_2(Si_2Al_2)O_{10}(OH)_2$	CV, Inc	
Maghemite	γ-Fe_2O_3 (tetragonal)	C, OC	r, w
Magnéli phases	Ti_5O_9, Ti_8O_{15}	C	
Magnesiochromite	$MgCr_2O_4$	many	
Magnesioferrite	$MgFe_2O_4$		r
Magnesiowüstite	(Mg,Fe)O		sh
Magnesite	$(Mg,Fe)CO_3$	C	r
Magnetite	Fe_3O_4	many, SNC	r, w
Majorite	$Mg_3(Mg,Si)Si_3O_{12}$		m, r, sh
Marcasite	FeS_2	C, SNC	r
Maricite	$NaFePO_4$	I	
Martensite	α_2-(Fe,Ni)	U	sh
Maskelynite	isotropic, glassy plagioclase	SNC	r, sh
Maucherite	$Ni_{11}As_8$	OC	r, w
Meionite	$Ca_4(Al_2Si_2O_8)CO_3$	OC	u
Melanterite	$FeSO_4 \cdot 7H_2O$	some	r, w
Melilite (Ak–Ge)	$Ca_2(Mg,Al)(Si,Al)_2O_7$	C, Inc	r
Merrihueite	$(K,Na)_2Fe_5Si_{12}O_{30}$	C, OC	m, r
Metahalloysite	$Al_4(Si_4O_{10})(OH)_8$	C	

continued

Table 16.7 *(continued)*

Mineral Name	Chemical Formula	Found in	Notes
Mica	$(K,Na,Ca)_2Al_4[Si_6Al_2O_{21}](OH,F)_4$	C	w
Millerite	NiS	CK	
Moissanite	SiC (hexagonal)	C, I	r
Molybdenite	MoS_2	CV, Inc	r
Molybdenum	Mo	Inc	
Molybdenum carbide	MoC, Mo_2C		m, pre
Molysite	$FeCl_3$	I	r, u, w
Monazite	$(Ce,La,Th)PO_4$	Inc	
Monticellite	$Ca(Mg,Fe)SiO_4$	OC, Inc	r
Montmorillonite	$(Na,Ca)_{0.3}(Al,Mg)_2(Si_4O_{10})(OH)_2 \cdot xH_2O$	C, OC	u
Muscovite	$KAl_2(AlSi_3O_{10})(OH,F)_2$	CV	
Nepheline	$NaAlSiO_4$	C, Inc	r
Nesquehonite	$Mg(HCO_3)(OH) \cdot 2H_2O$	CI	w
Ni-Cr-alloy	Ni-Cr	U	r
Nickel	Ni	OC	w
Nickel hydroxide	$Ni(OH)_2$		r, w
Nickeline	NiAs		r, w
Nierite	$\alpha\text{-}Si_3N_4$	EC	pre, r
Niningerite	(Mg,Fe)S	EC, U?	m, r
Niobium	Nb	Inc	
Nyerereite	$Na_2Ca(CO_3)_2$	CM	w
Oldhamite	CaS	EC, EA, U?	m, r
Oligoclase	$(K,Na)AlSi_3O_8 (Ab_{90-70})$	many	
Olivine (fo–fa)	$Mg_2SiO_4 - Fe_2SiO_4$	many	
Omeiite	$(Os,Ru)As_2$	CK	r
Opal	$SiO_2 \cdot xH_2O$	I	w
Orcelite	$Ni_{5-x}As_2$		r, w
Orthoclase (Or)	$KAlSi_3O_8$	many	
Orthopyroxene	$(Mg,Fe)_2Si_2O_6$	many	
Osbornite	TiN	EC, EA, CH	m, r
Osumulite	$(K,Na)(Fe,Mg)_2(Al,Fe)_3(Si,Al)_{12}O_{30}$	I, OC	r
Panethite	$(Ca,Na)_2(Mg,Fe)_2(PO_4)_2$	I	m, r
Pecoratite	$Ni_3Si_2O_5(OH)_4$		w
Pentlandite	$(Fe,Ni)_9S_8$	C, OC, Ach, Inc	r, w

continued

Table 16.7 *(continued)*

Mineral Name	Chemical Formula	Found in	Notes
Periclase	MgO	C, Inc	
Perovskite	$CaTiO_3$	C, Inc	r
Perryite	$(Ni,Fe)_5(Si,P)_2$	EC, EA, I	m, r
Pigeonite	$(Fe,Mg,Ca)SiO_3$	many	
Platinum	Pt	Inc	r
Pleonaste	$(Mg,Fe)Al_2O_4$	CK, Inc	r
Plessite	fine kamacite and taenite	I	
Polygorskite	$(Mg,Al)_2(Si_4O_{10})(OH)\cdot2H_2O$	C	u
Portlandite	$Ca(OH)_2$	EA	w
Powellite	$CaMoO_4$	CV, Inc	
Prehnite	$Ca_2Al_2Si_3O_{10}(OH)$	OC	u
Pseudobookite	Fe_2TiO_5		r, w
Pumpellyite	$Ca_2(Mg,Fe^{2+})Al_2(SiO_4)(Si_2O_7)(OH)_2\cdot H_2O$	CV	w
Pyrite	FeS_2	C, SNC	r, w
Pyrope	$Mg_3Al_2(SiO_4)_3$		sh
Pyrochlore	$(Na,Ca)_2Nb_2O_6(OH,F)$	Inc	
Pyrophanite	$MnTiO_3$	OC	r
Pyroxferroite	$(Fe,Mn,Ca)SiO_3$	lunar rock	
Pyroxene	$(Ca,Mg,Fe)SiO_3$	many	
Pyrrhotite	$Fe_{1-x}S$	C, Inc, SNC	r
Quartz	SiO_2 (trigonal)	many	w
Rammelsbergite	$NiAs_2$	OC	r, w
Reevesite	$Ni_6Fe_2CO_3(OH)_{16}\cdot4H_2O$	I	r, w
Rhenium	Re	Inc	
Rhodochrosite	$MnCO_3$	CI, SNC	
Rhodonite	$CaMn_4Si_5O_{15}$		r
Rhönite	$CaMg_2TiAl_2SiO_{10}$	CV, Inc	r
Ringwoodite	$\gamma\text{-}(Mg,Fe)_2SiO_4$		m, r, sh
Roaldite	$(Fe,Ni)_4N$	I	r
Roedderite	$(Na,K)_2(Mg,Fe)_5Si_{12}O_{30}$	EA, EC, I	m, r
Ruthenium	Ru	Inc	
Rutile	TiO_2 (tetragonal)	some, Inc, SNC	r
Safflorite	$CoAs_2$	OC	w

continued

Table 16.7 *(continued)*

Mineral Name	Chemical Formula	Found in	Notes
Sanidine	$KAlSi_3O_8$	SNC	
Sapphirine	$(Mg,Al)_7(Mg,Al)O_2(Al,Si)_6O_{18}$		r
Sarcopside	$(Fe,Mn)_3(PO_4)_2$	I	r
Scheelite	$CaWO_4$	CV, Inc	
Schöllhornite	$Na_{0.3}CrS_2 \cdot H_2O$	EA	w
Schreibersite, rhabdite	$(Fe,Ni)_3P$	many	m, r
Sepiolite	$Mg_4Si_6O_{15}(OH)_2 \cdot 6H_2O$	C	u
Serpentine	$(Mg,Fe)_6Si_4O_{10}(OH)_8$	CI, CM	w
Siderite	$FeCO_3$	CI, SNC	r, w
Silicon carbide (cubic)	SiC		m, pre
β-Silicon nitride	β-Si_3N_4		m, pre
Sinoite	Si_2N_2O	EC	m, r
Smythite	Fe_9S_{11}	EC	pre, r
Sodalite	$Na_8Al_6Si_6O_{24}Cl_2$	CV	r
Sperrylite	$PtAs_2$	CK	r
Sphalerite	α-$(Zn,Fe)S$	EA, EC, C, I, SNC	r, w
Spinel	$MgAl_2O_4$	C, Inc	r
Stanfieldite	$Ca_4(Mg,Fe)_5(PO_4)_6$	I, Pa	m, r
Starkeyite	$MgSO_4 \cdot 4H_2O$	C	w
Suessite	$(Fe,Ni)_3Si$	U	r, sh
Sulfur	α-S (orthorhombic)	CI	r, w
Sylvite	KCl	CM, U	r
Taenite	γ-FeNi (8–55 wt% Ni)	many	
Talc	$Mg_3(Si_4O_{10})(OH)_2$	CM, CV	w
Tetrataenite	FeNi	CM	
Thorianite	ThO_2	Inc	
Titanite	$CaTiSiO_5$	Euc	
Titanium carbide	TiC		m, pre
Titanium magnetite	$(Fe,Mg)(Al,Ti)_2O_4$	CV, Inc, SNC	
Tochilinite	$2(Fe,Mg,Cu,Ni)S$ $\cdot 1.57-1.85(Mg,Fe,Ni,Al,Ca)(OH)_2$	C	
Trevorite	$NiFe_2O_4$	C, I	r, w
Tridymite	SiO_2 (monoclinic, triclinic)	some	r
Troilite	FeS (hexagonal)	many, Inc, SNC	

continued

Table 16.7 *(continued)*

Mineral Name	Chemical Formula	Found in	Notes
Tungstenite	WS_2	CV, Inc	r
Ulvöspinel	Fe_2TiO_4	Inc	
Ureyite (kosmochlor)	$NaCrSi_2O_6$	I	m, r
Vanadium magnetite	$(Fe,Mg)(Al,V)_2O_4$	Inc	
Vaterite	$CaCO_3$	EA	w
Vermiculite	$(Mg,Fe^{2+},Al)_3(Al,Si)_4O_{10}(OH)_2 \cdot 4H_2O$	C, OC, SNC	w
Violarite	$FeNi_2S_4$	some	r
Vivianite	$Fe_3(PO_4)_2 \cdot 8H_2O$	some	r, w
Wadsleyite	$(Mg,Fe)_2SiO_4$		r
Wairauite	$FeCo-Fe_3Co_2$	CK	
Whewellite (Ca-oxalate)	$CaC_2O_4 \cdot H_2O$	CM	r
Whitlockite (merrillite)	$Ca_3(PO_4)_2$	many, Inc, SNC	
Wollastonite (wo)	$CaSiO_3$	CV	r
Wurtzite	β-ZnS		r
Wüstite	$(Fe,Ni)_{1-x}O$	OC, I	r, w
Yagiite	$(K,Na)_2(Mg,Al)_5(Si,Al)_{12}O_{30}$	I	m, r
Zaratite	$Ni_3CO_3(OH)_4 \cdot 4H_2O$	I	w
Zeolites	$(Na,K)_{0-2}(Ca,Mg)_{1-2}(Al,Si)_{5-10}O_{10-20} \cdot xH_2O$	Inc	w
Zircon	$ZrSiO_4$	I	r
Zirconium carbide	ZrC		m, pre
Zirconolite	$(Ca,Th,Ce)Zr(Ti,Nb)_2O_7$	Inc	

C: carbonaceous chondrites; CH, CI, CK, CM, CV: subtypes of carbonaceous chondrites; see Table 16.4. OC: ordinary chondrites; EC: enstatite chondrites; Ach: achondrites; EA: enstatite achondrites; Euc: eucrites; I: irons; Inc: in Ca, Al-rich inclusions, plagioclase-olivine inclusions, refractory inclusion rims, or fremdlinge; SNC: Shergottites, Nakhlites, and Chassignites; Pa: Pallasites; U: Ureilites.
m: mineral in meteorites only; pre: in presolar grains; r: relatively rare; sh: shock product; u: uncertain; w: aqueous alteration/weathering product.

Sources: Mason, B., 1972, *Meteoritics* 7, 309–326. Olsen, E., 1981, in *The encyclopedia of mineralogy,* (Frye, K., ed.) Hutchinson Ross Publ. Co., Stroudsburg, PA, pp. 240–246. Rubin, A. E., 1997, *Meteoritics & Planet. Sci.* 32, 231–247. Ulyanov, A. A., 1991, *The meteorite minerals,* Brown-Vernadsky Microsymp. on Comparative Planetology 14th, pp. 20.

Table 16.8   Comparison of Some Meteorite Ages (Myrs before present)

Name and Type	$^{87}Rb/^{86}Sr$ Age	Source	$^{147}Sm/^{143}Nd$ Age	Source	$^{207}Pb/^{206}Pb$ Age	Source
Chondrites						
Murray CM2	...		...		4511±42	TKA73
Allende CV3	...		...		4496±10	TKA73
					4548±25	TUD76
Tieschitz H3	4530±60	MA79	...		...	
Beardsley H5	4690±70	KW69	...		4574±12	TKA73
Plainview H5	4690±70	KW69	...		4529±10	TKA73
Richardton H5	4390±30	ECH79	...		4519±15	Til73,
	4400±90	MBA82			4519±15	TKA73
Guareña H6	4560±80	WPS69	...		...	
	4480±80	MBA82				
Modoc L6	...		4050±80	NT80	4530±15	TKA73
Chainpur LL3	4510±56	MA81	...		...	
Soko-Banja LL4	4452±20	MA81	...		...	
	4420±20	MBA82				
St. Sèverin LL6	4510±50	MBA82	4550±330	JW84	4543±19	MMA78
Achondrites						
Angra dos Reis	...		4564±37	JW84	4551±4	CW81
(ADOR)			4550±4	LM77	4557.80±0.42	LG92
					4555±5	TKA73
Acapulco	...		4600±30	PPW92	...	
Eucrites						
Bereba	4170±260	BA78	...		4521±0.4	TCB97
Ibitira	4520±250	BA78	4460±20	PPW92	4556±6	CW85
					4560±3	MGA87
Juvinas	4570±130	BA78	4560±80	Lug74	4539±4	MAP84
Moore County	...		4456±25	TCB97	4484±19	TCB97
Pasamonte	<2600	BA78	4580±120	UNT77	4530±30	UNT77
Sioux County	4190±140	BA78	...		4526±10	TKA73
Stannern	3300±500	BA78	4480±70	LS75	4128±16	TCB97
Y75011	4460±60	NTB86	4550±140	NTB86	...	

Note: Lead ages in Ca, Al-rich inclusions are 4553–4575 Myrs [MGA87, CW81, TUD76].

continued

Table 16.8 *(continued)*

Sources: **[BA78]** Birck, J. L., & Allègre, C. J., 1978, *Earth Planet Sci. Lett.* 39, 37–51. [CW81] Chen, J. H., & Wasserburg, G. J., 1981, *Earth Planet. Sci. Lett.* 52, 1–15. **[CW85]** Chen, J. H., & Wasserburg, G. J., 1985, *Lunar Planet. Sci. Conf.* XVI, 119–120. **[ECH79]** Evensen, N. M., Carter, S. R., Hamilton, P. J., O'Nions, R. K., & Ridley, W. I., 1979, *Earth Planet. Sci. Lett.* 42, 223–236. **[JW84]** Jacobsen, S. B., & Wasserburg, G. J., 1984, *Earth Planet Sci. Lett.* 67, 137–150. **[KW89]** Kausal, S. K., & Wetherill, G. W., 1969, *J. Geophys. Res. 74*, 2717–2726. **[LG92]** Lugmair, G. W., & Galer, S. J. G, 1992, *Geochim. Cosmochim. Acta* 56, 1673–1694. **[LM77]** Lugmair, G. W., & Marti, K., 1977, *Earth Planet Sci. Lett.* 35, 273–284. **[LS75]** Lugmair, G. W., & Scheinin, N. B., 1975, *Meteoritics* 10, 447–448. **[Lug74]** Lugmair, G. W., 1974, *Meteoritics* 9, 369. **[MA79]** Minster, J. F., & Allègre, C. J., 1979, *Earth Planet. Sci. Lett.* 42, 333–347. **[MA81]** Minster, J. F., & Allègre, C. J., 1981, *Earth Planet. Sci. Lett.* 56, 89–106. **[MAP84]** Manhes, G., Allègre, C. J., & Provost, A., 1984, *Geochim. Cosmochim. Acta* 48, 2247–2264. **[MBA82]** Minster, J. F., & Birck, J. L., & Allègre, C. J., 1982, *Nature* 300, 414–419. **[MGA87]** Manhes, G., Göpel, C, & Allègre, C. J., 1987, *Meteoritics* 22, 453–454. **[MMA78]** Manhes, G., Minster, J. F., & Allègre, C. J., 1978, *Earth Planet Sci. Lett.* 39, 14–24. **[NT80]** Nakamura, N., & Tatsumoto, M., 1980, *Meteoritics* 15, 334–335. **[NTB86]** Nyquist, L. E., Takeda, H., Bansal, B. M., Shih, C. Y., Wiesmann, H., & Wooden, J. L., 1986, *J. Geophys. Res.* 91, 8137–8150. **[PPW92]** Prinzhofer, A, Papanastassiou, D. A., & Wasserburg, G. J., 1989, *Geochim. Cosmochim. Acta*, 56, 797–815. **[TCB97]** Tera, F., Carlson, R. W., & Boctor, N. Z., 1997, *Geochim. Cosmochim. Acta* 61, 1713–1731. **[Til73]** Tilton, G. R., 1973, *Earth Planet Sci. Lett.* 19, 321–329. **[TKA73]** Tatsumoto, M., Knight, R. J., & Allègre, C. J., 1973, *Science* 180, 1279–1283. **[TUD76]** Tatsumoto, M., Unruh, D. M., & Desborough, G. A., 1976, *Geochim. Cosmochim. Acta* 40, 617–634. **[UNT77]** Unruh, D. M., Nakamura, N., & Tatsumoto, M. 1977, *Earth Planet Sci. Lett.* 37, 1–12. **[WPS69]** Wasserburg, G. J., Papanastassiou, D. A., & Sanz, H. G., 1969, *Earth Planet Sci. Lett.* 7, 33–43.

Table 16.9 Elemental Abundances in CI-Chondrites

Z	Element	Unit	Selected	[MS95]	[PB93]	[AG89]	[WK88]	[AE82]	[PSZ81]
1	H	wt%	2.02	...	2.02	2.02	2.00	2.02	...
2	He	nL/g	56	...	...	56	...	56	...
3	Li	ppm	1.5	1.5	1.49	1.50	1.57	1.59	1.45
4	Be	ppb	25	25	24.9	24.9	27	26.7	25
5	B	ppb	870	900	870	870	1200	1250	270
6	C	wt%	3.45	3.50	3.22	3.45	3.2	3.45	3.50
7	N	ppm	3180	3180	3180	3180	1500	3180	...
8	O	wt%	46.40	...	46.50	46.40	46.00	46.40	47.00
9	F	ppm	60	60	58.2	60.7	64	58.2	54
10	Ne	pL/g	203	...	...	203	...	203	...
11	Na	ppm	5000	5100	4982	5000	4900	4830	5020
12	Mg	wt%	9.70	9.65	9.61	9.89	9.70	9.55	93.60
13	Al	ppm	8650	8600	8650	8680	8600	8620	8200
14	Si	wt%	10.64	10.65	10.68	10.64	10.50	10.67	10.68
15	P	ppm	950 *	1080	1105	1220	1020	1180	1010
16	S	wt%	5.41[†]	5.4	5.25	6.25	5.90	5.25	5.80
17	Cl	ppm	700	680	698	704	680	698	678
18	Ar	pL/g	751	...	...	751	...	751	...
19	K	ppm	550	550	544	558	560	569	517
20	Ca	ppm	9260	9250	9510	9280	9200	9020	9000
21	Sc	ppm	5.9	5.92	5.90	5.82	5.8	5.76	5.9
22	Ti	ppm	440	440	441	436	420	436	440
23	V	ppm	55	56	54.3	56.5	55	56.7	55.6
24	Cr	ppm	2650	2650	2646	2660	2650	2650	2670
25	Mn	ppm	1940	1920	1933	1990	1900	1960	1820
26	Fe	wt%	18.20	18.10	18.23	19.04	18.20	18.51	18.30
27	Co	ppm	505	500	506	502	508	509	501
28	Ni	wt%	1.10	1.05	1.077	1.10	1.07	1.10	1.08
29	Cu	ppm	125	120	131	126	121	112	108
30	Zn	ppm	315	310	323	312	312	308	347
31	Ga	ppm	9.8	9.2	9.71	10	9.8	10.1	9.1
32	Ge	ppm	33	31	32.6	32.7	33	32.2	31.3
33	As	ppm	1.85	1.85	1.81	1.86	1.84	1.91	1.85
34	Se	ppm	21	21	21.3	18.6	19.6	18.2	18.9
35	Br	ppm	3.5	3.57	3.5	3.57	3.6	3.56	2.53

continued

Table 16.9 *(continued)*

Z	Element	Unit	Selected	[MS95]	[PB93]	[AG89]	[WK88]	[AE82]	[PSZ81]
36	Kr	pL/g	8.7	...	...	8.7	...	8.7	...
37	Rb	ppm	2.3	2.3	2.32	2.30	2.22	2.30	2.06
38	Sr	ppm	7.3	7.25	7.26	7.80	7.9	7.91	8.6
39	Y	ppm	1.56	1.57	1.57	1.56	1.44	1.50	1.44
40	Zr	ppm	3.9	3.82	3.87	3.94	3.8	3.69	3.82
41	Nb	ppb	250	240	246	246	270	250	300
42	Mo	ppb	920	900	928	928	920	920	920
44	Ru	ppb	710	710	714	712	710	714	690
45	Rh	ppb	140 ‡	130	134	134	134	134	130
46	Pd	ppb	560	550	556	560	560	557	530
47	Ag	ppb	200	200	197	199	208	220	210
48	Cd	ppb	690	710	680	686	650	673	770
49	In	ppb	80	80	77	80	80	77.8	80
50	Sn	ppm	1.7	1.65	1.68	1.720	1.72	1.680	1.75
51	Sb	ppb	135	140	133	142	153	155	130
52	Te	ppm	2.3	2.33	2.27	2.320	2.4	2.280	2.34
53	I	ppb	430	450	433	433	500	430	560
54	Xe	pL/g	8.6	...	...	8.6	...	8.6	...
55	Cs	ppb	190	190	188	187	183	186	190
56	Ba	ppm	2.35	2.41	2.41	2.340	2.3	2.270	2.2
57	La	ppb	235	237	245	234.7	236	236	245
59	Ce	ppb	620	613	638	603.2	616	619	638
59	Pr	ppb	94	92.8	96.4	89.1	92.9	90	96
60	Nd	ppb	460	457	474	452.4	457	462	474
62	Sm	ppb	150	148	154	147.1	149	142	154
63	Eu	ppb	57	56.3	58.0	56.0	56.0	54.3	58
64	Gd	ppb	200	199	204	196.6	197	196	204
65	Tb	ppb	37	36.1	37.5	36.3	35.5	35.3	37
66	Dy	ppb	250	246	254	242.7	245	242	254
67	Ho	ppb	56	54.6	56.7	55.6	54.7	54	57
68	Er	ppb	160	160	166	158.9	160	160	166
69	Tm	ppb	25	24.7	25.6	24.2	24.7	22	26
70	Yb	ppb	160	161	165	162.5	159	166	165
71	Lu	ppb	25	24.6	25.4	24.3	24.5	24.3	25
72	Hf	ppb	105	103	107	104	120	119	120

continued

Table 16.9 (continued)

Z	Element	Unit	Selected	[MS95]	[PB93]	[AG89]	[WK88]	[AE82]	[PSZ81]
73	Ta	ppb	14	13.6	14.0	14.2	16	17	14
74	W	ppb	93	93	95	92.6	100	89	89
75	Re	ppb	38	40	38.3	36.5	37	36.9	37
76	Os	ppb	490	490	486	486	490	699	490
77	Ir	ppb	465	455	459	481	460	473	480
78	Pt	ppm	1.0	1.01	0.994	0.990	0.990	0.953	1.05
79	Au	ppb	145	140	152	140	144	145	140
80	Hg	ppb	310	300	310	258	390	390	...
81	Tl	ppb	142	140	143	142	142	143	140
82	Pb	ppm	2.50	2.47	2.53	2.470	2.4	2.430	2.43
83	Bi	ppb	110	110	111	114	110	111	110
90	Th	ppb	29	29	29.8	29.4	29	28.6	29
92	U	ppb	8	7.4	7.8	8.1	8.2	8.1	8.2

* from [WP97]
† from [DPS95]
‡ from [Jo96]
ppm = 1 $\mu g/g$ = 10^3 ppb = 10^3 ng/g = 10^{-3} mg/g = 10^{-4} wt%

Sources: [AE82] Anders, E., & Ebihara, M., 1982, Geochim. Cosmochim. Acta 46, 2363–2380. [AG89] Anders, E., & Grevesse, N., 1989, Geochim. Cosmochim. Acta 53, 197–214. [DPS95] Dreibus, G., Palme, H., Spettel, B., Zipfel, J., & Wänke, H., 1995, Meteoritics 30, 439–445. [Jo96] Jochum, K. P., 1996, Geochim. Cosmochim. Acta 60, 3353–3357. [MS95] McDonough, W. F., & Sun, S. S., 1995, Chem. Geol. 120, 223–253. [PB93] Palme, H., & Beer, H., 1993, in Landolt-Börnstein, Group VI: Astronomy and Astrophysics, Vol. 3 (Voigt, H. H., ed.), Springer Verlag, Berlin, pp. 196–221. [PSZ81] Palme, H., Suess, H. E., & Zeh, H. D., 1981, in Landolt-Börnstein, Group VI: Astronomy and Astrophysics, Vol. 2, (Schaiffers, K., & Voigt, H. H., eds.), Springer Verlag, Heidelberg, pp. 257–272. [WK88] Wasson, J. T., & Kallemeyn G. W., 1988, Phil. Trans R. Soc. Lond. A325, 535–544. [WP97] Wolf, D., & Palme, H., 1997, Meteoritics & Planet. Sci. Suppl. A32, 141.

Table 16.10 Elemental Abundances in Carbonaceous Chondrites

Z	Element	Unit	CI *	CM	CV	CO	CK	CR	CH
1	H	wt%	2.02	1.4	0.28	0.07	...	...	...
3	Li	ppm	1.5	1.5	1.7	1.8	1.4	...	...
4	Be	ppm	0.025	0.04	0.05	...	...	...	...
5	B	ppm	0.87	0.48	0.3	...	...	...	...
6	C	wt%	3.45	2.2	0.53	0.44	0.22	2.0	0.78
7	N	ppm	3180	1520	80	90	...	620	190
8	O	wt%	46.4	43.2	37.0	37.0	...	...	...
9	F	ppm	60	38	24	30	20	...	...
11	Na	ppm	5000	3900	3400	4200	3100	3300	1800
12	Mg	wt%	9.70	11.5	14.3	14.5	14.7	13.7	11.3
13	Al	wt%	0.865	1.13	1.68	1.40	1.47	1.15	1.05
14	Si	wt%	10.64	12.7	15.7	15.8	15.8	15.0	13.5
15	P	ppm	950	1030	1120	1210	1100	1030	...
16	S	wt%	5.41	2.7	2.2	2.2	1.7	1.90	0.35
17	Cl	ppm	700	430	250	280	260	...	...
19	K	ppm	550	370	360	360	290	315	200
20	Ca	wt%	0.926	1.29	1.84	1.58	1.7	1.29	1.3
21	Sc	ppm	5.9	8.2	10.2	9.5	11	7.8	7.5
22	Ti	ppm	440	550	870	730	940	540	650
23	V	ppm	55	75	97	95	96	74	63
24	Cr	ppm	2650	3050	3480	3520	3530	3415	3100
25	Mn	ppm	1940	1650	1520	1620	1440	1660	1020
26	Fe	wt%	18.2	21.3	23.5	25.0	23.0	23.8	38.0
27	Co	ppm	505	560	640	680	620	640	1100
28	Ni	wt%	1.10	1.23	1.32	1.42	1.31	1.31	2.57
29	Cu	ppm	125	130	104	130	90	100	120
30	Zn	ppm	315	180	110	110	80	100	40
31	Ga	ppm	9.8	7.6	6.1	7.1	5.2	6.0	4.8
32	Ge	ppm	33	26	16	20	14	18	...
33	As	ppm	1.85	1.8	1.5	2.0	1.4	1.5	2.3

continued

Table 16.10 (*continued*)

Z	Element	Unit	CI *	CM	CV	CO	CK	CR	CH
34	Se	ppm	21	12	8.7	8.0	8.0	8.2	3.9
35	Br	ppm	3.5	3.0	1.6	1.4	0.6	1.0	1.4
37	Rb	ppm	2.3	1.6	1.2	1.3	...	1.1	...
38	Sr	ppm	7.3	10	14.8	13.0	15	10	...
39	Y	ppm	1.56	2.0	2.6	2.4	2.7	...	...
40	Zr	ppm	3.9	7.0	8.9	9.0	8	5.4	...
41	Nb	ppm	0.25	0.4	0.5	...	0.4	0.5	...
42	Mo	ppm	0.92	1.4	1.8	1.7	0.38	1.4	2
44	Ru	ppm	0.71	0.87	1.2	1.08	1.1	0.97	1.6
45	Rh	ppm	0.14	0.16	0.17	...	0.18	...	...
46	Pd	ppm	0.56	0.63	0.71	0.71	0.58	0.69	...
47	Ag	ppb	200	160	100	100	...	95	...
48	Cd	ppb	690	420	350	8	...	300	...
49	In	ppb	80	50	32	25	...	30	...
50	Sn	ppm	1.70	0.79	0.68	0.89	0.49	0.73	...
51	Sb	ppb	135	130	85	110	60	80	90
52	Te	ppm	2.3	1.3	1.0	0.95	0.8	1.0	...
53	I	ppm	0.43	0.27	0.16	0.2	0.2	...	...
55	Cs	ppm	0.19	0.11	0.09	0.08	...	0.084	...
56	Ba	ppm	2.35	3.1	4.55	4.3	4.7	3.4	3
57	La	ppm	0.235	0.320	0.469	0.38	0.46	0.31	0.29
59	Ce	ppm	0.620	0.940	1.190	1.14	1.27	0.75	0.87
59	Pr	ppm	0.094	0.137	0.174	0.14	...	...	...
60	Nd	ppm	0.460	0.626	0.919	0.85	0.99	0.79	...
62	Sm	ppm	0.150	0.204	0.294	0.25	0.29	0.23	0.185
63	Eu	ppm	0.057	0.078	0.105	0.096	0.11	0.08	0.076
64	Gd	ppm	0.200	0.290	0.405	0.39	0.44	0.32	0.29
65	Tb	ppm	0.037	0.051	0.071	0.06	...	0.05	0.05
66	Dy	ppm	0.250	0.332	0.454	0.42	0.49	0.28	0.31
67	Ho	ppm	0.056	0.077	0.097	0.096	0.10	0.10	0.07

continued

Table 16.10 *(continued)*

Z	Element	Unit	CI *	CM	CV	CO	CK	CR	CH
68	Er	ppm	0.160	0.221	0.277	0.305	0.35	...	...
69	Tm	ppm	0.025	0.035	0.048	0.04	...	...	0.04
70	Yb	ppm	0.160	0.215	0.312	0.27	0.32	0.22	0.21
71	Lu	ppm	0.025	0.033	0.046	0.039	0.046	0.032	0.03
72	Hf	ppm	0.105	0.18	0.23	0.22	0.25	0.15	0.14
73	Ta	ppb	14	19	...	...	...	...	...
74	W	ppb	93	160	160	150	180	110	150
75	Re	ppb	38	50	57	58	60	50	73
76	Os	ppb	490	670	800	805	815	710	1150
77	Ir	ppb	465	580	730	740	760	670	1070
78	Pt	ppm	1.0	1.1	1.25	1.24	1.3	0.98	1.7
79	Au	ppb	145	150	153	190	120	160	250
80	Hg	ppb	310	...	...	...	...	...	...
81	Tl	ppb	142	92	58	40	...	60	...
82	Pb	ppm	2.50	1.6	1.1	2.15	0.8	...	...
83	Bi	ppb	110	71	54	35	20	40	...
90	Th	ppb	29	41	58	80	58	42	...
92	U	ppb	8	12	17	18	15	13	...
	Fe#	mol%†	45	39	38	36	38	1–3	1–6
metal		wt%	0	<0.1	~0.2	~4.4	~0	~7	~37
sulfide		wt%	3	7.2	~6.5	~4	~3	~1	~1
H_2O		wt%	10.8	9.5	~0.15	~0.2	~3	~6	...
Density		g cm^{-3}	2.23	2.71	3.42	3.63	3.4	3.27	4.2

* Selected values from Table 16.9.
† Fe# = FeO/(FeO + MgO) in mol% of olivine. Equilibrium values for CI and CM chondrites.
For data sources, see text.

Table 16.11 Elemental Abundances in Ordinary and Enstatite Chondrites

Z	Element	Unit	H	L	LL	R	Acap.[a]	K	EH	EL
3	Li	ppm	1.7	1.85	1.8	...	...	...	1.9	0.70
4	Be	ppb	30	40	45	...	...	...	21	...
5	B	ppm	0.4	0.4	0.7	...	...	...	1	...
6	C	ppm	2100	2500	3100	580	...	...	3900	4300
7	N	ppm	48	43	70	...	...	...	420	240
8	O	wt%	35.70	37.70	40	...	...	...	28	31
9	F	ppm	125*	100*	70*	...	...	...	155	140
11	Na	ppm	6110	6900	6840	6630	6440	6800	6880	5770
12	Mg	wt%	14.1	14.9	15.3	12.9	15.6	15.4	10.73	13.75
13	Al	wt%	1.06	1.16	1.18	1.06	1.2	1.3	0.82	1.00
14	Si	wt%	17.1	18.6	18.9	18	17.7	16.9	16.6	18.8
15	P	ppm	1200	1030	910	...	1600	1400	2130	1250
16	S	wt%	2.0	2.2	2.1	4.07	2.7	5.5	5.6	3.1
17	Cl	ppm	140	270	200	<100	...	...	570	230
19	K	ppm	780	920	880	780	475	710	840	700
20	Ca	wt%	1.22	1.33	1.32	0.914	1.1	1.22	0.85	1.02
21	Sc	ppm	7.8	8.1	8.0	7.75	8.1	7.9	6.1	7.7
22	Ti	ppm	630	670	680	900	560	700	460	550
23	V	ppm	73	75	76	70	83	73	56	64
24	Cr	ppm	3500	3690	3680	3640	4160	3600	3300	3030
25	Mn	ppm	2340	2590	2600	2960	3000	2400	2120	1580
26	Fe	wt%	27.2	21.75	19.8	24.4	23.5	24.7	30.5	24.8
27	Co	ppm	830	580	480	610	850	750	870	720
28	Ni	wt%	1.71	1.24	1.06	1.44	1.5	1.46	1.84	1.47
29	Cu	ppm	94	90	85	...	110	...	215	120
30	Zn	ppm	47	57	56	150	230	145	290	18
31	Ga	ppm	6.0	5.4	5.3	8.1	9	8.2	16.7	11
32	Ge	ppm	10	10	10	...	16	...	38	30
33	As	ppm	2.2	1.36	1.3	1.9	2.3	2.4	3.5	2.2
34	Se	ppm	8.0	8.5	9	14.1	8.9	20	25	15
35	Br	ppm	0.01–1*	0.05–2*	1.0*	0.55	0.2	0.9	2.7	0.8

continued

Table 16.11 *(continued)*

Z	Element	Unit	H	L	LL	R	Acap.[a]	K	EH	EL
37	Rb	ppm	2.3	2.8	2.2	...	0.2	1.7	3.1	2.3
38	Sr	ppm	8.8	11	13	...	...	...	7.0	9.4
39	Y	ppm	2.0	1.8	2.0	...	...	...	1.2	...
40	Zr	ppm	7.3	6.4	7.4	...	...	...	6.6	7.2
41	Nb	ppm	0.4	0.4	...	...	...	...	...	...
42	Mo	ppm	1.4	1.2	1.1	0.9	...	...	...	...
44	Ru	ppb	1100	750	...	960	670	850	930	770
45	Rh	ppb	210	155	...	...	...	...	...	...
46	Pd	ppb	845	620	560	...	...	...	820	730
47	Ag	ppb	45*	50*	75*	...	50	...	280	85
48	Cd	ppb	1–10*	30*	40*	...	20	30	705	35
49	In	ppb	0.2–1.5*	0.1–20*	1–20*	...	4	3	85	4
50	Sn	ppb	350*	540*	...	...	...	...	1360	...
51	Sb	ppb	66	78	75	72	90	150	190	90
52	Te	ppb	520	460	380	...	1100	2000	2400	930
53	I	ppb	60*	70*	...	...	...	...	210	80
55	Cs	ppb	10–200*	20–500*	150*	...	...	50	210	125
56	Ba	ppm	4.4	4.1	4.0	...	...	...	2.4	2.8
57	La	ppb	301	318	330	310	585	320	240	196
59	Ce	ppb	763	970	880	830	...	...	650	580
59	Pr	ppb	120	140	130	...	...	...	100	70
60	Nd	ppb	581	700	650	...	...	...	440	370
62	Sm	ppb	194	203	205	180	240	200	140	149
63	Eu	ppb	74	80	78	72	100	80	52	54
64	Gd	ppb	275	317	290	...	...	...	210	196
65	Tb	ppb	49	59	54	...	60	...	34	32
66	Dy	ppb	305	372	360	29	...	...	230	245
67	Ho	ppb	74	89	82	59	...	...	50	51
68	Er	ppb	213	252	240	...	...	...	160	160
69	Tm	ppb	33	38	35	...	...	...	24	23
70	Yb	ppb	203	226	230	216	240	215	154	157

continued

Table 16.11 *(continued)*

Z Element	Unit	H	L	LL	R	Acap.[a]	K	EH	EL
71 Lu	ppb	33	34	34	32	32	33	25	25
72 Hf	ppb	150	170	170	150	130	...	140	210
73 Ta	ppb	21	21	...	...	...	...	...	...
74 W	ppb	164	138	115	<180	...	...	140	140
75 Re	ppb	78	47	32	43	60	...	55	57
76 Os	ppb	835	530	410	690	680	550	660	670
77 Ir	ppb	770	490	380	610	600	550	570	560
78 Pt	ppm	1.58	1.09	0.88	<1.0	1.3	...	1.29	1.25
79 Au	ppb	220	156	146	183	190	220	330	240
80 Hg	ppb	...	30*	22*		...	...	60	...
81 Tl	ppb	0.01–1*	0.1–5*	1–30*	...	20	3	100	7
82 Pb	ppb	240*	40	...	...	...	...	1500	240
83 Bi	ppb	0.1–10*	14	5–30*	...	27	25	90	13
90 Th	ppb	38	42	47	<50	...	...	30	38
92 U	ppb	13	15	15	<25	...	...	9.2	7.0
Fe#	mol%[†]	17	22	27	35	11	3.3	1.0	0.3
metal	wt%	16	7.7	2.5	<1	15	14	24	20
sulfide	wt%	5.5	6.0	5.8	11	7.4	15	15	8.5
H_2O	wt%	0.3	0.8	1.1	0	~0.1	0	0	0
Density	g cm^{-3}	3.8	3.6	3.55	3.6	3.7	3.7	3.67	3.58

[a] Acap.: Acapulcoites
* abundances are very variable; an approximate range is listed.
[†] Fe# = FeO/(FeO + MgO) in mol% of olivine. Approximate values.
1 wt% = 10000 ppm; 1 ppm = 1000 ppb
For data sources, see text.

Table 16.12 Structural Classification of Iron Meteorites

Structural Class	Symbol	Kamacite Bandwidth (mm)	Remarks	Chemical Class
Hexahedrites	H	>50	no octahedral orientation	IIA
Coarsest octahedrites	Ogg	>3.3	± taenite	IA, IIB
Coarse octahedrite	Og	1.3–3.3	—	IIIE, IIIF, (IIE)
Medium octahedrites	Om	0.5–1.3	—	IA, IB, IID, IIIA, IIIB, IIIF, (IIE)
Fine octahedrites	Of	0.2–0.5	—	IID, IIIC
Finest octahedrites	Off	0.05–0.2	distinct kamacite bands	IIIC, IIID
Plessitic octahedrites	Opl	0.03–0.05	kamacite sparks and spindels	IIC
Ataxites	D	0.006–0.03	well developed, slowly annealed plessite, kamacite spindels rare	IB, IIID, IVB
Anomalous	Anom	—	all irons that do not fit previous categories	IIE, (IIIF)

Table 16.13 Chemical Trends in Iron Meteorites

Correlation	Trend	For Group(s)
Ga-Ge	positive	all groups
Ga-Ni	positive	IIIA, IVB
Ga-Ni	negative	IA, IB, IIB, IIIB, IIIC
Ge-Ni	positive	(IIA), IIC, IID, IIIA, IVA, IVB
Ge-Ni	negative	IA, IB, IIB, IIIB, IIIC, IIID
Ge-Ni	absent	IIE, IIIC, IIIE, IIIF
Ir-Ni	negative	all groups
W-Ni	negative	IA, IB, IIA, IIB, IIIA, IIB, IVA

Sources: Buchwald, V. F., 1975, *Handbook of iron meteorites*, Vol. 1–3, Univ. of California Press, Berkeley, pp. 1426. Scott, E. R. D., 1972, *Geochim. Cosmochim. Acta* 36, 1205–1236. Wasson, J. T., 1985, *Meteorites*, W. H. Freeman & Co., New York, pp. 267.

Table 16.14 Chemical Classification of Iron Meteorites

Chem. Group	Ni (wt%)	Ga (ppm)	Ge (ppm)	Ir (ppb)	Structure[a]	Kamacite Bandwidth (mm)	Cooling Rates (K Myr⁻¹)	Frequency (%)	Examples
IA	6.5–8.5	55–100	190–520	10.6–5.5	Om–Ogg	1.0–3.1	1–5	17	Cañyon Diablo, Odessa
IB	8.5–25.0	11–55	25–190	0.3–2.0	D–Om	0.01–1.0	1–5	1.7	Colfax, Four Corners
IC	6.1–6.8	49–55	212–247	0.07–2.1	Anom, Og	<3	3–>100	2.1	Bendegó, Etosha
IIA	5.3–5.8	57–62	170–185	2–60	H	>50	2–10	8.1	Chesterville, Hex River
IIB	5.5–6.9	46–59	107–183	0.05–0.46	Ogg	5–15	2–10	2.7	El Burro, Sikhote Alin
IIC	9.0–12	37–39	88–114	4–10	Opl	0.06–0.07	100–500	1.4	Kumerina, Perryville
IID	9.9–11.4	70–83	82–98	3.5–18.5	Of–Om	0.40–0.85	1–2	2.7	Elbogen, Needles
IIE	7.5–9.7	21–28	62–75	1–8	Anom	0.7–2	0.2–400	2.5	Elga, Weekeroo Station
IIF	10.6–14.3	8.9–11.6	99–193	0.75–23	D–Of	0.05–0.21	...	1.0	Dorofeevka, Monahans
IIIA	7.1–8.9	17–23	32–47	0.1–20	Om	0.9–1.3	1–10	24.8	Cape York, Henbury
IIIB	8.6–10.6	16–21	27–46	0.01–1.6	Om	0.6–1.3	1–10	7.5	Bald Eagle, Turtle River
IIIC	10.5–13.3	11–92	8–280	0.07–0.55	Off–Ogg	0.2–0.5	1–5	1.4	Carlton, Havana
IIID	17.0–23.0	1.5–5.2	1.4–4.0	0.02–0.07	D–Off	0.01–0.05	1–5	1.0	Tazewell, Wedderburn
IIIE	8.3–8.8	17–19	34–37	0.05–0.6	Og	1.3–1.6	0.5–2	1.7	Rhine Villa, Willow Creek
IIIF	6.8–8.5	6.3–7.2	0.7–1.1	0.006–7.9	Om–Og	0.5–1.5	5–20	1.0	Moonbi, Nelson County
IVA	7.5–9.5	1.6–2.4	0.09–0.14	0.1–3.5	Of	0.25–0.45	3–200	8.3	Gibeon, Yanhuitlan
IVB	16–18	0.17–0.27	0.03–0.07	13–36	D	0.006–0.03	5–200	2.3	Hoba, Tawallah Valley

[a] see Table 16.12 for explanation of symbols.

Sources: Buchwald, V. F., 1975, *Handbook of iron meteorites*, Vol. 1–3, Univ. of California Press, Berkeley, pp. 1426. Kracher, A., Willis, J., & Wasson, J. T., 1980, *Geochim. Cosmochim. Acta* 44, 773–787. Scott, E. R. D., 1979, in *Asteroids*, (Gehrels, T., & Matthews, M. S., eds.), Univ. of Arizona Press, Tucson, pp. 892–925. Scott, E. R. D., & Wasson, J. T., 1975, *Rev. Geophys. Space Phys.* 13, 527–546.

Table 16.15 Lunar Meteorites

Meteorite	Recovered Mass	Rock Type, Lunar Source Region
Anorthositic breccias		
Allan Hills 81005	31 g	regolith breccia, highland
Dar Al Gani 262*	513 g	polymict breccia, highland
MacAlpine Hills 88104/88105	724 g	regolith/fragmental breccia, highland
QUE 93069/94269	21/3.2 g	regolith breccia, highland
Yamato 791197	52.4 g	regolith breccia, highland
Yamato 82192/82193	36.67/27.04 g	fragmental breccia, contains low-Ti mare basalt clasts
Yamato 86032	648.43 g	fragmental breccia
Calcalong Creek †	19 g	regolith breccia, KREEP-rich, highland
Basaltic breccias		
Elephant Moraine 87521	31 g	polymict fragmental breccia, mare
Yamato 793274	8.66 g	regolith breccia, mare
QUE 94281	23.4	basaltic breccia, mare
Mare gabbros (low to very low Ti content)		
Asuka 881757	442.12 g	igneous, small cumulate component, mare basalt
Yamato 793169	6.07 g	igneous, noncumulate, mare basalt

* found in the Sahara (see Bischoff & Weber, 1997)
† found in a desert region in Australia (see Hill *et al.*, 1991)
All other meteorites are found in Antarctica.

Some literature sources for lunar meteorites: Several articles in *Geochim. Cosmochim. Acta* 55 (no. 11) 1991, pp. 2999–3180. Bischoff, A., 1996, *Meteoritics and Planet. Sci.* 31, 849–855. Bischoff, A., & Weber, D., 1997, *Meteoritics and Planet. Sci.* 32, A13. Eugster, O., 1989, *Science* 245, 1197–1202. Hill, D. H., Boynton, W. V., & Haag, R. A., 1991, *Nature* 352, 614–617. Palme, H., Spettel, B., Weckwerth, G., & Wänke, H., 1983, *Geophys. Res. Lett.* 10, 817–820. Warren, P. H., 1994, *Icarus* 111, 338–363. Yanai, K., & Kojima, H., 1991, *Proc. NIPR Symp. Antarct. Meteor.* 4th, 70–90.

Table 16.16 SNC meteorites

Name	Recovered Mass	Recovery Location and Date	Major Minerals	Minor Minerals
Shergottites (basalts)				
EET79001	7.94 kg	Antarctica, 1979	pigeonite, augite, maskelynite	ilmenite, Ti-magnetite, whitlockite
Shergotty *	5 kg	India, 25 Aug. 1865		
QUE94201	12 g	Antarctica, 1994		
Zagami *	23 kg	Nigeria, 3 Oct. 1962		
Shergottites (lherzolites/harzburgites)				
ALH77005	480 g	Antarctica, 1977	cumulate olivine, chromite, pigeonite, augite, maskelynite	opx, plag, Ti-magnetite, ilmenite
LEW88516	13 g	Antarctica, 1988		
Y793605	18 g	Antarctica, 1979		
Nakhlites (clinopyroxenites/wehrlites)				
Governador Valadares	160 g	Brazil, 1958	cumulate augite, olivine	plagioclase, Ti-magnetite, ilmenite
Lafayette	600 g	USA, 1931		
Nakhla *	40 kg	Egypt, 28 June 1911		
Nakhlites (orthopyroxenites)				
ALH84001	1.93 kg	Antarctica, 1984	cumulate orthopyroxene	chromite, maske-lynite, augite, olivine
Chassignites (dunites)				
Chassigny *	4 kg	France, 3 Oct. 1815	cumulate olivine	augite, opx, plagio-clase, chromite, ilmenite

* observed fall

Some literature sources for SNC meteorites: Banin, A., Clark, B. C., & Wänke, H., 1992, in *Mars* (Kieffer, H. H., Jakosky, B. M., Snyder, C. W., & Matthews, M. S., eds.), Univ. of Arizona Press, Tucson, pp. 594–625. Jones, J. H., 1989, *Proc. Lunar Planet. Sci. Conf.* 19th, 465–474. Lodders, K., 1998, *Meteoritics & Planet. Sci.* 33, A183–A190. Lodders, K., & Fegley, B., 1997, *Icarus* 126, 373–394. McSween, H. Y., 1994, *Meteoritics* 29, 757–779. Shih, C. Y., Nyquist, L. E., Bogard, D. D., McKay, G. A., Wooden, J. L., Bansal, B. M., & Wiesmann, H., 1982, *Geochim. Cosmochim. Acta* 46, 2323–2344. Wood, A., & Ashwal, L. D., 1981, *Proc. Lunar Planet Sci. Conf.* 12B, 1359–1375.

Table 16.17 Compositions of Shergottites, Nakhlites, and Chassignites

Element	Unit	Shergotty	Zagami	ALH 84001	Nakhla	Lafayette	Chassigny
Li	ppm	4.5	2.9	...	3.9	...	1.4
C	wt%	0.053	...	0.058	0.03	0.01	0.085
F	ppm	46	41	...	57	...	15
Na	ppm	10300	9100	1000	3400	3000	920
Mg	wt%	5.58	6.8	15.1	7.3	7.8	19.2
Al	wt%	3.64	3.2	0.68	0.89	1.31	0.42
Si	wt%	24.0	23.6	24.7	22.7	21.9	17.5
P	ppm	3230	2200	60	500	1960	275
S	wt%	0.13	0.19	0.011	0.026	0.041	0.026
Cl	ppm	108	137	8.0	80	65	34
K	ppm	1440	1170	140	1070	900	300
Ca	wt%	6.86	7.5	1.30	10.5	9.6	0.47
Sc	ppm	52	55	13	51	58	5.3
Ti	ppm	4900	4720	1240	2020	2540	480
V	ppm	290	310	200	190	170	40
Cr	ppm	1350	2260	7760	1770	1280	5240
Mn	ppm	4010	3880	3560	3820	3880	4120
Fe	wt%	15.1	14.1	13.6	16.0	16.8	21.2
Co	ppm	40	36	47	48	43	123
Ni	ppm	79	48	5.8	90	96	500
Cu	ppm	16	...	...	12	12	2.6
Zn	ppm	69	60	92	54	78	72
Ga	ppm	16	14	2.9	3.0	3	0.7
Ge	ppm	0.73	0.8	1.1	3.0	2.5	0.01
As	ppm	0.025	0.05	<0.03	0.015	<0.15	0.008
Se	ppm	0.38	0.32	<0.16	0.08	0.07	0.04
Br	ppm	0.88	0.83	...	4.5	0.37	0.088
Rb	ppm	6.4	5.6	0.8	3.8	2.8	0.73
Sr	ppm	48	45	4.5	59	75	7.2
Y	ppm	19	...	1.6	3.3	4.4	0.6
Zr	ppm	57	73	5.9	8.8	9.4	2.1
Nb	ppm	4.6	5.5	0.42	1.6	1.46	0.34
Mo	ppm	0.37	...	...	0.086	...	...
Pd	ppb	1.7	1.8	...	30	1.7	0.15
Ag	ppb	11	27	...	40	58	2.6
Cd	ppb	28	66	77	93	95	14

continued

Table 16.17 *(continued)*

Element	Unit	Shergotty	Zagami	ALH 84001	Nakhla	Lafayette	Chassigny
In	ppb	26	24	...	20	20	4
Sn	ppm	0.01	...	...	0.6	...	...
Sb	ppb	5.2	9	...	40	...	0.9
Te	ppb	3.3	1.6	...	<4.3	<5.2	50
I	ppb	43	<5	...	180	100	<10
Cs	ppb	440	360	43	390	320	37
Ba	ppm	34	26	4.0	29	27	7.6
La	ppm	2.16	1.51	0.19	2.06	1.86	0.53
Ce	ppm	5.45	3.75	0.59	5.87	4.82	1.12
Pr	ppm	0.81	...	0.06	0.67	0.80	0.13
Nd	ppm	4.2	2.75	0.265	3.23	3.09	0.62
Sm	ppm	1.47	1.11	0.12	0.77	0.84	0.14
Eu	ppm	0.60	0.475	0.035	0.24	0.22	0.045
Gd	ppm	2.54	...	0.14	0.86	0.92	0.11
Tb	ppm	0.48	0.28	0.038	0.12	0.12	0.03
Dy	ppm	3.50	2.20	0.28	0.77	0.95	0.20
Ho	ppm	0.71	...	0.076	0.155	0.17	0.044
Er	ppm	1.88	1.6	0.21	0.37	0.4	0.09
Tm	ppm	0.300	...	0.036	0.047	0.057	...
Yb	ppm	1.63	1.27	0.29	0.39	0.33	0.11
Lu	ppm	0.26	0.195	0.049	0.055	0.052	0.015
Hf	ppm	2.0	1.7	0.14	0.27	0.28	<0.1
Ta	ppm	0.25	0.20	32	0.09	80	<0.02
W	ppb	460	610	80	120	400	46
Re	ppb	0.04	0.035	0.002	0.036	0.03	0.06
Os	ppb	0.4	0.12	0.01	0.007	0.6	1.6
Ir	ppb	0.06	0.03	0.08	0.22	0.09	2.1
Au	ppb	0.92	2.0	0.009	0.72	...	0.73
Tl	ppb	13	12	...	3.5	7.0	3.7
Bi	ppb	1.0	1.4	...	0.5	5.3	0.4
Th	ppb	380	370	35	200	150	57
U	ppb	105	120	11	52	46	18

Note: mean values from literature data survey
Sources: Lodders, K., 1998, *Meteoritics & Planet. Sci.* 33, A183–A190 and references
      therein.

Table 16.18 Compositions of Eucrites, Howardites, and Diogenites

Element	Unit	Juvinas (noncumulate eucrite)	Serra de Magé (cumulate eucrite)	Binda (howardite)	Johnstown (diogenite)
Li	ppm	7.8	3	3.5	...
C	%	...	0.05	...	0.3
F	ppm	19	...	...	...
Na	ppm	3200	2430	1150	3150
Mg	%	4.22	5.88	4.4	15.9
Al	%	6.9	8.91	7.04	0.79
Si	%	22.97	22.33	23.12	24.44
S	%	0.15	0.13	0.34	0.27
P	ppm	400	200	430	115
Cl	ppm	29	16	19	...
K	ppm	328	71	67	25
Ca	%	7.66	7.90	7.30	1.07
Sc	ppm	28	18.3	19	27
Ti	ppm	3710	875	5000	710
V	ppm	85	116	...	...
Cr	ppm	1780	2770	5300	5300
Mn	ppm	4040	3230	3900	4000
Fe	%	13.9	9.6	12.5	13.1
Co	ppm	4.7	7.4	18	16
Ni	ppm	4	7	25	25
Cu	ppm	2.3	2.7	8	...
Zn	ppm	1.81	0.67	26	0.71
Ga	ppm	1.7	1.4	0.7	...
Ge	ppm	0.036	0.003	0.14	...
As	ppm	0.239	0.154	...	...
Se	ppm	0.14	0.31	0.25	0.4
Br	ppm	0.16	0.03	0.21	...
Rb	ppm	0.2	0.05	0.32	0.10
Sr	ppm	74.9	56.9	31	1.7
Y	ppm	16.5	19	...	1.2
Zr	ppm	45	14	17	3
Nb	ppm	2.7	...	...	...
Mo	ppm	0.015	...	...	...
Pd	ppb	0.4	...	10	2

continued

Table 16.18 *(continued)*

Element	Unit	Juvinas	Serra de Magé	Binda	Johnstown
Ag	ppb	58	1.9	...	11
Cd	ppb	20	6	...	21
In	ppb	1.45	0.38	...	3.2
Sb	ppb	13.8	0.54	62	11
Te	ppm	0.009	0.0015	...	0.005
I	ppb	40	...	97	25
Cs	ppb	6.3	1.39	20	1.1
Ba	ppm	30	10.2	8.5	...
La	ppb	2580	390	840	44
Ce	ppb	6930	730	1900	130
Pr	ppb	970	...	280	...
Nd	ppb	4960	...	1400	110
Sm	ppb	1620	230	440	80
Eu	ppb	620	410	220	9
Gd	ppb	2290	...	530	...
Tb	ppb	400	40	120	21
Dy	ppb	2860	600	760	140
Ho	ppb	530	...	180	36
Er	ppb	1740	...	590	140
Tm	ppb	280	...	1000	20
Yb	ppb	1600	235	590	175
Lu	ppb	250	50	96	36
Hf	ppm	1.3	0.13	6.5	0.1
Ta	ppb	150	80	60	8
W	ppb	30	...	...	6.5
Re	ppb	0.0097	0.0008	...	0.06
Os	ppb	0.008	...	...	0.7
Ir	ppb	0.096	0.084	...	0.7
Au	ppb	4	1.3	...	0.9
Tl	ppb	0.7	0.08	...	...
Bi	ppb	3	1.1	...	0.2
Th	ppb	297	53	...	...
U	ppb	123	13	27	6

Note: mean compositions from literature survey
Sources: Kitts, K., & Lodders, K., 1998, *Meteoritics & Planet. Sci.* 33, A197–A213 and
   references therein.

Table 16.19   Model Elemental Abundances in the Silicate Portion of the Eucrite Parent Body

Element	Unit	[Jo84]	[DW80]	[MHT78]	[CD77]	[HVA77]
Li	ppm	...	2.7	...	...	...
F	ppm	...	4.8	...	...	...
Na	ppm	820	830	340	450	400
Mg	mass%	16.7	19.0	17.2	16.9	17.7
Al	mass%	1.69	1.73	1.3	1.35	1.27
Si	mass%	18.8	21.6	18.8	19.3	19.3
P	ppm	96	97	...	...	...
Cl	ppm	...	4.6	...	...	...
K	ppm	83	76	40	...	36
Ca	mass%	1.90	1.84	1.5	1.32	1.4
Sc	ppm	...	12.4	...	...	...
Ti	ppm	960	960	800	...	720
V	ppm	...	91	...	...	...
Cr	ppm	2300	5980	...	...	...
Mn	ppm	4900	3290	...	...	3500
Fe	mass%	19.60	11.50	20.7	20.40	18.7
Co	ppm	...	13	...	...	...
Ni	ppm	30–80	38	...	...	...
Ga	ppm	...	0.67	...	...	...
Br	ppm	...	0.03	...	...	...
Sr	ppm	...	19.1	...	...	...
Zr	ppm	...	9.9	...	...	...
Ba	ppm	...	6.0	...	...	...
La	ppm	...	0.65	...	...	...
Sm	ppm	...	0.38	...	...	...
Eu	ppm	...	0.14	...	...	...
Yb	ppm	...	0.43	...	...	...
Hf	ppm	...	0.29	...	...	...
Ta	ppb	...	31	...	...	...
W	ppb	...	14	...	...	...
U	ppb	22	22	...	...	...

Sources: **[CD77]** Consolmagno, G. J., & Drake, M. J., 1977, *Geochim. Cosmochim. Acta* 41, 1271–1282. **[DW80]** Dreibus, G., & Wänke, H., 1980, *Z. Naturforsch.* 35a, 204–216. **[HVA77]** Hertogen J., Vizgirda, J., & Anders, E., 1977, *BAAS* 9, 458–459. **[Jo84]** Jones, J. H., 1984, *Geochim. Cosmochim. Acta* 48, 641–648. **[MHT78]** Morgan, J. W., Higuchi, H., Takahashi, H., & Hertogen, J., 1978, *Geochim. Cosmochim. Acta* 42, 27–38.

Table 16.20 Model Compositions of the Eucrite Parent Body

Compound	[Jo84]	[DW80]	[MHT78]	[CD77]	[HVA77]
Mantle & crust					
MgO	27.7	31.55	28.5	29.7	29.3
Al_2O_3	3.2	3.27	2.5	1.8	2.4
CaO	2.6	2.58	2.06	1.2	2.0
SiO_2	40.3	46.26	40.2	39.0	41.2
TiO_2	0.16	0.16	0.13	...	0.12
Na_2O	0.11	0.11	0.05	0.04	0.05
K_2O	0.01	0.0092	0.004	...	0.004
FeO	25.2	14.82	26.6	28.3	24.0
Cr_2O_3	0.34	0.78	...	...	...
MnO	0.63	0.47	...	...	0.46
P_2O_5	0.022	0.022	...	...	...
Core					
Fe	...	91.78	86.8	...	82.5
Ni	...	7.86	13.2	...	17.5
Co	...	0.36	...	...	...
Relative masses (%)					
Mantle & crust	...	78.3	87.1	...	92
Core	...	21.7	12.9	...	8

Note: All data in mass%

Sources: **[CD77]** Consolmagno, G. J., & Drake, M. J., 1977, *Geochim. Cosmochim. Acta* 41, 1271–1282. **[DW80]** Dreibus, G., & Wänke, H., 1980, *Z. Naturforsch.* 35a, 204–216, and Dreibus, G., Brückner, J., & Wänke, H., 1997, *Meteoritics & Planet. Sci. 32 Suppl.*, A36. **[HVA77]** Hertogen J., Vizgirda, J., & Anders, E., 1977, *BAAS* 9, 458–459. **[Jo84]** Jones, J. H., 1984, *Geochim. Cosmochim. Acta* 48, 641–648. **[MHT78]** Morgan, J. W., Higuchi, H., Takahashi, H., & Hertogen, J., 1978, *Geochim. Cosmochim. Acta* 42, 27–38.

Table 16.21   Approximate Mean Oxygen, Nitrogen, Carbon, and Hydrogen Isotopic Compositions of Meteorites and of the Earth

Object	$\delta^{18}O$ ‰ rel. SMOW [a]	$\delta^{17}O$ ‰ rel. SMOW	$\Delta^{17}O$ ‰	$\delta^{15}N$ ‰ rel. air	$\delta^{13}C$ ‰ rel. PDB [b]	δD ‰ rel. SMOW
Chondrites						
CI	16.35	8.81	0.39	+42	~ −10	~ +220
CM	7.3	1.2	−2.6	+20 to +50	~ −8	~ +50
CO	−1.13	−5.09	−4.5	+100 to −30	~ −17	~ −180
CV	2.12	−2.23	−3.3	+20 to −50	~ −19	~ +240
CK	−2.2	−5.6	−4.5	2	~ −24	+14
CR	4.3	0.9	−1.3	+170	~ −10	+500 to +1000
CH	0.8	−0.9	−1.3	+700 to +850	0.5	~ −110
H	4.08	2.85	0.75	+22	...	...
L	4.70	3.52	1.10	+4.7	...	...
LL	5.04	3.88	1.28	+13	...	~ 3000
R	4.97	5.28	2.72	...	...	...
K	2.73	−0.13	−1.54	+10 to −20	...	...
EH	5.31	2.74	0.01	−20	~ −10	~ −110
EL	5.57	2.96	0.09	−35	~ +2	~ +180
Acapulcoites	3.46	0.76	−1.04	...	~ −24	...
Lodranites	3.38	0.57	−1.18	...	...	...
Achondrites, stony-irons, irons, and others						
Angrites	3.69	1.77	−0.15	...	...	...
Aubrites	5.26	2.75	0.02	...	~ −24	~ −40
Brachinites	4.00	1.81	−0.26	...	...	...
EHD	3.68	1.70	−0.25	+5.4	...	~ −70
Mesosiderites	3.41	1.53	−0.24	...	...	...
Pallasites	2.91	1.23	−0.28	...	...	...
SNC	4.31	2.51	0.27	...	...	~ −90
Ureilites	6.9	2.4	−1.2	12	~ −3	...
Winonaites	5.25	2.22	−0.5	...	~ −4	...
IAB-silicates	4.94	2.07	−0.48	...	...	...
IIE-silicates	4.26	2.81	0.59	...	...	...
IIIAB-silicates	2.28	0.98	−0.21	...	...	...
IVA-silicates	4.49	3.50	1.17	...	...	...
Earth	5.38	2.78	=0.0	...	−6.4	~ −90
Moon	5.54	2.89	0	...	...	...

continued

Table 16.21 *(continued)*

[a] SMOW: Standard Mean Ocean Water.
[b] PDB: Peedee Belemnite.

Conversions and notation:

$$\delta^x M = [(^xM/^{ref}M)_{sample}/(^xM/^{ref}M)_{standard} - 1] \times 1000$$

where $^xM/^{ref}M$ is the isotopic ratio of the element in the sample or standard.

$\Delta^{17}O = \delta^{17}O - 0.52 \times \delta^{18}O$; describes offset from terrestrial fractionation line.

$^{16}O_{excess} = \frac{0.52}{(1-0.52)}\delta^{18}O - \frac{1}{(1-0.52)}\delta^{17}O$; describes ^{16}O excess.

Sources:

O-isotopes: Clayton, R. N., 1993, *Annu. Rev. Earth Planet. Sci.* 21, 115–149 (general summary). Clayton, R. N., & Mayeda, T. K., 1978, *Earth Planet. Sci. Lett.* 40, 168–174 (irons & stony meteorites). Clayton, R. N., & Mayeda, T. K., 1983, *Earth Planet. Sci. Lett.* 62, 1–6 (achondrites). Clayton, R. N., Mayeda, T. K., Olsen, E. J., & Prinz, M., 1983, *Earth Planet. Sci. Lett.* 65, 229–232 (irons). Clayton, R. N., & Mayeda, T. K., 1984, *Earth Planet. Sci. Lett.* 67, 151–161 (C-chondrites). Clayton, R. N., & Mayeda, T. K., 1996, *Geochim. Cosmochim. Acta* 60, 1999–2018 (achondrites, lunar). Clayton, R. N., Onuma, N., & Mayeda, T. K., 1976, *Earth Planet. Sci. Lett.* 30, 10–18 (chondrites). Clayton, R. N., Mayeda, T. K., & Rubin, A.E., 1984, *Proc. 15th Lunar Planet. Sci. Conf., J. Geophys. Res.* 89 Suppl., C245–C249 (enstatite meteorites). Clayton, R. N., Mayeda, T. K., Goswami, J. N., & Olsen, E. J., 1991, *Geochim. Cosmochim. Acta* 55, 2317–2337 (ordinary chondrites). Mayeda, T. K., & Clayton, R. N., 1980, *Proc. Lunar Planet. Sci. Conf.* 11th, 1145–1151 (aubrites). Robert, F., Rejou-Michel, A., & Javoy, M., 1992, Earth Planet Sci. Lett. 108, 1–9 (terrestrial and lunar). Weisberg, M. K., Prinz, M., Clayton, R. N., Mayeda, T. K., Grady, M. M., & Pillinger, C. T., 1995, *Proc. NIPR Symp. Antarctic Meteorites* 8, 11–32 (CH).

H, C, & N-isotopes: Chaussidon, M. Sheppard, M. F., & Michard, A., 1991, in *Stable isotope geochemistry: A tribute to Samuel Epstein* (Taylor, H. P., O'Neil, J. R., & Kaplan, I. R. (eds), The Geochemical Society, San Antonio, pp. 325–337. Hashizume, K., & Sigura, N., 1995, *Geochim. Cosmochim. Acta* 59, 4057–4070. Kung, C. C., & Clayton, R. N., 1978, *Earth Planet. Sci. Lett.* 38, 421–435. Kerridge, J. F., 1985, *Geochim. Cosmochim. Acta* 49, 1707–1714. McNaughton, N. J., Borthwick, J., Fallick, A. E., & Pillinger, C. T., 1981, *Nature* 294, 639–641.

17

BEYOND THE SOLAR SYSTEM

Table 17.1 Constellations

Constellation Name	Genitive Form	Abbr.	Meaning
Andromeda	Andromedae	And	princess of Ethiopia
Antila	Antliae	Ant	air pump
Apus	Apodis	Aps	bird of paradise
Aquarius	Aquarii	Aqr	water bearer
Aquila	Aquilae	Aql	eagle
Ara	Arae	Ara	altar
Aries	Arietis	Ari	ram
Auriga	Aurigae	Aur	charioteer
Boötes	Boötis	Boo	herdsman
Caelum	Caeli	Cae	chisel
Camelopardalis	Camelopardis	Cam	giraffe
Cancer	Cancri	Cnc	crab
Canes Venatici	Canum Venaticorum	CVn	hunting dogs
Canis Major	Canis Majoris	CMa	big dog
Canis Minor	Canis Minoris	CMi	little dog
Capricornus	Capricorni	Cap	goat
Carina	Carinae	Car	ship's keel
Cassiopeia	Cassiopeiae	Cas	queen of Ethiopia
Centaurus	Centauri	Cen	centaur
Cepheus	Cephei	Cep	king of Ethiopia
Cetus	Ceti	Cet	whale, sea monster
Chamaeleon	Chamaeleontis	Cha	chameleon
Circinus	Circini	Cir	compasses
Columba	Columbae	Co	dove
Coma Berenices	Comae Berenices	Com	Berenice's hair
Corona Australis	Coronae Australis	CrA	southern crown
Corona Borealis	Coronae Borealis	CrB	northern crown

continued

Table 17.1 *(continued)*

Constellation Name	Genitive Form	Abbr.	Meaning
Corvus	Corvi	Crv	crow
Crater	Crateris	Crt	cup
Crux	Crucis	Cru	southern cross
Cygnus	Cygni	Cyg	swan
Delphinus	Delphini	Del	dolphin, porpoise
Dorado	Doradus	Dor	dorado fish, swordfish
Draco	Draconis	Dra	dragon
Equuleus	Equulei	Equ	little horse
Eridanus	Eridani	Eri	river Eridanus
Fronax	Fornacis	For	furnace
Gemini	Geminorum	Gem	twins
Grus	Gruis	Gru	crane
Hercules	Herculis	Her	son of Zeus
Horologium	Horologii	Hor	clock
Hydra	Hydrae	Hya	water snake
Hydrus	Hydri	Hyi	sea serpent
Indus	Indi	Ind	indian
Lacerta	Lacertae	Lac	lizard
Leo	Leonis	Leo	lion
Leo Minor	Leonis Minoris	LMi	little lion
Lepus	Leporis	Lep	hare
Libra	Librae	Lib	scales, balance
Lupus	Lupi	Lup	wolf
Lynx	Lyncis	Lyn	lynx
Lyra	Lyrae	Lyr	lyre
Mensa	Mensae	Men	table (mountain)
Microscopium	Microscopii	Mic	microscope
Monoceros	Monocerotis	Mon	unicorn
Musca	Muscae	Mus	fly
Norma	Normae	Nor	square (level)
Octans	Octantis	Oct	octant
Ophiuchus	Ophiuchi	Oph	serpent bearer

continued

Table 17.1 *(continued)*

Constellation Name	Genitive Form	Abbr.	Meaning
Orion	Orionis	Ori	hunter
Pavo	Pavonis	Pav	peacock
Pegasus	Pegasi	Peg	winged horse
Perseus	Persei	Per	rescuer of Andromeda
Phoenix	Phoenicis	Phe	phoenix
Pictor	Pictoris	Pic	painter, easel
Pisces	Piscium	Psc	fishes
Piscis Austrinus	Picis Austrini	PsA	southern fish
Puppis	Puppis	Pup	poop (stern)
Pyxis (= malus)	Pyxidis	Pyx	ship's compass
Reticulum	Reticuli	Ret	net
Sagitta	Sagittae	Sge	arrow
Sagittarius	Sagittarii	Sgr	archer
Scorpius	Scorpii	Sco	scorpion
Sculptor	Sculptoris	Scl	sculptor
Scutum	Scuti	Sct	shield
Serpens (caput/cauda)	Serpentis	Ser	serpent (head/tail)
Sextans	Sextantis	Sex	sextant
Taurus	Tauri	Tau	bull
Telescopium	Telescopii	Tel	telescope
Triangulum	Trianguli	Tri	triangle
Triangulum Australe	Trianguli Australis	TrA	southern triangle
Tucana	Tucanae	Tuc	toucan
Ursa Major	Ursae Majoris	UMa	big bear
Ursa Minor	Ursae Minoris	UMi	little bear
Vela	Velorum	Vel	ship's sail
Virgo	Virginis	Vir	virgin
Volans	Volantis	Vol	flying fish
Vulpecula	Vulpeculae	Vul	little fox

17.1 Some Definitions and Practical Equations for Stellar Parameters

Stellar Classification

Table 17.2 Stellar Spectral Classes

Spectral Class	Color	Effective Temperature (K) [a]	Characteristic Lines or Bands [b]
O	blue-white	30,000	He II
B	blue-white	11,000–25,000	He I, (H)
A	blue-white	7,500–11,000	H, (Ca II)
F	blue-white; white	6,000–7,500	Ca II, H, (Fe II)
G	white; yellow-white	5,000–6,000	Ca II, FeII, (H)
K	yellow-orange	3,500–5,000	metallic lines, CH, CN
M	reddish	3,500	TiO, VO
S	reddish	3,500	ZrO
C (R)	reddish	>2,500	CN, C_2
C (N)	reddish	<2,500	C_2, CN

[a] approximate temperatures
[b] decreasing strength from subclass 0–9

Table 17.3 Stellar Luminosity Classes

Class	Star Type	Class	Star Type
Ia	bright supergiants	IV	subgiants
Ib	supergiants	V	main sequence (dwarfs)
II	bright giants	VI	sub-dwarfs
III	giants	VII	white dwarfs

Table 17.4 Other Notations Associated with Spectral Classification

Symbol	Meaning	Symbol	Meaning
e	emission lines present	p	peculiar spectrum
k	interstellar lines present	s	sharp lines
m	metallic lines present	v, var	variable
n	diffuse lines	wk	weak lines

Effective Temperature (T_{eff}) Measure of total energy radiated from the stellar surface, normally taken as equal to that of a blackbody emitting the same total energy. The effective temperature is related to stellar luminosity L and radius R, via

$$T_{eff}^4 = L/(4\pi\sigma R^2) \tag{1}$$

where σ = 5.6705×10⁻⁸ Wm⁻²K⁻⁴ is the Stefan-Boltzmann constant. In practice, effective temperatures are derived from stellar line spectra.

Magnitude The magnitude scale describes the stellar brightness scale. The brightest stars are of low magnitudes, whereas the faintest stars are of higher magnitudes. A one-magnitude difference corresponds to a factor of $100^{1/5} \approx 2.512$ in brightness; an interval of five magnitudes equals a factor of one hundred in brightness.

Magnitudes are measured in different wavelength band systems, for example; the U (ultraviolet, centered at ~365 nm), B (blue, centered at ~440 nm), or V (visual, centered at ~550 nm) bands. Other bands used are the J, H, K, L, M, and N bands in the infrared, as well as the R and I bands in the far red.

Table 17.5 Wavelength Band Systems (center wavelengths)

Band	λ	Band	λ	Band	λ
U	365 nm = 0.365 μm	I	0.9 μm	L	3.40 μm
B	440 nm = 0.44 μm	J	1.25 μm	M	5.0 μm
V	550 nm = 0.55 μm	H	1.6 μm	N	10.2 μm
R	700 nm = 0.7 μm	K	2.2 μm	Q	21 μm

Apparent magnitudes are commonly designated by either the capital band letter (e.g., B, V) or by using the notation m_B, m_V, etc. Absolute magnitudes are denoted by a subscript to M (e.g., M_B, M_V).

Apparent Magnitude (m) The magnitude difference of two stars (x–y) is related to their brightness (flux) ratio l_y/l_x via:

$$x - y = 2.5 \log(l_y/l_x) \tag{2}$$

Absolute Magnitude (M) The apparent magnitude (m) of a star depends on the star's distance (d in pc). The absolute magnitude is the apparent magnitude normalized to a distance of D = 10 pc from the sun

$$M = m - 5 \log (d/D) \tag{3a}$$

Stellar brightness (l) and luminosity (L) are related to the apparent and absolute magnitude via

$$L/l = (d/D)^2 \text{ and } m - M = 2.5 \log (L/l) = 5 \log (d/D) \tag{3b}$$

Using equation (3a) and D = 10 pc, we obtain for the absolute magnitude

$$M = m + 5 - 5 \log (d) - A \tag{4a}$$

and with equation (10)

$$M = m + 5 + 5 \log (\pi) - A \tag{4b}$$

In equations (4a) and (4b), A is a term to correct for interstellar extinction.

Color Index The color index is given by the magnitude difference at two wavelengths (e.g., $B - V = M_B - M_V$) and is independent of distance of the star. The color index gives a measure of stellar temperature; for example, a reddish star has a positive $B - V$ color index because it is brighter in V than in B.

Bolometric Flux and Bolometric Magnitude The integrated brightness over all wavelengths is called bolometric flux (l_{bol}, in W m^{-2}). The bolometric magnitude (m_{bol}) follows from

$$m_{bol} = -2.5 \log l_{bol} + \text{constant} \tag{5}$$

For practical purposes, the bolometric correction (BC) is introduced, e.g., for the V band

$$BC_V = m_{bol} - m_V = M_{bol} - M_V = 2.5 \log (l_V/l_{bol}) \tag{6}$$

Luminosity The luminosity (L) can be obtained from the absolute magnitude (M_{bol}) via

$$M_{bol\odot} - M_{bol} = 2.5 \log (L/L_\odot) \tag{7}$$

For the sun, $M_{bol\odot} = +4.75$ so that

$$\log (L/L_\odot) = 1.90 - 0.4 M_{bol} \tag{8}$$

Absolute luminosities are obtained from this equation by using the solar luminosity of $L_\odot = 3.826 \times 10^{26}$ W.

Stellar Distances The distance (d) of a stellar object from the earth is related to the semimajor axis (a) of the earth's orbit and the parallax angle (π, in arcsec) by

$$\sin \pi = a/d \tag{9}$$

Typically, $\pi < 1''$ and setting $a = 1$ AU, equation (9) simplifies to

$$d \text{ (pc)} = 1/\pi \text{ (arcsec)} \tag{10}$$

From Earth, distance determinations up to 100 pc ($\pi > 0.01''$) are practical.

Stellar Radii The Stefan-Boltzmann relation (eq. 1) relates luminosity (L, in W), effective temperature (T_{eff}, in K), and stellar radius (R, in m), so radius determinations using equation (1) require the determination of T_{eff} and stellar luminosity, which is often not practical. Although more feasible, direct determinations of stellar radii are difficult because the disks of distant stars have small angular diameters (on the order of 10^{-3} arcsec). Radii determinations are possible during lunar occulations, which allow high angular resolution measurements. From the angular diameter (α, in arcsec) and the distance (d, in pc), the stellar diameter (D, in km) is

$$D \text{ (km)} = \alpha \text{ (arcsec)} \cdot d \text{ (pc)} \cdot 1.496 \times 10^{8} \tag{11}$$

The stellar radius in terms of solar radii is then

$$R/R_{\odot} = \alpha \text{ (arcsec)} \cdot d \text{ (pc)} \cdot 107.5 \tag{12}$$

Surface gravity The surface gravity (g) can be obtained from the Stefan-Boltzmann relation once the effective temperature (T_{eff}) and mass (M) are known by substituting the radius (R) in the Stefan-Boltzmann relation with $R^2 = GM/g$ (where G is the universal gravitational constant), so that

$$\log T_{eff} = 0.25 \, [\log g + \log L - \log M - \log (4\pi\sigma G)] \tag{13}$$

In terms of solar values ($M_{bol\odot}$; $L_{\odot}$; $M_{\odot}$), this relation can be transformed into

$$\log g = 4 \, (\log T_{eff} + 0.1 \, M_{bol} + 0.25 \log M/M_{\odot} - 3.626) \tag{14}$$

Table 17.6 Stars Within 5 Parsecs of the Sun

Names Gl/GJ[a]	LHS[a]	Alternate Name(s)	Spectral Type	d (pc)	m_v	B–V	V–R	R–I	V–I	Notes
551	49	Proxima Cen	M5 Ve	1.29	+11.05	1.83	1.62	2.001	3.623	flare star, planet. comp.?
559A	50	α Cen A, Rigil Centaurus	G2 V	1.35	+0.01	0.633	0.362	0.331	0.693	
559B	51	α Cen B	K1 V	1.35	+1.34	0.84	0.474	0.404	0.878	
699	57	Barnard's Star, BD+04°3561A	M4 V	1.82	+9.57	1.745	1.216	1.568	2.783	unseen companion?
411	37	BD+36°2147, G119–052, HD 95735	M2 V	2.55	+7.47	1.51	1.01	1.14	2.15	planetary companion
244A	219	α CMa A, Sirius	A1 V	2.64	−1.44	0.01	...	...	...	
244B		α CMa B	DA2	2.64	+8.44	−0.03	...	...	...	
729	3414	HIP 92403	M3.5 Ve	2.97	+10.44	1.745	1.233	1.550	2.785	flare star
144	1557	ε Eri, BS 1084, HD 22049	K2 V	3.22	+3.73	0.88	0.505	0.44	0.945	
887	70	HD 217987	M2 V	3.29	+7.34	1.50	0.975	1.045	2.020	
447	315	FI Vir, Ross 128, G10–050	M4.5 Ve	3.34	+11.16	1.758	1.31	1.678	2.988	flare star
15A	3	GQ And, G171–047	M1 V	3.45	+8.08	1.55	...	...	2.14	flare star
15B	4	GX And, G171–048	M6 Ve	3.45	+11.07	1.79	1.24	1.24	2.82	flare star
820A	62	61 Cyg A	K5 V	3.48	+5.21	1.18	...	...	...	
820B	63	61 Cyg B	K7 V	3.50	+6.05	1.31	...	...	1.62	unseen companion
280A	233	α CMi A, Procyon, HD 61421	F5 IV	3.50	+0.37	0.42	0.245	0.245	0.49	
280B		α CMi B	DA	3.50	+10.70					

continued

Table 17.6 *(continued)*

Gl/GJ [a]	LHS [a]	Alternate Name(s)	Spectral Type	d (pc)	m_v	B–V	V–R	R–I	V–I	Notes
725B	59	G227–047, HD 173740	M3.5 V	3.52	+9.68	1.59	1.12	1.43	2.55	flare star
725A	58	G227–046, HD 173739	M3 V	3.57	+8.90	1.54	1.07	1.39	2.46	flare star
845	67	ε Ind, HD 209100	K5 V	3.63	+4.68	1.05	0.62	0.525	1.145	
71	146	τ Cet, HD 10700	G8 V	3.65	+3.49	0.72	0.43	0.385	0.815	unseen companion
54.1	138	YZ Cet, G268–135	M5.5 Ve	3.72	+12.10	1.85	1.374	1.771	3.145	flare star
273	33	Luyten's Star, BD+5°1668, G089–019	M3.5 V	3.80	+9.85	1.57	1.145	1.537	2.69	flare star
191	29	Kapteyn's Star, HD 33793	M0 V	3.92	+8.85	1.56	0.955	0.995	1.95	
825	66	AX Mic, HD 202560	M0 V	3.95	+6.67	1.42	0.900	0.865	1.765	flare star
860A	3814	Kruger 60, G232–075	M2 Ve	4.01	+9.59	1.66	1.19	1.49	2.68	
860B	3815	DO Cep	M6 Ve	4.01	+11.30	1.80	...	...	...	
		HIP 30920	M4.5 Ve	4.12	+11.12	1.69	...	...	...	
563.2A	380	CD–25°10553	M3 V	4.25	+11.72	1.48	...	...	...	
628	419	Wolf 1061, BD–12°4523, G153–058	M3.5 V	4.26	+10.10	1.58	1.165	1.515	2.680	
1	1	G267–025, HD 225213	M2 V	4.36	+8.54	1.48	0.967	1.159	2.126	
		HIP 15689	...	4.40	+12.16	...	...	...	...	
		HIP 3829	DG	4.41	+12.37	0.55	...	...	...	
		HIP 72509	M	4.51	+12.07	1.52	...	...	...	

continued

Table 17.6 (continued)

| Names | | Alternate | Spectral | d | | | | | | |
GI/GJ [a]	LHS [a]	Name(s)	Type	(pc)	m_V	B–V	V–R	R–I	V–I	Notes
687	450	BD+68°0946, G240–063	M3.5 V	4.53	+9.22	1.50	1.09	1.41	2.50	
		HIP 85523	K5	4.54	+9.38	1.553	...	...	...	
		HIP 114110	...	4.62	+12.24	...	...	...	...	
		HIP 57367	DC:	4.62	+11.5	0.96	...	...	...	
876	530	Ross 780, G156–057	M5 V	4.70	+10.18	1.58	1.183	1.570	2.753	
412A	38	BD+44°2051	M2 Ve	4.83	+8.82	1.491	1.00	1.02	2.02	
380	280	HD 88230	K7 V	4.87	+6.59	1.36	0.85	0.77	1.62	suspected binary
		HIP 82725	...	4.93	+11.72	...	...	...	...	
		HIP 85605	...	4.93	+11.39	1.10	...	...	...	
832	3685	CD–49°13515, HD 204961	M1 V	4.94	+8.66	1.515	1.00	1.185	2.185	

[a] Catalogue identifications: Gl = Gliese, GJ = Gliese & Jahreiss, LHS = Luyten Half Second, BD = Bonner Durchmusterung, CD: Cordoba Durchmusterung, G = Giclas, HD = Henry Draper, HIP = Hipparcos.
Spectral type: luminosity class V indicates main sequence stars; e indicates H-α emission. m_V: apparent visual magnitude.

Sources: The Hipparcos catalogue, European Space Agency, ESA SP-1200. Bessel, M. S., 1990, *Astron. & Astrophys. Suppl. Ser.* 83, 357–378. Leggett, S. K., 1992, *ApJ. Suppl. Ser.* 82, 351–394. Wood, B. E., Brown, A., Linsky, J. L., Kellett, B. J., Bromage, G. E., Hodkin, S. T., & Pye, J. P., 1994, *ApJ. Suppl. Ser.* 93, 287–307.

Table 17.7   Properties of Low-mass Substellar Objects (Extrasolar Planets and Brown Dwarfs) and Comparison with Jupiter

Name	a (AU)	e	$P_{Orbital}$	M (M_{Jup})	T_{eff} (K)	Notes	Sources
51 Peg B	0.05	0.015	4.23 d	0.47	1240	$R=1.2–1.4\ R_{Jup}$	BM97, GBH96, MQ95
υ And B	0.054	0.109	4.61 d	0.77	1300		BMW97, Gon97
55 ρ^1 Cnc C	0.11	0.050	14.64 d	0.84	1000	in binary sys.	BM97, BMW97
Gl 411 B	2.33	...	5.8 y	0.9	150		Gat96
Gl 411 C (?)	6.9	...	30 y	~1.1	70*		Gat96
ρ CrB B	0.23	0.028	39.65 d	1.1	610*		NJK97
16 Cyg C	1.6	0.63	800 d	1.5	230*	in binary sys.	CHB97
47 UMa B	2.11	0.01	1108 d	2.39	175		BM96, BM97
CM Dra C	...	...	2.2 y	5	...	$R=0.94\ R_{Jup}$ in binary sys.	GMW97
70 Vir B	0.45	0.404	116.7 d	6.6	360		MB96, BM97
τ Boo B	0.046	0.018	3.31 d	4.1, 6.7	1500		BMW97, Gon97
HD 114762 B	0.34	0.35	83.90 d	9.4 40–50	450	brown dwarf or M-dwarf?	Hal95, MLS96
GD 165B	123	...	~1900 y	~13–80	2100	primary is DA4	BZ88, JLA96
HD 110833 B	~0.8	0.69	270.04 d	18	...		MQU97
BD–04°0782B	~0.7	0.28	240.92 d	22	...		MQU97
HD 112758 B	~0.35	0.16	103.22 d	37	...		MQU97
HD 98230 B	~0.06	0.00	3.98 d	39	...	quadruple sys.	MD91, MQU97
HD 18445 E	~1.3	0.54	554.7 d	41	...	quadruple sys.	MD91, MQU97
HD 29587 B	2.6	0.33 0.37	1472 d 1481 d	42 58	...		MLS96, MQU97
Gl 229 B	~40	...	~200 y	42–58	960	$\log g = 5.3$ $L/L_\odot=6.4\times10^{-6}$	AHB95, NOK95
HD 140913 B	0.58	0.61	147.9 d	48	...		MLS96, MQU97
BD+26°0730B	~0.03	0.02	1.79 d	52	...		MQU97
Calar 3	dM8; $L/L_\odot=7.8\times10^{-4}$			55	2600	single, Pleiades	RMB96
HD 89707 B	0.93	0.95	298.25 d	55(–75)	...		MD91, MQU97
HD 217580 B	1.05	0.52	454.66 d	63	...		TDH94, MQU97

continued

Table 17.7 *(continued)*

Name	a (AU)	e	$P_{Orbital}$	M (M_{Jup})	T_{eff} (K)	Notes	Sources
Denis-PJ1228.2-1547	brown dwarf?			≤ 68	<1800	single obj.	DTF97
Denis-PJ1058.7-1548	brown dwarf?			≤ 68	~1800	single obj.	DTF97
Denis-PJ0205.4-1159	brown dwarf			≤ 68	~1500	single obj.	DTF97
PC0025+0447	dM9.5, brown dwarf?			≤ 84	~1900	single obj.	GMG92
Kelu-1	log g = 5.0–5.5, 10 pc			≤ 79	~1900	single obj.	RLA97
PPl 15	dM8; $L/L_\odot$=6.6×10^{-4}			55	~2600	single; Pleiades	RMB96
Teide 1	dM6,5; $L/L_\odot$ =1.6×10^{-3}			80	~2800	single; Pleiades	RMB96
BD+31°0643C	accretion disk, R = 6600 AU					in binary sys.	KJ97
1257+12 B	0.19	= 0.0	25.34 d	0.015$M_\oplus$	...	around pulsar	Wol94
B0329+54 C	2.3	...	3.3 y	0.3 $M_\oplus$	...	around pulsar	Sha95
B0329+54 B	7.3	0.23	16.9 y	2.2 $M_\oplus$	...	around pulsar	Sha95
1257+12 C	0.36	0.0182	66.54 d	3.4 $M_\oplus$	...	around pulsar	Wol94
1257+12 D	0.47	0.0264	98.22 d	2.8 $M_\oplus$	...	around pulsar	Wol94
B1620–26	38	...	~100 y	10 M_{Jup} 0.25–13	...	around pulsar	BFS93, JR97
Jupiter	5.20	0.048	11.86 y	1.0	165		

* temperature estimated from log (T/T_{eff^*}) = $-1.3 - 0.5$ log a; where T_{eff^*} is the effective temperature of the primary star in K and a is the semimajor axis of the planet's orbit in AU. *Notes: a:* semimajor axis; *e:* orbital eccentricity; *?* uncertain. Objects with ~13 to 80 M_{Jup} are brown dwarf candidates. Masses listed are minimum masses for objects orbiting stars.
1 $M_\oplus$ = 5.9736×10^{24} kg = 3.15×10^{-3} M_{Jup}; 1 M_{Jup} = 1.8986×10^{27} kg = 9.545×10^{-4} $M_\odot$

Sources: **[AHB95]** Allard, F., Hauschildt, P. H., Baraffe, I., & Chabrier, G., 1995, *ApJ.* 465, L123–L127. **[BFS93]** Backer, D. C., Foster, R. S., & Sallen, S., 1993, *Nature* 365, 817–819. **[BM96]** Butler, R. P., & Marcy, G. W., 1996, *ApJ.* 464, L153–L156. **[BM97]** Butler, R. P., & Marcy, G. W., 1997, in *Astronomical and biochemical origins and the search for life in the universe* (Cosmovici, C. B., Bowyer, S., & Werthimer, D., eds.) Proc. 5th Intern. Conf. on Bioastronomy, IAU Colloq. 161, Capri. **[BMW97]** Butler, R. P., Marcy, G. W., Williams E., Hauser, H., & Shirts, P., 1997, *ApJ.* 474, L115–L118. **[BZ88]** Becklin, E. E., & Zuckerman, B., 1988, *Nature* 336, 656–658. **[CHB97]** Cochran, W. D., Hatzes, A. P., Butler, R. P., & Marcy, G. W., 1997, *ApJ.* 483, 457–463. **[DTF97]** Delfosse, X., Tinney, C. G., Forveille, T, Epchtein, N., Bertin, E., Borsenberger, J., Copet, E., de Batz, B., Fouqué, P., Kimeswenger, S., Le Bertre, T., Lacombe, F., Rouan, D., Tiphène, 1997, *ApJ.* 327, L25–L28. **[GMG92]** Graham, J. R., Matthews, K., Greenstein, J. L., Neugebauer, G., Tinney, C. G., & **[Gat96]** Gatewood, G., 1996, *BAAS* 28, 885. **[GBH96]** Guillot, T., Burrows, A., Hubbard, W. B., Lunine, J. I., Saumon, D., 1996, *ApJ.* 459, L35–L38.

Persson, S. E., 1992, *Astron. J.*, 104, 2016–2021. **[GMW97]** Guinan, E. F., McCook, G. P., Wright, S. R., & Bradstreet, D. H., 1997, *BAAS* 29, 810. **[Gon97]** Gonzales, G., 1997, *Mon. Not. R. Astron. Soc.* 285, 403–412. **[Hal95]** Hale, A., 1995, *Publ. Astron. Soc. Pac.* 107, 22–26. **[JLA96]** Jones, H. R. A., Longmore, A. J., Allard F., & Hauschildt, P. H., 1996, *Mon. Not. R. Astron. Soc.* 280, 77–94. **[JR97]** Joshi, K. J., & Rasio, F. A., 1997, *ApJ.* 479, 948–959. **[KJ97]** Kalas, P. & Jewitt, D., 1997, *Nature* 386, 52–54. **[MB96]** Marcy, G. W., & Butler, R. P., 1996, *ApJ.* 464, L147–L151. **[MD91]** Mathioudakis, M., & Doyle, J. G., 1991, *Astron. & Astrophys.* 244, 409–418. **[MLS96]** Mazeh, T., Latham, D. W., & Stefanik, R. P., 1996, *ApJ.* 466, 415–426. **[MQ95]** Mayor, M., & Queloz, D., 1995, *Nature* 378, 355–359. **[MQU97]** Mayor, M., Queloz, D., Udry, S., & Halbwachs, J. L., 1997, in *Astronomical and biochemical origins and the search for life in the universe* (Cosmovici, C. B., Bowyer, S., & Werthimer, D., eds) Proc. 5th Intern. Conf. on Bioastronomy, IAU Colloq. 161, Capri. **[NJK97]** Noyes, R. W., Jha, S., Korzennik, S. G., Krockenberger, M., Nisenson, P., Brown, T. M., Kennelly, E. J., & Horner, S. D., 1997, *ApJ.* 483, L111–L114. **[NOK95]** Nakajima, T., Oppenheimer, B. R., Kulkarni, S. R., Golimowski, D. A., Matthews K., & Durrance, S. T., 1995, *Nature* 378, 463–465. **[RLA97]** Ruiz, M. T., Leggett, S. K., & Allard, F., 1997, *ApJ.* 491, L107–L110. **[RMB96]** Rebollo, R., Martin, E. L., Basri, G., Marcy, G. W., & Zapatero-Osorio, M. R., 1996, *ApJ.* 469, L53–L56. **[Sha95]** Shananova, T. V., 1995, *ApJ.* 453, 779–782. **[TDH94]** Tokovinin, A. A., Duquennoy, A., Halbwachs, J. L., & Mayor, M., 1994, *Astron. & Astrophys.* 282, 831–834. **[Wol94]** Wolszczan, A., 1994, *Science* 264, 538–542.

Table 17.8 Properties of Some Primary Stars with Low-mass Companions and Comparison to the Sun

Name	Spec. Type	Dist. (pc)	$P_{Rot.}$ (d)	Age (Ga)	m_V (mag)	log g (cgs)	[Fe/H]	T_{eff} (K)	Mass ($M_\odot$)	Radius ($R_\odot$)
16 Cyg A	G1.5 V	21.4	26.9	...	5.96	4.28	+0.05	5785	1.05	1.23
16 Cyg B	G2.5 V	21.4	29.1	...	6.22	4.35	+0.05	5760	1.00	1.11
47 UMa	G1 V	14.1	16	6.9	5.05	4.31	+0.01	5860	1.05	1.19
51 Peg	G5 V	15.4	37.1	10	4.60	4.32	+0.19	5770	1.00	1.29
55 ρ^1 Cnc	G8 V	12.5	41.7	5	6.82	4.50	+0.30	5196	0.85	0.86
70 Vir	G4V	18.1	36.3	9	4.98	4.00	–0.11	5480	0.92	0.86
CM Dra A	M4.5Ve	14.7	1.27	...	12.77	5.00	...	...	0.24	0.25
CM Dra B	M4.5Ve	14.7	...	...	12.92	5.03	...	...	0.21	0.235
Gl 229 A	M1 Ve	5.77	...	...	8.12	5.0	0.0	3400	1.74	0.69
Gl 411 A	M2 V	2.55	53	...	7.48	4.8	–0.20	3830	0.37	0.40
HD 18445	G5 V	25.7	...	...	8.5	...	...	...	0.92	...
HD 29587	G2 V	28.3	...	...	7.29	4.5	0.0	6000	1.04	0.95
HD 89707	F8 V	34.6	...	8.1	7.18	4.42	–0.42	5989	1.19	1.11
HD 98230	G0 Ve	7.87	3.98	...	4.87	...	–0.12	5600	~1.8	...

continued

Table 17.8 *(continued)*

Name	Spec. Type	Dist. (pc)	P_Rot. (d)	Age (Ga)	m_V (mag)	log g (cgs)	[Fe/H]	T_eff (K)	Mass (M_☉)	Radius (R_☉)
HD 110833	K3 V	15.1	...	...	6.98	...	...	...	~0.94	...
HD 112758	K0 V	21.0	...	...	7.54	4.5	−0.29	5116	~0.54	...
HD 114762	F9 V	40.6	12.0	5	7.31	4.5	−0.79	5750	0.73	0.80
HD 140913	G0 V	48.0	10	...	8.08	4.5	0.0	5250	1.19	1.02
HD 217580	K4 V	16.9	...	...	7.46	...	...	...	0.74	...
BD+26°0730A	K5 Ve	17.9	1.79	...	8.16	...	...	4540	0.8	
BD−04°0782	K5 Ve	20.4	1.85	...	7.9	3.0	...	4450	0.65	0.77
GD 165 A	DA4	29	...	...	14.32	7.86	...	14618	0.507	0.0014
1257+12	pulsar	300	6.2 ms	...	...	...	...	...	~1.4	...
B0329+54	pulsar	780	714.5 ms	...	...	...	...	...	1.4	...
15ρ CrB A	G0 V	17.4	17	11.7	5.41	4.24	−0.26	5810	1.1	1.31
τ Boo	F7 V	15.6	3.3	1.3–2	4.50	4.5	+0.34	6600	1.55	1.16
υ And	F8 V	13.5	12	3–4	4.09	4.3	+0.17	6250	1.34	1.36
Sun	G2 V	—	25.4,26.1	4.6	4.79	4.45	= 0.0	5780	1.0	1.0

Notes: $M_☉ = 1.989 \times 10^{30}$ kg; $R_☉ = 695950$ km

Sources: [BMW97, BZ88, CHB97, Gon97, MLS96, NJK97, Sha95, TDH94, Wol94] as identified in Table 17.7, and Alonso, A., Arribas, S., & Martinez-Roger, C., 1996, *Astron. & Astrophys. Suppl. Ser.* 117, 227–254. Baliunas, S, L., Henry, G. W., Donahue, R. A., Fekel, F. C., & Soon, W. H., 1997, *ApJ.* 474, L119–L122. Blackwell, D. E., & Lynas-Gray, A. E., 1994, *Astron. & Astrophys.* 282, 899–910. Baliunas, S., Sokoloff, D., & Soon, W., 1996, *ApJ.* 457, L99–L102. Barrado Y Navascués, Fernández-Figueroa, M. J., Garcia López, R. J., De Castro, E., & Cornide, M., 1997, *Astron. & Astrophys.* 326, 780–792. Bragalia, A., Renzini, A., & Bergeron, P., 1995, *ApJ.* 443, 735–752. Byrne, P. B., Doyle, J. G., & Menzies, J. W., 1985, *Mon. Not. R. Astron. Soc.* 214, 119–130. Cayrel de Strobel, G., Hauck, B., Francois, P., Thevenin, F., Friel, E., Mermilliod, M., & Borde, S., 1992, *Astron. & Astrophys. Suppl. Ser.* 95, 273–336. Doyle, J. G., & Butler, C. J., 1990, *Astron. & Astrophys.* 235, 335–339. Doyle, J. G., 1987, *Mon. Not. R. Astron. Soc.* 224, 1–6. Edvardsson, B., Andersen, J., Gustafsson, B., Lambert, D. L., Nissen, P. E., & Tomkin, J., 1993, *Astron. & Astrophys.* 275, 101–152. Friel, E., Cayrel de Strobel, G., Chmielewski, Y., Spite, M., Lèbre, A., & Bentolila, C., 1993, *Astron. & Astrophys.* 274, 825–837. Henry, G. W., Baliunas, S. L., Donahue, R. A., Soon, W. H., & Saar, S. H., 1997, *ApJ.* 474, 503–510. Henry, T. J., Kirkpatrick, J. D., & Simons, D. A., 1994, *Astron. J.* 108, 1437–1444. Keenan, P. C., & Pitts, R. E., 1980, *ApJ. Suppl. Ser.* 42, 541–563. Kirkpatrick, J. D., & McCarthy, D. W., 1994, *Astron. J.* 107, 333–349. Leggett, S. K., 1992, *ApJ. Suppl. Ser.* 82, 351–394. Leggett, S. K., Allard, F., Berriman, G., Dahn, C. C., & Hauschildt, P. H., 1996, *ApJ. Suppl. Ser.* 104, 117–143. Metcalfe, T. S., Mathieu, R. D., Latham, D. W., & Torres, G., 1996, *ApJ.*

456, 356–364. Mould, J. R., 1978, *ApJ.* 226, 923–930. Peterson, R. C., & Carney, B. W., 1979, *ApJ.* 231, 762–780. Pettersen, B. R., 1989, *Astron. & Astrophys.* 209, 279–295. Soderblom, D. R., 1985, *Astron. J.* 90, 2103–2115. Strassmeier, K. G., Hall, D. S., Zeilik, M., Nelson, E., Eker, Z., & Fekel, F. C., 1988, *Astron. & Astrophys. Suppl. Ser.* 72, 291–345. Wood, B. E., Brown, A., Linsky, J. L., Kellett, B. J., Bromage, G. E., Hodkin, S. T., & Pye, J. P., 1994, *ApJ. Suppl. Ser.* 93, 287–307.

Table 17.9 Other Primary Star Designations

Name	HD	Gl/GJ	BS/HR	BD (DM)	LHS	GC or Giclas
50υ And	9826	Gl 61	458	BD+40°0332	—	GC 1948
—	18445	GJ 120.1 C	—	CD–25°1168	—	GC 3558
—	29587	—	—	BD+41°0931	1696	G081–030
—	42581	Gl 229 A	—	BD–21°1377	1827	—
55 ρ¹Cnc A	75732	GJ 324 A	3522	BD+28°1660	2062	GC 12244
—	89707	Gl 388.2	—	BD–14°3093	—	GC 14202
47 UMa	95128	Gl 407	4277	BD+41°2147	—	GC 15087
Lal. 21185	95735	Gl 411	—	BD+36°2147	37	G119–052
53 ξ UMa B	98230/1	Gl 423 B	4374	BD+32°2132B	2391	—
—	110833	GJ 483	—	BD+52°1650	—	G199–036
—	112758	GJ 491 A	—	BD–09°3595	2656	GC 17617
—	114762	—	—	BD+18°2700	2693	G063–009
70 Vir	117176	GJ 512.1	5072	BD+14°2621	2740	GC 18212
τ Boo	120136	GJ 527 A	5185	BD+18°2782	—	GC 18637
—	140913	—	—	BD+28°2469	—	—
15 ρ CrB	143761	GJ 606.2	5968	BD+33°2663	3145	GC 21527
16 Cyg B	186427	GJ 765.1 B	7504	BD+50°2848	—	GC 27285
51 Peg	217014	Gl 882	8729	BD+19°5036	—	GC 32003
—	217580	GJ 886	—	BD–04°5804	6415	G156–075
—	283750	Gl 171.2 A	—	BD+26°0730	—	G039–028
CM Dra	—	Gl 630.1	—	—	421	G225–067
V833 Tau	—	GJ 1069	—	BD–04°0782	—	—

Table 17.10 Some Properties of Interstellar Molecular Clouds

	T (K)	Mass ($M_\odot$)	Number Density (cm^{-3})	Mean Diameter (pc)	ΔV [a] (km s^{-1})	Object Examples
Diffuse H I clouds						
	20–100	1–100	20–1,000	1–5	0.5–3	ζ Oph
Translucent molecular clouds						
	15–50	3–100	200–5,000	0.5–5	0.5–3	HD 169454
Cold, dark nebulae						
complex	~10	1,000–10^4	100–1,000	3–20	1–3	Taurus-Auriga
clouds	~(5–30)	10–1,000	100–10^4	0.2–4	0.5–2	B1, B5, B227
cores	~10	0.3–10	10^4–10^5	0.05–0.4	0.2–0.4	TMC-1, B335
Giant molecular clouds (GMC)						
complex	15–20	10^5–10^6	100–300	20–80	6–15	Orion, M17
clouds	15–40	1,000–10^5	100–10^4	3–50	3–12	OMC-1, W 33, W3A
warm clumps	25–100	10–1,000	10^4–10^7	0.05–3	1–3	M17 clumps
hot cores	100–200	10–1,000	10^7–10^9	0.05–1	1–15	Orion hot core

[a] CO emission line width

Sources: Blitz, L., 1993, in *Protostars and planets III* (Levy, E. H., & Lunine, J. I., eds.), Univ. Arizona Press, Tucson, pp. 125–161, and references therein. Goldsmith, P. F., 1987, in *Interstellar processes* (Hollenbach, D. J., & Thronson, H. A., eds.), D. Reidel, Dortrecht, The Netherlands, pp. 51–70.

Table 17.11 Some Observed Interstellar Molecules

No. of Atoms	Linear, Open Molecules
1	H, C
2	H_2, C_2, CH, CN, CO, CS, NH, NO, NS, CP*, PN, PO, OH, SO, HCl, NaCl*, KCl*, AlF*, AlCl*, SiC*, SiN*, SiO, SiS, CH^+, SO^+
3	C_3*, CH_2, C_2H, C_2D, C_2O, C_2S, HCO, HNO, HCN, DCN, HNC, DNC, H_2O, HDO, H_2S, OCS, SO_2, SiC_2*, HCO^+, DCO^+, N_2H^+, N_2D^+, HOC^+, HCS^+
4	C_3H, C_2H_2, C_3O, H_2CO, HDCO, D_2CO, NH_3, NH_2D, C_3N, HC_2N, HNCO, C_3S, H_2CS, HNCS, $HCNH^+$, H_3O^+, $HOCO^+$
5	C_5*, C_3H_2, C_3HD, C_4H, C_4D, CH_4, CH_3D, H_2CCC, CH_2CO, HOCHO, HCOOH, HC_3N, DC_3N, CH_2NH, NH_2CN, CH_2CN, HCCNC, SiH_4*
6	C_5*, C_5H, C_2H_4, H_2CCCC, C_5O, CH_3OH, CH_3OD, HC_2CHO, CH_3SH, CH_3NC, CH_3CN, CH_2DCN, H_3CNC, NH_2CHO
7	C_7*, C_6H, CH_3CCH, CH_2DCCH, DC_5N, CH_3NH_2, CH_2CHCN, CH_3CHO, HC_5N,
≥8	C_9, CH_3C_4H, HC_7N, HC_9N, $HC_{11}N$, CH_3C_2CN, CH_3C_4CN, CH_3CH_2CN, CH_3CH_2OH, CH_3OCHO, CH_3OCH_3, CH_3COCH_3, $HCOOCH_3$

Cyclic Molecules

SiC_2, C_3H, C_3H_2, SiC_4*

---

* Molecules only observed in circumstellar outflows of late-type giant stars (e.g., IRC +10°216); see Glassgold, 1996.

Sources: Blake, G. A., Sutton, E. C., Masson, C. R., & Phillips, T. G., 1987, *ApJ.* 315, 621–645. Glassgold, A. E., 1996, *Annu. Rev. Astron. Astrophys.* 34, 241–277. Irvine, W. M., Schloerb, F. P., Hjalmarson, A., & Herbst, E., 1985, in *Protostars & Planets II* (Black, D. C., & Matthews, M. S., eds.), Univ. Arizona Press, Tucson, pp. 579–620. van Dishoeck, E. F., Blake, G. A., Draine, B. T., & Lunine, J. I., 1993, in *Protostars and planets III*, (Levy, E. H., & Lunine, J. I., eds.) Univ. of Arizona Press, Tucson, pp. 163–241.

GLOSSARY

Ab: Albite

Achilles: 588 Achilles was the first Trojan asteroid discovered in 1906. It orbits 60° ahead of Jupiter

Accretion: Formation of planetesimals and planets by accumulation of smaller objects from the solar nebula

Achondrite: Stony meteorite without chondrules, mainly igneous in origin

Adonis: 2101 Adonis is an Apollo asteroid discovered in 1936 when it passed within 0.015 AU of Earth. It has a diameter of ~1 km and q ~ 0.51 AU.

ADOR: Initial $^{87}Sr/^{86}Sr$ (= 0.69883±0.00002) for the Angra dos Reis achondrite

Adrastea: Second satellite from Jupiter; discovered in 1979 by E. Danielson and D. Jewitt.

Alba Patera: A large volcano north of the Tharis Ridge on Mars

Albedo: The fraction of incident light reflected from a planetary surface. It can be calculated in several different ways. Also see Bond albedo and Geometric albedo.

Albite: $NaAlSi_3O_8$; Na-endmember of feldspar solid solutions

Alkali elements: Li, Na, K, Rb, Cs, Fr; first group of elements in the periodic table, excluding H. Francium (Fr) is a radioactive element not found naturally.

Alkali feldspar: Na- and K-rich feldspars (albite, orthoclase, sanidine)

ALL: Lowest $^{87}Sr/^{86}Sr$ ratio (= 0.69877±0.00002) measured in a CAI from the Allende (CV3) meteorite

Alpha Regio: A 1300 km highland plateau on Venus. The first surface feature on Venus that was discovered by Earth-based radar.

Amalthea: Third satellite from Jupiter; discovered in 1892 by E. Barnard

Amor asteroids: Asteroids with perihelion distances 1.017 AU < q < 1.3 AU

Amphoterites: Old name for the LL-chondrites

An: Anorthite

AMU: Atomic mass unit; 1 AMU = u = 1/16 of the mass of an O atom

Ananke: Thirteenth satellite from Jupiter, one of the four retrograde outer satellites; discovered by S. Nicholson in 1951

Angular momentum: A property of any revolving or rotating system given by the vector product of the moment of inertia (I) and angular velocity (ω) of a body rotating about a point. Angular momentum is conserved in a closed system.

Angular momentum density: Angular momentum normalized to mass

Anomalistic month: The time (27.55455 days) between two successive perigee passages of the moon

Anomalistic year: The time (365.25964 days) between two successive perihelion passages of the earth in its orbit. It is longer than the sidereal and tropical years because of the advance in Earth's perihelion as a result of planetary perturbations.

Anorthite: $CaAl_2Si_2O_8$; Ca-rich endmember of feldspar solid solutions

Anorthosite: Igneous rock made almost entirely of plagioclase feldspar

ANT: Acronym for lunar highland rocks anorthosite, norite, and troctolite

Aphelion: Most distant orbital point from the sun of a body gravitationally bound to the sun

Aphrodite Terra: Second largest highland region on Venus

Apocenter: The point in an orbit of a planet, satellite, etc., that is farthest from the center of mass of the system

Apogee: The point in the orbit of the moon or of an artificial Earth satellite that is farthest from Earth. The maximum orbital velocity is reached at apogee.

Apollo asteroids: Asteroids with semimajor axis a > 1.0 AU and perihelion distances q < 1.017 AU. Well-known Apollo asteroids include 1566 Icarus, 1862 Apollo, and 2201 Oljato. The group is named after 1862 Apollo, which was discovered in 1932 when it passed within 0.07 AU of Earth.

Argyre Basin: A Martian impact basin 1200 km in diameter and ~3 km deep

Ariel: Twelfth satellite from Uranus. Discovered by W. Lassell in 1851, it is the fourth largest Uranian satellite. *Voyager 2* imaging revealed smooth-floored valleys and a cratered surface

Assimilation: Incorporation of surrounding rock into magma by melting and dissolution

Asthenosphere: Part of the earth's mantle located directly under the plates with lower viscosity than the lithosphere.

Asteroids: Objects (>10,000) between the orbits of Mars and Jupiter at 1.7–4 AU, ranging in diameter from less than one to about 1000 km

Ataxites: Iron meteorites with little or no visible structures containing >16 wt% Ni

Aten asteroids: Asteroids with a < 1 AU. Named after 2062 Aten discovered in 1976. Other Aten asteroids are 2100 Ra-Shalom (the largest Aten) and 1986 TO (no. 3753), which is in an overlapping horseshoe orbit with Earth (Wiegert, P. A., Innanen, K. A., & Mikkola, S., 1997, *Nature* 387, 685–686).

Atlas: Second satellite from Saturn; discovered by R. Terrile in 1980

AU: Astronomical unit, the mean Earth-Sun distance = 1.496×10^8 km

Aubrite: Differentiated stone meteorite mainly consisting of almost pure enstatite. These highly reduced meteorites are also called enstatite achondrites.

Augite: Ca- and Mg-rich clinopyroxene

Autumnal equinox: The date in the autumn, around September 21st, when the sun crosses from north to south of the earth's equator. The sun's declination is $0°$ at the autumnal and vernal equinoxes. The autumnal and vernal equinoxes are formally defined as the two points on the celestial sphere where the sun crosses the celestial equator in its apparent annual motion. The equinoxes are not fixed in position but are moving retrograde (westward) at 50.28 arc seconds per year (precession of the equinoxes). Also see Tropical year and Vernal equinox.

BABI: Basaltic Achondrite Best Initial. Best estimate for the initial $^{87}Sr/^{86}Sr$ ratio in basaltic achondrites, taken as representative for the solar nebula. $^{87}Sr/^{86}Sr = 0.69898$

Barycenter: The position of the center of mass of a system of bodies such as the earth and moon. The barycenter of the Earth-Moon system is 4671 km from the center of the earth and is given by $M_m a_m/(M_e + M_m)$, where M_m and M_e are the relative masses of the moon and the earth, and a_m is the semimajor axis of the moon.

Basalt: Dark, fine-grained, mafic igneous rock consisting mainly of plagioclase feldspar and pyroxene. Minor constituents include olivine and ilmenite.

Basaltic achondrites: Stony meteorites formed from basaltic magmas

Belinda: Ninth satellite from Uranus; discovered by *Voyager 2* in 1986

Bianca: Third satellite from Uranus; discovered by *Voyager 2* in 1986

Blackbody: An idealized body perfectly absorbing all incident radiation at all wavelengths

Blackbody radiation: Continuous spectrum emitted by a blackbody at a given temperature. The energy radiated per unit area per unit wavelength interval ($B_\lambda(T)$) is described by Planck's law $B_\lambda(T) = (2\pi c^2/\lambda^5)/(e^{hc/\lambda kT} - 1)$, where c, h, k, and l are the speed of light in vacuum, the Planck constant, the Boltzmann constant, and wavelength, respectively.

Bond albedo: The fraction of total incident spectral energy (over all wavelengths) that a body radiates back to space in all directions.

Bonner Durchmusterung: The Bonn survey, a general star catalog. Catalogued stars are identified by the prefix BD and a number giving the declination (with 1° resolution), and a number giving its right ascension. Also see Córdoba Durchmusterung.

Breccia: Rock containing coarse rock and mineral fragments (clasts) in a finer grained matrix. Breccias are further subdivided into genomict (clasts are of same class but differ petrographically), monomict (clasts and matrix are of the same class and type), and polymict (clasts and/or matrix are compositionally different) breccias.

Brown Dwarf: An object of ~80 solar masses, or larger, that is too small to sustain nuclear fusion, but instead followed the Hayashi track for ~100 Ma before cooling off. Gliese 229B is the first unambiguous brown dwarf that has been identified and imaged.

CAI: Calcium- and aluminum-rich inclusions often found in CV, CO, and CM chondrites

Callisto: Fourth Galilean satellite, apparently undifferentiated ice-rock body

Calypso: Eleventh satellite from Saturn; discovered by B. Smith in 1980. Calypso is co-orbital with Telesto and Tethys.

Cape Photographic Durchmusterung: A general star catalog of the southern sky listing 455,000 stars with declinations of −19° to −90°. Stars in this catalog have the prefix CPD. The CPD catalog should not be confused with the CD catalog.

Carbon star: Red giant stars with spectra showing strong C_2, CN, or other C-bearing molecular lines.

Carbonaceous chondrites: Primitive stony meteorites containing up to about 3–4 mass% carbonaceous material. Subgroups are CI, CM, CO, CV, CR, and CH, depending on composition and mineralogy. Most of the carbonaceous chondrites are highly oxidized and have chemical compositions similar to that of the solar photosphere (except for very volatile elements).

Carme: Fourteenth satellite from Jupiter, one of the four retrograde outer satellites; discovered by S. Nicholson in 1938

Ceres: 1 Ceres was the first asteroid discovered. It has a diameter of 914 km, a mass of 1.2×10^{21} kg, and a density of 2.7 g cm^{-3}. Ceres has an albedo of 0.10 and cannot be seen with the naked eye.

Chalcophile elements: Sulfide-loving elements or elements that preferentially enter sulfide phases (e.g., S, Se, Cd, Cu, Tl)

Chandler wobble: The wobble of Earth's rotational axis about its mean position with periods of 12 and 14 months. The 12 month period is due to seasonal variations in ice, snow, and atmospheric mass; the 14 month period is due to mass movements within the earth.

Charon: The satellite of Pluto; discovered by J. Christy in 1978

Chassignite: Rare type of achondrites consisting mainly of olivine and minor pyroxene, plagioclase, chromite, and sulfide, related to shergottites and nakhlites

Chiron: 2060 Chiron was the first Centaur object discovered in 1977.

Chondrites: Most abundant class of stony meteorites containing chondrules. The term chondrite is also applied to all meteorites that have bulk compositions close to that of the solar photosphere (except for volatile elements) even if no chondrules are present in the meteorite (e.g., CI chondrites).

Chondrules; Millimeter-size, generally spherical objects consisting mainly of olivine and/or low-Ca pyroxene found in chondritic meteorites

Chromosphere: Transparent, intermediate temperature region between photosphere and corona of main sequence stars, such as the sun. Temperatures in the solar chromosphere range from ~4000 K at the base to ~50,000 K at its top.

Chryse Planitia: A Martian plains area 1600 km across and 2.5 below the Mars datum radius; the landing site for the *Viking 1* spacecraft

CHUR: Chondritic Uniform Reservoir obtained from average chondrite Sm/Nd ratio and present ^{143}Nd/^{144}Nd ratios. Allows calculation of the Nd isotopic composition of the reservoir as a function of time.

Clinopyroxene: Minerals of the pyroxene group crystallizing in the monoclinic form (e.g., augite, pigeonite).

Column density: Number of atoms or molecules per unit area. Often used to express the total atmospheric abundance of a species or the absorption by a species along the line of sight.

Compatible elements: Elements that readily enter crystalline phases rather than silicate melts during igneous differentiation

Continental drift: Relative motion of the continents on the surface of the earth as a result of plate tectonics

Cordelia: Innermost satellite of Uranus; discovered by *Voyager 2* in 1986. Cordelia and Ophelia are shepherd satellites to the Epsilon ring (the outermost ring) of Uranus.

Córdoba Durchmusterung: A catalog of stars visible at the Córdoba, Argentina, observatory in the southern hemisphere. The prefix CD is followed by numbers giving the declination (to the nearest 1°) and the right ascension. A companion to the Bonner Durchmusterung.

Corona: Outermost hot and extended region of main-sequence stars, such as the sun. Temperatures in the solar corona reach 2×10^6 K.

Cosmic-ray exposure age: Time interval a meteorite was exposed to cosmic radiation, i.e., the time between ejection from its parent body and its arrival on Earth

Cosmic rays: Highly energetic particles produced by various processes, such as supernova explosions and pulsars, that continuously bombard the earth in all directions. Discovered in 1912 by V. F. Hess. Cosmic ray energies range from 10^8 to 10^{19} electron volts. Cosmic rays are composed of all atomic nuclei.

Cosmogenic nuclides: Nuclides (e.g., ^{21}Ne) produced by interaction with cosmic rays (e.g., spallation reactions)

Cressida: Fourth satellite from Uranus; discovered by *Voyager 2* in 1986

Cumulate: Plutonic igneous rock that accumulated crystals by sinking or floating from magma

Declination: The declination of a celestial body is measured in degrees ($0-90°$) north (positive) or south (negative) of the celestial equator.

Deimos: One of the two Martian satellites; discovered by A. Hall in 1877

Desdemona: Fifth satellite from Uranus; discovered by *Voyager 2* in 1986

Despina: Third satellite from Neptune; discovered by *Voyager 2* in 1989

Diogenite: Achondrite consisting mainly of Mg-rich orthopyroxene, related to eucrites and howardites

Dione: Twelfth satellite from Saturn; discovered by G. Cassini in 1684. The small satellite Helene is co-orbital with Dione.

Diopside: $CaMgSi_2O_6$, one endmember of the pyroxene group

Distribution coefficient: See Partition coefficient

Dunite: Rock consisting mainly of olivine

Eccentricity: A measure describing the deviation of an orbit from circularity. Defined as $e = c/2a$, where c is the distance between the foci of the ellipse and 2a is the length of the major axis. Eccentricity = 0 for a circular orbit.

Ecliptic: The mean plane of the earth's orbit around the sun

Eclogite: Mantle-derived rock consisting of garnet and pyroxene, chemically similar in composition to basalts

EHD: Eucrites, Howardites, and Diogenites, genetically related groups of basaltic achondrites

Elara: Twelfth satellite from Jupiter, one of the four prograde outer satellites; discovered by C. Perrine in 1905

Elysium Planitia: A volcanic province in the eastern hemisphere of Mars. The largest volcano in this region is Elysium Mons, 250 km in diameter at its base and 15 km high.

En: Enstatite

Enceladus: Eighth satellite from Saturn; discovered by W. Herschel in 1789

Enstatite: $MgSiO_3$; Mg-endmember of the pyroxene group

Enstatite chondrites: Highly reduced chondritic meteorites containing almost Fe-free pyroxene (enstatite), iron-nickel metal, and troilite (FeS). Subgroups are EH

(Enstatite chondrite, High metallic iron) and EL (Enstatite chondrite, Low metallic iron).

Eos family: A Hirayama family of asteroids located at ~3 AU

E-process: Equilibrium nucleosynthetic process. Process in which photodisintegration of previously synthesized nuclides leads to a population of nuclides approaching local statistical equilibrium.

Epimetheus: Fifth satellite from Saturn; discovered by R. Walker in 1966. It is co-orbital with Janus.

Eros: 433 Eros is a near Earth Amor asteroid that is the target of the NEAR mission. It was discovered in 1898.

Eucrite: Basaltic achondrite consisting mainly of plagioclase and Ca-pyroxene; related to howardites and diogenites

Europa: Second Galilean satellite, thought to have an internal ocean

Fa: Fayalite

Fayalite: Fe_2SiO_4; Fe-endmember of the olivine group

Feldspar: Solid solution of the aluminous silicate minerals anorthite, albite, and/or orthoclase/sanidine

Ferrosilite: $FeSiO_3$; Fe-endmember of the pyroxene group

Fixed stars: Stars in general, which until ~200 years ago, were thought to be fixed in the sky. Also see Proper motion.

Fo: Forsterite

Forsterite: Mg_2SiO_4; Mg-endmember of the olivine group

Fraunhofer lines: Absorption lines in the photospheric spectrum of the sun. At visible wavelengths, the most prominent lines are due to singly ionized Ca, neutral H, Na, and Mg. Many weaker lines arise from Fe.

Fs: Ferrosilite

Fugacity: Thermodynamic property describing the chemical potential of gaseous species. Fugacity (f) is used instead of partial pressure (p_i) to describe nonideal behavior of gases. The relationship between fugacity and pressure is $f_i = \gamma_i p_i$, where γ_i is the fugacity coefficient of species i.

FUN: Isotopic anomalies in CAIs caused by Fractionation and Unknown Nuclear effects

Gabbro: Coarse-grained igneous rock consisting mainly of augite and plagioclase; forms from basaltic magma at depth

Galatea: Fourth satellite from Neptune; discovered by *Voyager 2* in 1989

Ganymede: Third Galilean satellite and largest satellite in the solar system

Gardening: Process of turning over soil or regolith by micrometeorite bombardment; leads to fragmentation of surface constituents

Garnet: Minerals with the general formula $A_3Y_2(SiO_4)_3$, where A= Ca, Mg, Fe^{2+}, Mn, and B = Al, Fe^{3+}, Cr^{3+}. Endmembers of the garnet solid solution series are almandine ($Fe_3Al_3(SiO_4)_3$) and pyrope ($Mg_3Al_2(SiO_4)_3$).

Geomagnetic tail: Part of the terrestrial magnetic field pushed back by the solar wind plasma to form a tail

Geometric albedo: The ratio of light reflected at zero phase angle (opposition) to that which would be reflected by a perfectly diffusing disk of the same size

Granite: Igneous rock consisting mainly of alkali feldspar and quartz

Gregorian calendar: The calendar now in use throughout most of the world. It was instituted by Pope Gregory XIII in 1582 when the Julian calendar was corrected for accumulated errors of 10.4 days. The corrections involved removing 10 days from the calendar (so 15 October 1582 immediately followed 4 October 1582) and designating years divisible by 4 as leap years (e.g., 1588, but not 1589). Only centesimal years evenly divisible by 400 would henceforth be leap years (e.g., 1600 and 2000 are leap years, but not 1700, 1800, or 1900). The Gregorian calendar was adopted at various times by different nations, with France, Britain, Russia, and Greece adopting the calendar in 1582, 1752, 1918, and 1923, respectively. There is a small discrepancy between the Gregorian year and the tropical year amounting to 3 days per 10,000 years.

Half-life: Time interval during which the number of atoms of a radioactive nuclide is decreased to half the initial value

Hayashi Track: Evolutionary path calculated by C. Hayashi for convective protostars on the Hertzsprung-Russell (HR) diagram

Heat flow: Rate of heat energy leaving a planet's surface per unit area

Helene: Thirteenth satellite from Saturn; discovered by P. Laques and J. Lecacheux in 1980. Helene is co-orbital with Dione.

Helioseismology: The study of the solar interior by observations of solar oscillations. Leighton and colleagues first detected solar oscillations in 1960 by measuring the Doppler shifts of Fraunhofer lines.

Henry Draper Catalog: A stellar catalog compiled at the Harvard College Observatory. Stars in this catalog have the prefix HD and are classified according to the Harvard classification of spectral types. The extension to this catalog is the Henry Draper Extension (HDE) catalog. About 359,000 stars are in the HD and HDE catalogs.

Hertzsprung-Russell Diagram: A two-dimensional graph relating spectral type (i.e., temperature) to absolute stellar magnitude (i.e., luminosity)

Hidalgo: 944 Hidalgo was discovered in 1920 and is interesting because it has an orbit ranging from 9.64 AU to 2.00 AU. It has an albedo of 0.03 and may be an extinct cometary nucleus.

Himalia: Tenth satellite from Jupiter, one of the four prograde outer satellites; discovered by C. Perrine in 1904

Hot spot: Volcanic source, probably originating from deep mantle plumes, causing age-progressive lines of volcanoes, typically not located near plate boundaries. The Hawaiian island chain and Iceland are two examples of hot spot volcanism.

Howardite: Polymict brecciated basaltic achondrite consisting of fragments of eucritic or diogenitic compositions

Hyperion: Sixteenth satellite from Saturn; discovered by W. Bond in 1971

Iapetus: Seventeenth satellite from Saturn; discovered by G. Cassini in 1671. It has a dark leading hemisphere and a bright trailing hemisphere.

Icarus: 1566 Icarus is an Apollo asteroid. It was discovered in 1949; in 1968, it passed 0.04 AU from Earth.

IDP: Interplanetary Dust Particle, micrometeoroids or micrometeorites
Igneous rock: Rock formed by melting and subsequent solidification
Ilmenite: $FeTiO_3$, trace opaque mineral in basalts
Impact melting: Melting of target rock by large meteorite, comet, or planetesimal impact
Impactite: Glassy rocks formed by fusion of target rock during heating by impact
Inclination: The angle between the plane of a planetary orbit and the ecliptic or between the plane of a satellite's orbit and its planet's equator
Incompatible elements: Minor or trace elements that partition preferentially into a silicate melt during igneous differentiation
Io: Innermost Galilean satellite; volcanically active because of tidal heating
IRAS: Infrared Astronomical Satellite to conduct a high sensitivity survey of the sky in four infrared wavelength bands (12, 25, 60, and 100 μm). Joint project by the United States, the United Kingdom, and the Netherlands. IRAS was launched on 26 January 1983 and operated until 22 November 1983.
Ishtar Terra: A large highland region in the northern hemisphere of Venus
Isochron: Line of equal age for a suite of samples when the daughter nuclide is plotted against the radioactive parent nuclide
Isostasy: Balancing of topography by underlying density, e.g., the lithosphere floats on the weaker asthenosphere. At some uniform depth, the pressure is everywhere constant, and beneath this depth hydrostatic equilibrium exists.
Iron meteorite: Meteorite mainly consisting of metallic iron or iron-nickel
Jansky: A unit of flux density equal to 10^{-26} W m^{-2} Hz^{-1}
Janus: Sixth satellite from Saturn; reported by A. Dolfus in 1966, but not confirmed. Later a satellite, named Janus, was discovered at about the same orbital distance given by Dolfus. It is co-orbital with Epimetheus.
JCMT: The James Clerk Maxwell Telescope, a 15 meter-diameter radio telescope on Mauna Kea, Hawaii. The JCMT operates at $\lambda = 350$ μm to 2 mm.
Julian calendar: The calendar authorized by Julius Caesar in 46 B.C. and based on the assumption that the year contained 365.25 days. Every fourth year was a leap year containing 366 days. However, leap years were not correctly inserted until A.D. 8. The Julian year was 11 minutes 15 seconds longer than the tropical year and one extra day accumulated every 128 years. The Julian calendar was replaced by Pope Gregory XIII in 1582, but remained in use in some countries until 1923.
Julian date: The number of days elapsed since noon Greenwich mean time (GMT) on 1 January 4173 B.C. (the Julian day number), plus the decimal fraction of the day elapsed since the preceding noon. Modified Julian dates (MJD) are calculated by subtracting 2,400,000.5 from the Julian date (JD). The Julian date at noon (universal time) on 1 March 2000 will be 2,451,605. The Julian date is not to be confused with the Julian calendar.
Juliet: Sixth satellite from Uranus; discovered by *Voyager 2* in 1986
Kamacite: Fe metal or alloy (e.g., Fe-Ni with <7% Ni) with body-centered cubic structure
Koronis family: A Hirayama asteroid family located at 2.9 AU

KREEP: Lunar highland rock rich in potassium (K), rare earth elements (REE), phosphorous (P), and other incompatible elements

K-T boundary: The boundary between the Cretaceous and Tertiary, 65 Ma ago

Larissa: Fifth satellite from Neptune, discovered by *Voyager 2* in 1989

Leda: Ninth satellite from Jupiter, one of the four prograde outer satellites; discovered by C. Kowal in 1974

LIL: Large Ion Lithophile elements; incompatible elements with large ionic radii (e.g., K., Rb, U, Th, REE)

Lithophile element: Silicate-rock–loving elements; elements that concentrate in silicate and oxide minerals

Lithosphere: Outer layer of the silicate earth consisting of rigid plates

Lunar day: The moon rotates synchronously as it revolves around Earth, so the lunar day is 27.322 Earth days, equal to the sidereal month. Also see Synodic month and Lunar month

Lunar month: The interval (29.53059 days) between two successive new moons

Lunar year: Twelve synodic (i.e., lunar) months equal to 354.3672 days. A lunar calendar has 354 days and a leap year of 355 days. It has 12 months of either 29 or 30 days.

LUNI: Best estimate for initial $^{87}Sr/^{86}Sr$ (= 0.69903) of the moon, obtained from lunar anorthosites

Lysithea: Eleventh satellite from Jupiter, one of the four prograde outer satellites; discovered in 1938 by S. Nicholson

Mg, Mg*, Mg#: Magnesium number; molar ratio of $MgO/(MgO+FeO)$

Mare basalt: Basalts forming the dark-colored areas (the maria) on the moon

Maxwell Montes: The highest mountain range (10.4 km above the modal radius) on Venus, located in eastern Ishtar Terra

Mesosiderite: Stony iron meteorite consisting of silicates and Fe-Ni; believed to be related to EHD meteorites

Metis: Innermost satellite of Jupiter; discovered in 1979 by S. Synott

Mimas: Seventh satellite from Saturn; discovered by W. Herschel in 1789

Miranda: Eleventh satellite from Uranus; discovered by G. Kuiper in 1948. About half of Miranda's surface was imaged by *Voyager 2*, revealing complex surface geology and three major terrain types (old grooved plains, regions of enclosed grooved areas, and brighter regions with cliffs and scarps).

Moment of inertia: A property (I) of any rotating body that expresses the resistance to stopping or changing velocity. The kinetic energy of a body rotating with angular velocity ω is $\frac{1}{2}I\omega^2$, and the angular momentum (L) is $I\omega$. The dimensionless moment of inertia is a measure of the density distribution within a planet describing the increase of density with depth. A uniformly dense sphere has a dimensionless moment of inertia of $I/MR^2 = 0.4$, where M is mass and R is radius. Differentiated bodies with cores have $I/MR^2 < 0.4$.

Naiad: Innermost satellite of Neptune; discovered by *Voyager 2* in 1989

Nakhlite: Rare achondrite consisting of Ca-pyroxene (augite) and olivine. Related to shergottites and chassignites.

Nereid: The outermost (eighth) satellite of Neptune; discovered by G. Kuiper in 1949. Its orbit is highly eccentric.

Neumann lines: Lines visible upon etching of kamacite, resulting from mechanical twinning by mild shock

Neutrino: An elementary particle with zero charge and spin equal to one-half. Neutrinos either are massless or have very small mass. Experiments to measure solar neutrinos have been under way for three decades, but the observed solar neutrino flux is less than that predicted theoretically.

Noble gases: Inert gases He, Ne, Ar, Kr, Xe, and Rn. Radon is a radioactive element produced by the decay of radium.

Norite: Plagioclase-bearing rock similar to gabbro but with orthopyroxene dominant over clinopyroxene

NRM: Natural Remnant Magnetization; portion of the magnetization that is permanent in a rock; acquired by cooling of ferromagnetic minerals through the Curie temperature.

Nutation: Periodic but irregular movement caused by gravitational interactions of the moon and sun with Earth. Lunar notation leads to a wander of ±9 arc seconds in the celestial pole over an 18.6 year period. Nutation is superimposed on precession of Earth's rotational axis. Also see Precession and Obliquity.

Oberon: The outermost (fifteenth) and the second largest satellite of Uranus; discovered by W. Herschel in 1787. *Voyager 2* images showed large mountains up to 6 km high and several large craters.

Obliquity: The tilt angle between a planet's axis of rotation and the axis perpendicular to the orbital plane (the pole of the orbit). Earth's obliquity is about 23°26' and varies between 22.1° and 24.54° because of precession and nutation. Earth's obliquity is currently decreasing by 0.47" per year.

Olivine: Major rock-forming mineral, a solid solution series between forsterite and fayalite

Oort cloud: Spherical cloud believed to be the source region of comets with semimajor axis >20,000 AU. Comets are expelled from this shell by perturbation from nearby stars or giant molecular clouds so that some comets acquire orbits that bring them within the orbits of Jupiter or Saturn.

Ophelia: Second satellite from Uranus; discovered by *Voyager 2* in 1986. Ophelia and Cordelia are shepherd satellites to the Epsilon ring (the outermost ring) of Uranus.

Optical depth: An e-folding distance for absorption of light or radiation

opx: Orthopyroxene

Ordinary chondrite: Most common type of chondritic meteorites. Subdivisions are H (High metal), L (Low metal), and LL (Low metal, Low iron).

Orthoclase: $KAlSi_3O_8$; K-endmember of alkali feldspars

Orthopyroxene: Orthorhombic member of the pyroxene minerals. Most orthopyroxenes are low in Ca.

Pallasite: Stony iron meteorites mainly consisting of Fe-Ni with inclusions of olivine

Pan: Innermost satellite of Saturn; discovered by M. Showalter in 1990

Pandora: Fourth satellite from Saturn; discovered by S. Collins in 1980

Parsec: The distance at which the semimajor axis of Earth's orbit (1 AU) subtends an angle of one arc second. One parsec is 3.2616 light years.

Partition coefficient: Weight concentration ratio of a trace element between two equilibrated mineral phases (Nernstian partition coefficient). Often the abbreviation "D" is used.

Pasiphae: Fifteenth satellite from Jupiter, one of the four retrograde outer satellites; discovered by P. Melotte in 1908

PDB: Pee Dee Belemnite, fossil carbonate used as reference standard for carbon isotopic measurements

Peridotite: Plutonic igneous rock mainly consisting of olivine and pyroxene

Pericenter: The point in an orbit of a planet, satellite, etc., that is closest to the center of mass of the system

Perigee: The point in the orbit of the moon or an artificial earth satellite that is closest to Earth. The minimum orbital velocity is reached at perigee.

Perihelion: Closest orbital point from the sun of a body gravitationally bound to the sun. The earth is at perihelion around January 3rd each year.

PGE: Platinum Group Elements; the noble metals Ru, Rh, Pd, Os, Ir, and Pt

Phoebe: The outermost (eighteenth) satellite from Saturn; discovered by W. Pickering in 1898

Photosphere: Visible region of a main-sequence star, such as the sun. Solar abundance data are mainly obtained from this region. Temperatures in the solar photosphere range from 6000 K to 4000 K.

Pigeonite: Low-Ca, monoclinic pyroxene solid solution of enstatite and ferrosilite endmembers; minor amounts of wollastonite can be accommodated

Plagioclase feldspar: Solid solution series between albite and anorthite

Planetesimal: Bodies whose sizes range from meters to several hundred km in diameter that formed during planetary formation. Most of them are thought to have formed the planets.

Plessite: Fine-grained intergrowth of kamacite and taenite

Plutonic rocks: Igneous rocks formed at large depth

Portia: Seventh satellite from Uranus; discovered by *Voyager 2* in 1986

ppb: parts per billion; concentration measure by weight frequently used in geochemistry (equals ng/g). It is used in atmospheric chemistry to describe molecular concentration by volume (ppbv).

ppm: parts per million; concentration measure by weight frequently used in geochemistry (equals µg/g). It is used in atmospheric chemistry to describe molecular concentration by volume (ppmv)

p-process: Nucleosynthetic process producing rare heavy proton-rich nuclides; believed to operate in supernova envelopes

Precession: The periodic change in the rotational axis of Earth (and other objects) under the influence of an external torque. Earth's axis is precessing with a period of 25,800 years. The equinoxes are also precessing westward, with the same period, because of precession of the earth's rotational axis.

Prometheus: Third satellite from Saturn; discovered by S. Collins in 1980

Proper motion: The apparent angular motion of a star on the celestial sphere; a combination of the star's motion in space, the sun's motion in space, and the star's motion relative to the solar system

Proteus: Sixth satellite from Neptune; discovered by *Voyager 2* in 1989

Puck: Tenth satellite from Uranus; discovered by *Voyager 2* in 1986

Red giant star: High luminosity, late-type (K or M) star with very large radius occupying the upper right portion of the Hertzsprung-Russel (HR) diagram

REE: Rare Earth Elements; elements of the lanthanide series of the periodic table

Refractory element: Elements that condense or vaporize at high temperatures (e.g., Al, Ca, Os, Re, Ti, U, W, REE)

Regolith: Loose surface material consisting of soil and rock fragments

Rhea: Fourteenth satellite from Saturn; discovered by G. Cassini in 1672

Rosalind: Eighth satellite from Uranus; discovered by *Voyager 2* in 1986

Rossby number: The nondimensional ratio (R_0) of inertial to Coriolis acceleration in a planetary atmosphere or stellar interior. It is given by $R_0 = U/(f \cdot L)$, where U is a characteristic horizontal velocity, L is a characteristic horizontal scale length, and $f = 2\Omega \sin\lambda$ is the vorticity.

r-process: Nucleosynthetic process producing nuclides by neutron capture on a rapid time scale

Seafloor spreading: Creation of new igneous oceanic crust by volcanism at mid-ocean ridges, causing movement of the plates away from the ridges

Shergottite: Rare achondrite, consisting of pigeonite and maskelynite; related to nakhlites and chassignites

Sidereal month: The time (27.32166 days) for the moon to revolve once around the earth, with respect to the fixed stars.

Sidereal period: The time for a planet, asteroid, or satellite to complete one revolution around the sun or its primary, or to complete one rotation, with respect to the fixed stars

Sidereal year: The time (365.25636 days) for earth to complete one revolution around the sun, with respect to the fixed stars. The sidereal year is 20 minutes longer than the tropical year.

Siderophile elements: Metal-loving elements; elements that preferentially partition into metal phases (e.g., Ni, Co, As, Ge, Ga, Au, PGE).

Sinope: Outermost satellite of Jupiter, one of the four retrograde outer satellites; discovered by S. Nicholson in 1914

SMOW: Standard Mean Ocean Water, reference standard for H and O isotopic measurements

SNC: Shergottites, Nakhlites, and Chassignites, a group of achondrites believed to originate from Mars

s-process: Nucleosynthetic process producing nuclides by slow neutron capture; nuclei beyond ^{209}Bi cannot be synthesized by the s-process because of rapid alpha decay of ^{209}Bi.

Subduction: Sinking of an oceanic plate below another plate into the mantle

Summer solstice: The date, around June 22nd each year, when the sun's declination is +23.45° and the sun passes directly overhead at noon at 23.45° n. latitude, the Tropic of Cancer. At the summer solstice, the sun does not set for 24 hours at 66.5° N. latitude, the Arctic Circle.

Synodic month: The interval (29.53059 days) between two successive new moons

Taenite: Iron metal or alloy (e.g., Fe-Ni) with face centered cubic structure

Tektite: Centimeter-size, silica-rich, glassy, rounded objects produced from terrestrial surface material during cometary or asteroidal impacts. Tektites occur in strewn fields often far away from the original impact sites. Well known strewn fields (their ages and special tektite names) are in Australasia (0.7–0.8 Myr, australites, phillipinites), in Czechoslovakia (14.8 Myr, moldavites), at the Ivory Coast (1.1–1.3 Myr), and in Texas, U. S. A. (35 Myr, bediasites).

Telesto: Tenth satellite from Saturn; discovered by B. Smith in 1980. Telesto is co-orbital with Calypso and Tethys.

Terrestrial age: Time period since the fall of a meteorite

Tethys: Ninth satellite from Saturn; discovered by G. Cassini in 1684. Tethys is co-orbital with Telesto and Calypso.

Thalassa: Second satellite from Neptune; discovered by *Voyager 2* in 1989

Thebe: Fourth satellite from Jupiter; discovered in 1979 by S. Synott

Themis family: A Hirayama family of asteroids at 3.2 AU

Tisserand parameter or Tisserand invariant (T) describes perturbations of asteroidal and cometary orbits by Jupiter

$$T = \frac{a_J}{a} + 2\sqrt{\frac{a(1-e^2)}{a_J}} \cos i = \frac{2a_J}{q+Q} + 2\sqrt{\frac{2qQ}{a_J(q+Q)}} \cos i$$

where a, e, i, q, and Q are the objects semimajor axis, eccentricity, inclination, perihelion and aphelion, respectively, and a_J is the semimajor axis of Jupiter's orbit.

Titan: Largest satellite of Saturn, and fifteenth outward; discovered by C. Huygens in 1655. Titan has a 1.5 bar N_2, Ar, and CH_4 atmosphere.

Titania: Fourteenth satellite from Uranus; discovered by W. Herschel in 1787. The largest satellite of Uranus. *Voyager 2* images showed an extensive fault network, many small craters, and some large impact basins.

Titus-Bode rule: An empirical relationship (not a physical law) describing the distances (d in AU) of the planets and asteroids from the sun: $d = 0.4 + 0.3 \cdot 2^n$, where n is $-\infty$ (Mercury), 0 (Venus), and 1 to 8 (Earth to Pluto).

Triton: Seventh satellite from Neptune; discovered by W. Lassell in 1846. The largest satellite of Neptune. Triton is in a retrograde orbit and is spiraling in toward Neptune. *Voyager 2* images showed 8 km high cryovolcanic plumes.

Troctolite: Rock similar to gabbro with olivine dominant over pyroxene

Tropical month: The time (27.32158 days) for the moon to revolve around the earth, measured with respect to the vernal equinox

Tropical year: The time (365.24219 days) between two successive passages of the sun through the vernal equinox. The tropical year is about 20 minutes shorter than the sidereal year because precession produces an annual net retrograde motion of 50.28 arc seconds of the equinoxes relative to fixed stars.

Umbriel: Thirteenth satellite from Uranus; discovered by W. Lassell in 1851. The third largest satellite of Uranus. Very dark with an albedo ~0.18. *Voyager 2* images showed a heavily cratered surface.

Universal time: The basis of civil time keeping (abbreviated as UT). Defined by a mathematical formula relating UT to sidereal time

Ureilite: Carbon-rich achondrites mainly consisting of olivine and pigeonite

Van Allen radiation belts: Two regions in the earth's magnetosphere where charged particles are trapped and oscillate between the two magnetic poles. Discovered by James Van Allen in 1958 by his analysis of observations from *Explorer* Earth satellites.

Vernal equinox: The date in the spring, around March 21st, when the sun crosses from south to north of the earth's equator. The sun's declination is 0° at the autumnal and vernal equinoxes. The autumnal and vernal equinoxes are formally defined as the two points on the celestial sphere where the sun crosses the celestial equator in its apparent annual motion. The equinoxes are not fixed in position but are moving retrograde (westward) at 50.28 arc seconds per year (precession of the equinoxes). Also see Tropical year and Autumnal equinox.

Volatile element: Elements that condense or vaporize at relatively low temperatures (e.g., K, Rb, Cs, Tl, Pb)

Volcanic arc: Set of active volcanoes within the upper plate in a subduction zone

Widmanstätten pattern: Structures in iron meteorite consisting of large octahedral kamacite and small areas of taenite; made visible by etching

Wien's displacement law: Describes the wavelength at which the maximum power is radiated from a blackbody at given temperature:
$$\lambda_{max} \cdot T = 2.87756 \times 10^{-3} \text{ m K}$$

Winter solstice: The date, around December 22nd each year, when the sun's declination is $-23.45°$ and the sun passes directly overhead at noon at 23.45° S. latitude, the Tropic of Capricorn. The sun does not set for 24 hours at 66.5° S. latitude, the Antarctic Circle.

Wollastonite: $CaSiO_3$; Ca-endmember of the pyroxene group

Ws: Wollastonite